SYMBOLIC LOGIC

A FIRST COURSE

THIRD EDITION

GARY M. HARDEGREE
University of Massachusetts–Amherst

The McGraw-Hill Companies, Inc.
Primis Custom Publishing

New York St. Louis San Francisco Auckland Bogotá Caracas Lisbon London Madrid Mexico Milan Montreal New Delhi Paris San Juan Singapore Sydney Tokyo Toronto

McGraw-Hill Higher Education

*A Division of The **McGraw-Hill** Companies*

Symbolic Logic: A First Course

McGraw-Hill's Primis Custom Series consists of products that are produced from camera-ready copy. Peer review, class testing, and accuracy are primarily the responsibility of the author(s).

4 5 6 7 8 9 BKM BKM 0 9 8 7 6 5 4 3

ISBN 0-07-239027-1

Editor: Debbie Quigley
Cover Designer: Mark Anderson
Printer/Binder: Bookmart Press

TABLE OF CONTENTS

1. BASIC CONCEPTS OF LOGIC

2. TRUTH-FUNCTIONAL CONNECTIVES

3. VALIDITY IN SENTENTIAL LOGIC

4. TRANSLATIONS IN SENTENTIAL LOGIC

5. DERIVATIONS IN SENTENTIAL LOGIC

6. TRANSLATIONS IN MONADIC PREDICATE LOGIC

7. TRANSLATIONS IN POLYADIC PREDICATE LOGIC

8. DERIVATIONS IN PREDICATE LOGIC

PICTORIAL SUMMARY OF RULES OF DERIVATION

1 BASIC CONCEPTS OF LOGIC

1. WHAT IS LOGIC?

Logic may be defined as the science of reasoning. However, this is not to suggest that logic is an empirical (i.e., experimental or observational) science like physics, biology, or psychology. Rather, logic is a non-empirical science like mathematics. Also, in saying that logic is the science of reasoning, we do not mean that it is concerned with the *actual* mental (or physical) process employed by a thinking entity when it is reasoning. The investigation of the actual reasoning *process* falls more appropriately within the province of psychology, neurophysiology, or cybernetics.

Even if these empirical disciplines were considerably more advanced than they presently are, the most they could disclose is the exact process that goes on in a being's head when he or she (or it) is reasoning. They could not, however, tell us whether the being is reasoning correctly or incorrectly.

Distinguishing correct reasoning from incorrect reasoning is the task of logic.

2. INFERENCES AND ARGUMENTS

Reasoning is a special mental activity called *inferring*, what can also be called making (or performing) *inferences*. The following is a useful and simple definition of the word 'infer'.

To ***infer*** is to draw conclusions from premises.

In place of word 'premises', you can also put: 'data', 'information', 'facts'.

Examples of Inferences:

(1) You see smoke and *infer* that there is a fire.

(2) You count 19 persons in a group that originally had 20, and you *infer* that someone is missing.

Note carefully the difference between 'infer' and 'imply', which are sometimes confused. We *infer* the fire on the basis of the smoke, but we do not *imply* the fire. On the other hand, the smoke *implies* the fire, but it does not *infer* the fire. The word 'infer' is not equivalent to the word 'imply', nor is it equivalent to 'insinuate'.

The reasoning process may be thought of as beginning with *input* (premises, data, etc.) and producing *output* (conclusions). In each specific case of drawing (inferring) a conclusion C from premises P_1, P_2, P_3, ..., the details of the actual mental process (how the "gears" work) is not the proper concern of logic, but of psychology or neurophysiology. The proper concern of logic is whether the inference of C on the basis of P_1, P_2, P_3, ... is warranted (correct).

Inferences are made on the basis of various sorts of things – data, facts, information, states of affairs. In order to simplify the investigation of reasoning, logic treats all of these things in terms of a single sort of thing – *statements*. Logic correspondingly treats inferences in terms of collections of statements, which are called *arguments*. The word 'argument' has a number of meanings in ordinary English. The definition of 'argument' that is relevant to logic is given as follows.

> An ***argument*** is a collection of **statements**, one of which is designated as the **conclusion**, and the remainder of which are designated as the **premises**.

Note that this is *not* a definition of a *good* argument. Also note that, in the context of ordinary discourse, an argument has an additional trait, described as follows.

> Usually, the premises of an argument are intended to **support** (**justify**) the conclusion of the argument.

Before giving some concrete examples of arguments, it might be best to clarify a term in the definition. The word 'statement' is intended to mean *declarative sentence*. In addition to declarative sentences, there are also interrogative, imperative, and exclamatory sentences. The sentences that make up an argument are all declarative sentences; that is, they are all statements. The following may be taken as the official definition of 'statement'.

> A ***statement*** is a **declarative sentence**, which is to say a sentence that is capable of being **true** or **false**.

The following are examples of statements.

it is raining
I am hungry
2+2 = 4
God exists

On the other hand the following are examples of sentences that are *not* statements.

are you hungry?
shut the door, please
#$%@!!! (replace '#$%@!!!' by your favorite expletive)

Observe that whereas a *statement* is capable of being true or false, a *question*, or a *command*, or an *exclamation* is not capable of being true or false.

Note that in saying that a statement is capable of being true or false, we are not saying that we *know for sure* which of the two (true, false) it is. Thus, for a sentence to be a statement, it is not necessary that humankind knows for sure whether it is true, or whether it is false. An example is the statement 'God exists'.

Now let us get back to inferences and arguments. Earlier, we discussed two examples of inferences. Let us see how these can be represented as arguments. In the case of the smoke-fire inference, the corresponding argument is given as follows.

(a1)	there is smoke	(premise)
	therefore, there is fire	(conclusion)

Here the argument consists of two statements, 'there is smoke' and 'there is fire'. The term 'therefore' is not strictly speaking part of the argument; it rather serves to designate the conclusion ('there is fire'), setting it off from the premise ('there is smoke'). In this argument, there is just one premise.

In the case of the missing-person inference, the corresponding argument is given as follows.

(a2)	there were 20 persons originally	(premise)
	there are 19 persons currently	(premise)
	therefore, someone is missing	(conclusion)

Here the argument consists of three statements – 'there were 20 persons originally', 'there are 19 persons currently', and 'someone is missing'. Once again, 'therefore' sets off the conclusion from the premises.

In principle, any collection of statements can be treated as an argument simply by designating which statement in particular is the conclusion. However, not every collection of statements is *intended to be* an argument. We accordingly need criteria by which to distinguish arguments from other collections of statements.

There are no hard and fast rules for telling when a collection of statements is intended to be an argument, but there are a few rules of thumb. Often an argument can be identified as such because its conclusion is marked. We have already seen one conclusion-marker – the word 'therefore'. Besides 'therefore', there are other words that are commonly used to mark conclusions of arguments, including 'consequently', 'hence', 'thus', 'so', and 'ergo'. Usually, such words indicate that what follows is the conclusion of an argument.

Other times an argument can be identified as such because its premises are marked. Words that are used for this purpose include: 'for', 'because', and 'since'. For example, using the word 'for', the smoke-fire argument (a1) earlier can be rephrased as follows.

(a1′) there is fire
for there is smoke

Note that in (a1′) the conclusion comes *before* the premise.

Other times neither the conclusion nor the premises of an argument are marked, so it is harder to tell that the collection of statements is intended to be an argument. A general rule of thumb applies in this case, as well as in previous cases.

> In an argument, the premises are intended to support (justify) the conclusion.

To state things somewhat differently, when a person (speaking or writing) advances an argument, he(she) expresses a statement he(she) believes to be true (the conclusion), and he(she) cites other statements as a reason for believing that statement (the premises).

3. DEDUCTIVE LOGIC VERSUS INDUCTIVE LOGIC

Let us go back to the two arguments from the previous section.

(a1) there is smoke;
therefore, there is fire.

(a2) there were 20 people originally;
there are 19 persons currently;
therefore, someone is missing.

There is an important difference between these two inferences, which corresponds to a division of logic into two branches.

On the one hand, we know that the existence of smoke does not *guarantee* (ensure) the existence of fire; it only makes the existence of fire *likely* or *probable*. Thus, although inferring fire on the basis of smoke is reasonable, it is nevertheless *fallible*. Insofar as it is *possible* for there to be smoke without there being fire, we *may* be wrong in asserting that there is a fire.

The investigation of inferences of this sort is traditionally called *inductive logic*. Inductive logic investigates the process of drawing *probable* (likely, plausible) *though fallible* conclusions from premises. Another way of stating this: inductive logic investigates arguments in which the truth of the premises *makes likely* the truth of the conclusion.

Inductive logic is a very difficult and intricate subject, partly because the practitioners (experts) of this discipline are not in complete agreement concerning what constitutes correct inductive reasoning.

Inductive logic is *not* the subject of this book. If you want to learn about inductive logic, it is probably best to take a course on probability and statistics. Inductive reasoning is often called statistical (or probabilistic) reasoning, and forms the basis of experimental science.

Inductive reasoning is important to science, but so is deductive reasoning, which *is* the subject of this book.

Consider argument (a2) above. In this argument, *if* the premises are in fact true, *then* the conclusion is *certainly* also true; or, to state things in the subjunctive mood, *if* the premises *were* true, *then* the conclusion *would certainly* also be true.

Still another way of stating things: the truth of the premises *necessitates* the truth of the conclusion.

The investigation of these sorts of arguments is called *deductive logic*.

The following should be noted. suppose that you have an argument and suppose that the truth of the premises *necessitates* (guarantees) the truth of the conclusion. Then it follows (logically!) that the truth of the premises *makes likely* the truth of the conclusion. In other words, if an argument is judged to be *deductively correct*, then it is also judged to be *inductively correct* as well. The converse is not true: not every inductively correct argument is also deductively correct; the smoke-fire argument is an example of an inductively correct argument that is not deductively correct. For whereas the existence of smoke *makes likely* the existence of fire it does not *guarantee* the existence of fire.

In deductive logic, the task is to distinguish deductively correct arguments from deductively incorrect arguments. Nevertheless, we should keep in mind that, although an argument may be judged to be *deductively incorrect*, it may still be *reasonable*, that is, it may still be *inductively correct*.

Some arguments are not inductively correct, and therefore are not deductively correct either; they are just plain unreasonable. Suppose you flunk intro logic, and suppose that *on the basis of this* you conclude that it will be a breeze to get into law school. Under these circumstances, it seems that your reasoning is faulty.

4. STATEMENTS VERSUS PROPOSITIONS

Henceforth, by 'logic' I mean deductive logic.

Logic investigates inferences in terms of the arguments that represent them. Recall that an argument is a collection of statements (declarative sentences), one of which is designated as the conclusion, and the remainder of which are designated as the premises. Also recall that usually in an argument the premises are offered to support or justify the conclusions.

Statements, and sentences in general, are linguistic objects, like words. They consist of strings (sequences) of sounds (spoken language) or strings of symbols (written language). Statements must be carefully distinguished from the *propositions* they express (assert) when they are uttered. Intuitively, statements stand in the same relation to propositions as nouns stand to the objects they denote. Just as the word 'water' denotes a substance that is liquid under normal circumstances, the *sentence* (statement) 'water is wet' denotes the *proposition* that water is wet; equivalently, the sentence denotes the state of affairs the wetness of water.

The difference between the five letter word 'water' in English and the liquid substance it denotes should be obvious enough, and no one is apt to confuse the word and the substance. Whereas 'water' consists of letters, water consists of molecules. The distinction between a statement and the proposition it expresses is very much like the distinction between the word 'water' and the substance water.

There is another difference between statements and propositions. Whereas statements are always part of a particular language (e.g., English), propositions are not peculiar to any particular language in which they might be expressed. Thus, for example, the following are different statements in different languages, yet they all express the same proposition – namely, the whiteness of snow.

snow is white
der Schnee ist weiss
la neige est blanche

In this case, quite clearly different sentences may be used to express the same proposition. The opposite can also happen: the same sentence may be used in different contexts, or under different circumstances, to express different propositions, to denote different states of affairs. For example, the statement 'I am hungry' expresses a different proposition for each person who utters it. When I utter it, the proposition expressed pertains to my stomach; when you utter it, the proposition pertains to your stomach; when the president utters it, the proposition pertains to his(her) stomach.

5. FORM VERSUS CONTENT

Although propositions (or the meanings of statements) are always lurking behind the scenes, logic is primarily concerned with statements. The reason is that statements are in some sense easier to point at, easier to work with; for example, we can write a statement on the blackboard and examine it. By contrast, since they are essentially abstract in nature, propositions cannot be brought into the classroom, or anywhere. Propositions are unwieldy and uncooperative. What is worse, no one quite knows exactly what they are!

There is another important reason for concentrating on statements rather than propositions. Logic analyzes and classifies arguments according to their *form*, as opposed to their *content* (this distinction will be explained later). Whereas the form of a statement is fairly easily understood, the form of a proposition is not so easily understood. Whereas it is easy to say what a statement consists of, it is not so easy to say what a proposition consists of.

A statement consists of words arranged in a particular order. Thus, the form of a statement may be analyzed in terms of the arrangement of its constituent words. To be more precise, a statement consists of *terms*, which include simple terms and compound terms. A simple term is just a single word together with a specific grammatical role (being a noun, or being a verb, etc.). A compound term is a string of words that act as a grammatical unit within statements. Examples of compound terms include noun phrases, such as 'the president of the U.S.', and predicate phrases, such as 'is a Democrat'.

For the purposes of logic, terms divide into two important categories – *descriptive terms* and *logical terms*. One must carefully note, however, that this distinction is *not* absolute. Rather, the distinction between descriptive and logical terms depends upon the level (depth) of logical analysis we are pursuing.

Let us pursue an analogy for a moment. Recall first of all that the core meaning of the word 'analyze' is *to break down a complex whole into its constituent parts*. In physics, matter can be broken down (analyzed) at different levels; it can be analyzed into molecules, into atoms, into elementary particles (electrons, protons, etc.); still deeper levels of analysis are available (e.g., quarks). The basic idea in breaking down matter is that in order to go deeper and deeper one needs ever increasing amounts of energy, and one needs ever increasing sophistication.

The same may be said about logic and the analysis of language. There are many levels at which we can analyze language, and the deeper levels require more logical sophistication than the shallower levels (they also require more energy on the part of the logician!)

In the present text, we consider three different levels of logical analysis. Each of these levels is given a name – Syllogistic Logic, Sentential Logic, and Predicate Logic. Whereas syllogistic logic and sentential logic represent relatively superficial (shallow) levels of logical analysis, predicate logic represents a relatively deep level of analysis. Deeper levels of analysis are available.

Each level of analysis – syllogistic logic, sentential logic, and predicate logic – has associated with it a special class of logical terms. In the case of syllogistic logic, the logical terms include only the following: 'all', 'some', 'no', 'not', and 'is/are'. In the case of sentential logic, the logical terms include only sentential connectives (e.g., 'and', 'or', 'if...then', 'only if'). In the case of predicate logic, the logical terms include the logical terms of both syllogistic logic and sentential logic.

As noted earlier, logic analyzes and classifies arguments according to their *form*. The (logical) form of an argument is a function of the forms of the individual statements that constitute the argument. The logical form of a statement, in turn, is a function of the arrangement of its terms, where the logical terms are regarded as more important than the descriptive terms. Whereas the logical terms have to do with the form of a statement, the descriptive terms have to do with its content.

Note, however, that since the distinction between logical terms and descriptive terms is relative to the particular level of analysis we are pursuing, the notion of logical form is likewise relative in this way. In particular, for each of the different logics listed above, there is a corresponding notion of logical form.

The distinction between form and content is difficult to understand in the abstract. It is best to consider some actual examples. In a later section, we examine this distinction in the context of syllogistic logic.

As soon as we can get a clear idea about form and content, then we can discuss how to classify arguments into those that are deductively correct and those that are not deductively correct.

6. PRELIMINARY DEFINITIONS

In the present section we examine some of the basic ideas in logic which will be made considerably clearer in subsequent chapters.

As we saw in the previous section there is a distinction in logic between *form* and *content*. There is likewise a distinction in logic between arguments that are *good in form* and arguments that are *good in content*. This distinction is best understood by way of an example or two. Consider the following arguments.

(a1) all cats are dogs
all dogs are reptiles
therefore, all cats are reptiles

(a2) all cats are vertebrates
all mammals are vertebrates
therefore, all cats are mammals

Neither of these arguments is good, but they are bad for different reasons. Consider first their content. Whereas all the statements in (a1) are false, all the statements in (a2) are true. Since the premises of (a1) are not all true this is not a good argument *as far as content goes*, whereas (a2) is a good argument *as far as content goes*.

Now consider their forms. This will be explained more fully in a later section. The question is this: do the premises *support* the conclusion? Does the conclusion *follow from* the premises?

In the case of (a1), the premises do in fact support the conclusion, the conclusion does in fact follow from the premises. Although the premises are not true, if they *were* true then the conclusion *would* also be true, of necessity.

In the case of (a2), the premises are all true, and so is the conclusion, but nevertheless the truth of the conclusion is *not* conclusively supported by the premises; in (a2), the conclusion does not follow from the premises. To see that the conclusion does not follow from the premises, we need merely substitute the term 'reptiles' for 'mammals'. Then the premises are both true but the conclusion is false.

All of this is meant to be at an intuitive level. The details will be presented later. For the moment, however we give some rough definitions to help us get started in understanding the ways of classifying various arguments.

In examining an argument there are basically two questions one should ask.

Question 1:	Are all of the premises true?

Question 2:	Does the conclusion ***follow from*** the premises?

The classification of a given argument is based on the answers to these two questions. In particular, we have the following definitions.

> An argument is ***factually correct***
> if and only if
> all of its *premises* are true.

> An argument is ***valid***
> if and only if
> its conclusion *follows from* its premises.

> An argument is ***sound***
> if and only if
> it is *both* factually correct *and* valid.

Basically, a factually correct argument has *good content*, and a valid argument has *good form*, and a sound argument has *both* good content *and* good form.

Note that a factually correct argument *may* have a false conclusion; the definition only refers to the premises.

Whether an argument is valid is sometimes difficult to decide. Sometimes it is hard to know whether or not the conclusion follows from the premises. Part of the problem has to do with knowing what 'follows from' means. In studying logic we are attempting to understand the meaning of 'follows from'; more importantly perhaps, we are attempting to learn how to distinguish between valid and invalid arguments.

Although logic can teach us something about validity and invalidity, it can teach us very little about factual correctness. The question of the truth or falsity of individual statements is primarily the subject matter of the sciences, broadly construed.

As a rough-and-ready definition of validity, the following is offered.

> An argument is ***valid***
> if and only if
> it is **impossible** for
> the conclusion to be **false**
> while the premises are **all true**.

An alternative definition might be helpful in understanding validity.

> To say that an argument is ***valid***
> is to say that
> if the premises **were** true,
> then the conclusion **would necessarily** also be true.

These will become clearer as you read further, and as you study particular examples.

7. FORM AND CONTENT IN SYLLOGISTIC LOGIC

In order to understand more fully the notion of logical form, we will briefly examine syllogistic logic, which was invented by Aristotle (384-322 B.C.).

The arguments studied in syllogistic logic are called syllogisms (more precisely, categorical syllogisms). Syllogisms have a couple of distinguishing characteristics, which make them peculiar as arguments. First of all, every syllogism has exactly two premises, whereas in general an argument can have any number of premises. Secondly, the statements that constitute a syllogism (two premises, one conclusion) come in very few models, so to speak; more precisely, all such statements have forms similar to the following statements.

(1)	all **Lutherans** are **Protestants**	all **dogs** are **collies**
(2)	some **Lutherans** are **Republicans**	some **dogs** are **cats**
(3)	no **Lutherans** are **Methodists**	no **dogs** are **pets**
(4)	some **Lutherans** are not **Democrats**	some **dogs** are not **mammals**

In these examples, the words written in bold-face letters are descriptive terms, and the remaining words are logical terms, relative to syllogistic logic.

In syllogistic logic, the descriptive terms all refer to *classes*, for example, the class of cats, or the class of mammals. On the other hand, in syllogistic logic, the logical terms are all used to express relations among classes. For example, the statements on line (1) state that a certain class (Lutherans/dogs) is entirely contained in another class (Protestants/collies).

Note the following about the four pairs of statements above. In each case, the pair contains both a true statement (on the left) and a false statement (on the right). Also, in each case, the statements are about different things. Thus, we can say that the two statements differ in *content*. Note, however, that in each pair above, the two statements have the same *form*. Thus, although 'all Lutherans are Protestants' *differs in content* from 'all dogs are collies', these two statements *have the same form*.

The sentences (1)-(4) are what we call *concrete sentences*; they are all actual sentences of a particular actual language (English). Concrete sentences are to be distinguished from *sentence forms*. Basically, a sentence form may be obtained from a concrete sentence by replacing all the descriptive terms by letters, which serve as place holders. For example, sentences (1)-(4) yield the following sentence forms.

(f1) all X are Y
(f2) some X are Y
(f3) no X are Y
(f4) some X are not Y

The process can also be reversed: concrete sentences may be obtained from sentence forms by *uniformly* substituting descriptive terms for the letters. Any concrete sentence obtained from a sentence form in this way is called a *substitution instance* of that form. For example, 'all cows are mammals' and 'all cats are felines' are both substitution instances of sentence form (f1).

Just as there is a distinction between concrete statements and statement forms, there is also a distinction between concrete arguments and argument forms. A *concrete argument* is an argument consisting entirely of concrete statements; an *argument form* is an argument consisting entirely of statement forms. The following are examples of concrete arguments.

(a1) all Lutherans are Protestants
some Lutherans are Republicans
/ some Protestants are Republicans

(a2) all Lutherans are Protestants
some Protestants are Republicans
/ some Lutherans are Republicans

Note: henceforth, we use the slash symbol (/) to abbreviate 'therefore'.

In order to obtain the argument form associated with (a1), we can simply replace each descriptive term by its initial letter; we can do this because the descriptive terms in (a1) all have different initial letters. this yields the following argument form. An alternative version of the form, using X,Y,Z, is given to the right.

(f1)	all L are P	all X are Y
	some L are R	some X are Z
	/ some P are R	/ some Y are Z

By a similar procedure we can convert concrete argument (a2) into an associated argument form.

(f2)	all L are P	all X are Y
	some P are R	some Y are Z
	/ some L are R	/ some X are Z

Observe that argument (a2) is obtained from argument (a1) simply by interchanging the conclusion and the second premise. In other words, these two arguments which are different, consist of precisely the same statements. They are different because their conclusions are different. As we will later see, they are different in that one is a *valid* argument, and the other is an *invalid* argument. Do you know which one is which? In which one does the truth of the premises guarantee the truth of the conclusion?

In deriving an argument form from a concrete argument care must be taken in assigning letters to the descriptive terms. First of all different letters must be assigned to different terms: we cannot use 'L' for both 'Lutherans' and 'Protestants'.

Secondly, we cannot use two different letters for the same term: we cannot use 'L' for Lutherans in one statement, and use 'Z' in another statement.

8. DEMONSTRATING INVALIDITY USING THE METHOD OF COUNTEREXAMPLES

Earlier we discussed some of the basic ideas of logic, including the notions of validity and invalidity. In the present section, we attempt to get a better idea about these notions.

We begin by making precise definitions concerning statement forms and argument forms.

> A ***substitution instance*** of an argument/statement form is a concrete argument/statement that is obtained from that form by substituting appropriate descriptive terms for the letters, in such a way that each occurrence of the same letter is replaced by the same term.

> A ***uniform substitution instance*** of an argument/statement form is a substitution instance with the additional property that distinct letters are replaced by distinct (non-equivalent) descriptive terms.

In order to understand these definitions let us look at a very simple argument form (since it has just one premise it is not a syllogistic argument form):

(F) all X are Y
/ some Y are Z

Now consider the following concrete arguments.

(1) all cats are dogs
/ some cats are cows

(2) all cats are dogs
/ some dogs are cats

(3) all cats are dogs
/ some dogs are cows

These examples are not chosen because of their intrinsic interest, but merely to illustrate the concepts of substitution instance and uniform substitution instance.

First of all, (1) is not a substitution instance of (F), and so it is not a uniform substitution instance either (why is this?). In order for (1) to be a substitution instance to (F), it is required that each occurrence of the same letter is replaced by the same term. This is not the case in (1): in the premise, Y is replaced by 'dogs', but in the conclusion, Y is replaced by 'cats'. It is accordingly not a substitution instance.

Next, (2) is a substitution instance of (F), but it is *not* a uniform substitution instance. There is only one letter that appears twice (or more) in (F) – namely, Y. In each occurrence, it is replaced by the same term – namely, 'dogs'. Therefore, (2) is a substitution instance of (F). On the other hand, (2) is not a *uniform* substitution instance since distinct letters – namely, X and Z – are replaced by the same descriptive term – namely, 'cats'.

Finally, (3) is a *uniform* substitution instance and hence a substitution instance, of (F). Y is the only letter that is repeated; in each occurrence, it is replaced by the same term – namely, 'dogs'. So (3) is a substitution instance of (F). To see whether it is a uniform substitution instance, we check to see that the same descriptive term is not used to replace different letters. The only descriptive term that is repeated is 'dogs', and in each case, it replaces Y. Thus, (3) is a uniform substitution instance.

The following is an argument form followed by three concrete arguments, one of which is not a substitution instance, one of which is a non-uniform substitution instance, and one of which is a uniform substitution instance, in that order.

(F) no X are Y
no Y are Z
/ no X are Z

(1) no cats are dogs
no cats are cows
/ no dogs are cows

(2) no cats are dogs
no dogs are cats
/ no cats are cats

(3) no cats are dogs
no dogs are cows
/ no cats are cows

Check to make sure you agree with this classification.

Having defined (uniform) substitution instance, we now define the notion of *having the same form.*

Two arguments/statements have the ***same form***
if and only if
they are both **uniform substitution instances** of the
same argument/statement form.

For example, the following arguments have the same form, because they can both be obtained from the argument form that follows as uniform substitution instances.

(a1) all Lutherans are Republicans
some Lutherans are Democrats
/ some Republicans are Democrats

(a2) all cab drivers are maniacs
some cab drivers are Democrats
/ some maniacs are Democrats

The form common to (a1) and (a2) is:

(F) all X are Y
some X are Z
/ some Y are Z

As an example of two arguments that do not have the same form consider arguments (2) and (3) above. They cannot be obtained from a common argument form by *uniform* substitution.

Earlier, we gave two intuitive definitions of *validity*. Let us look at them again.

> An argument is **valid**
> if and only if
> it is **impossible** for
> the conclusion to be **false**
> while the premises are **all true**.

> To say that an argument is ***valid***
> is to say that
> if the premises **were** true,
> then the conclusion **would necessarily** also be true.

Although these definitions may give us a general idea concerning what 'valid' means in logic, they are difficult to apply to specific instances. It would be nice if we had some methods that could be applied to specific arguments by which to decide whether they are valid or invalid.

In the remainder of the present section, we examine a method for showing that an argument is invalid (if it is indeed invalid) – the method of *counterexamples*. Note however, that this method cannot be used to prove that a *valid* argument is in fact valid.

In order to understand the method of counterexamples, we begin with the following fundamental principle of logic.

FUNDAMENTAL PRINCIPLE OF LOGIC

Whether an argument is valid or invalid is determined entirely by its form; in other words:

VALIDITY IS A FUNCTION OF FORM.

This principle can be rendered somewhat more specific, as follows.

FUNDAMENTAL PRINCIPLE OF LOGIC
(REWRITTEN)

If an argument is valid, then every argument *with the same form* is also valid.

If an argument is invalid, then every argument *with the same form* is also invalid.

There is one more principle that we need to add before describing the method of counterexamples. Since the principle almost doesn't need to be stated, we call it the Trivial Principle, which is stated in two forms.

THE TRIVIAL PRINCIPLE

No argument with all true premises but a false conclusion is valid.

If an argument has all true premises but has a false conclusion, then it is invalid.

The Trivial Principle follows from the definition of validity given earlier: an argument is valid if and only if it is impossible for the conclusion to be false while the premises are all true. Now, if the premises are all true, and the conclusion is in fact false, then it is possible for the conclusion to be false while the premises are all true. Therefore, if the premises are all true, and the conclusion is in fact false, then the argument is not valid that is, it is invalid.

Now let's put all these ideas together. Consider the following concrete argument, and the corresponding argument form to its right.

(A)	all cats are mammals	(F)	all X are Y
	some mammals are dogs		some Y are Z
	/ some cats are dogs		/ some X are Z

First notice that whereas the premises of (A) are both true, the conclusion is false. Therefore, in virtue of the Trivial Principle, argument (A) is invalid. But if (A) is invalid, then in virtue of the Fundamental Principle (rewritten), *every* argument with the *same form* as (A) is also invalid.

In other words, every argument with form (F) is invalid. For example, the following arguments are invalid.

(a2) all cats are mammals
some mammals are pets
/ some cats are pets

(a3) all Lutherans are Protestants
some Protestants are Democrats
/ some Lutherans are Democrats

Notice that the premises are both true and the conclusion is true, in both arguments (a2) and (a3). Nevertheless, both these arguments are *invalid*.

To say that (a2) (or (a3)) is invalid is to say that the truth of the premises does not guarantee the truth of the conclusion – the premises do not support the conclusion. For example, it is possible for the conclusion to be false even while the premises are both true. Can't we imagine a world in which all cats are mammals, some mammals are pets, but no cats are pets. Such a world could in fact be easily brought about by a dastardly dictator, who passed an edict prohibiting cats to be kept as pets. In this world, all cats are mammals (that hasn't changed!), some mammals are pets (e.g., dogs), yet no cats are pets (in virtue of the edict proclaimed by the dictator).

Thus, in argument (a2), it is *possible* for the conclusion to be false while the premises are both true, which is to say that (a2) is *invalid*.

In demonstrating that a particular argument is invalid, it may be difficult to *imagine* a world in which the premises are true but the conclusion is false. An easier method, which does not require one to imagine unusual worlds, is the method of counterexamples, which is based on the following definition and principle, each stated in two forms.

A. A ***counterexample*** to an **argument form** is any substitution instance (not necessarily uniform) of that form having true premises but a false conclusion.

B. A ***counterexample*** to a **concrete argument** $\mathcal{A}$ is any concrete argument that

(1) has the **same form** as $\mathcal{A}$
(2) has all true premises
(3) has a false conclusion

PRINCIPLE OF COUNTEREXAMPLES

A. An argument (form) is **invalid** if it admits a counterexample.

> B. An argument (form) is **valid** only if it does not admit any counterexamples.

The Principle of Counterexamples follows our earlier principles and the definition of the term 'counterexample'. One might reason as follows:

> Suppose argument $\mathcal{A}$ admits a counterexample. Then there is another argument $\mathcal{A}$* such that:
>
> (1) $\mathcal{A}$* has the same form as $\mathcal{A}$,
>
> (2) $\mathcal{A}$* has all true premises, and
>
> (3) $\mathcal{A}$* has a false conclusion.
>
> Since $\mathcal{A}$* has all true premises but a false conclusion, $\mathcal{A}$* is invalid, in virtue of the Trivial Principle. But $\mathcal{A}$ and $\mathcal{A}$* have the same form, so in virtue of the Fundamental Principle, $\mathcal{A}$ is invalid also.

According to the Principle of Counterexamples, one can demonstrate that an argument is *invalid* by showing that it admits a counterexample. As an example, consider the earlier arguments (a2) and (a3). These are both invalid. To see this, we merely look at the earlier argument (A), and note that it is a counterexample to both (a2) and (a3). Specifically, (A) has the same form as (a2) and (a3), it has all true premises, and it has a false conclusion. Thus, the existence of (A) demonstrates that (a2) and (a3) are *invalid.*

Let us consider two more examples. In each of the following, an invalid argument is given, and a counterexample is given to its right.

(a4)	no cats are dogs no dogs are apes / no cats are apes	(c4)	no men are women no women are fathers / no men are fathers
(a5)	all humans are mammals no humans are reptiles / no mammals are reptiles	(c5)	all men are humans no men are mothers / no humans are mothers

In each case, the argument to the right has the same form as the argument to the left; it also has all true premises and a false conclusion. Thus, it demonstrates the *invalidity* of the argument to the left.

In (a4), as well as in (a5), the premises are true, and so is the conclusion; nevertheless, the conclusion does *not follow* from the premises, and so the argument is invalid. For example, if (a4) were valid, then (c4) would be valid also, since they have exactly the same form. But (c4) is not valid, because it has a false conclusion and all true premises. So, (c4) is not valid either. The same applies to (a5) and (c5).

If all we know about an argument is whether its premises and conclusion are true or false, then usually we cannot say whether the argument is valid or invalid. In fact, there is only one case in which we can say: when the premises are all true, and the conclusion is false, the argument is definitely *invalid* (by the Trivial Principle). However, in all other cases, we cannot say, one way or the other; we need *additional* information about the *form* of the argument.

This is summarized in the following table.

PREMISES	CONCLUSION	VALID OR INVALID?
all true	true	can't tell; need more info
all true	false	**definitely invalid**
not all true	true	can't tell; need more info
not all true	false	can't tell; need more info

9. EXAMPLES OF VALID ARGUMENTS IN SYLLOGISTIC LOGIC

In the previous section, we examined a few examples of invalid arguments in syllogistic logic. In each case of an invalid argument we found a counterexample, which is an argument with the same form, having all true premises but a false conclusion.

In the present section, we examine a few examples of valid syllogistic arguments (also called valid *syllogisms*). At present we have no method to demonstrate that these arguments are in fact valid; this will come in later sections of this chapter.

Note carefully: if we cannot find a counterexample to an argument, it does not mean that no counterexample exists; it might simply mean that we have not looked hard enough. Failure to find a counterexample is not proof that an argument is valid.

Analogously, if I claimed "all swans are white", you could refute me simply by finding a swan that isn't white; this swan would be a counterexample to my claim. On the other hand, if you could not find a non-white swan, I could not thereby say that my claim was proved, only that it was *not disproved yet*.

Thus, although we are going to examine some examples of valid syllogisms, we do not presently have a technique to prove this. For the moment, these merely serve as examples.

The following are all valid syllogistic argument *forms*.

(f1) all X are Y
all Y are Z
/ all X are Z

(f2) all X are Y
some X are Z
/ some Y are Z

(f3) all X are Z
no Y are Z
/ no X are Y

(f4) no X are Y
some Y are Z
/ some Z are not X

To say that (f1)-(f4) are valid argument forms is to say that every argument obtained from them by substitution is a valid argument.

Let us examine the first argument form (f1), since it is by far the simplest to comprehend. Since (f1) is valid, every substitution instance is valid. For example the following arguments are all valid.

(1a)	all cats are mammals	T
	all mammals are vertebrates	T
	/ all cats are vertebrates	T
(1b)	all cats are reptiles	F
	all reptiles are vertebrates	T
	/ all cats are vertebrates	T
(1c)	all cats are animals	T
	all animals are mammals	F
	/ all cats are mammals	T
(1d)	all cats are reptiles	F
	all reptiles are mammals	F
	/ all cats are mammals	T
(1e)	all cats are mammals	T
	all mammals are reptiles	F
	/ all cats are reptiles	F
(1f)	all cats are reptiles	F
	all reptiles are cold-blooded	T
	/ all cats are cold-blooded	F
(1g)	all cats are dogs	F
	all dogs are reptiles	F
	/ all cats are reptiles	F
(1h)	all Martians are reptiles	?
	all reptiles are vertebrates	T
	/ all Martians are vertebrates	?

In the above examples, a number of possibilities are exemplified. It is possible for a valid argument to have all true premises and a true conclusion – (1a); it is possible for a valid argument to have some false premises and a true conclusion – (1b)-(1c); it is possible for a valid argument to have all false premises and a true conclusion – (1d); it is possible for a valid argument to have all false premises and a false conclusion – (1g).

On the other hand, it is not possible for a valid argument to have all true premises and a false conclusion – no example of this.

In the case of argument (1h), we don't know whether the first premise is true or whether it is false. Nonetheless, the argument is valid; that is, if the first premise *were* true, then the conclusion *would necessarily* also be true, since the second premise is true.

The truth or falsity of the premises and conclusion of an argument is not crucial to the validity of the argument. To say that an argument is valid is simply to say that the conclusion *follows from* the premises.

The truth or falsity of the premises and conclusion may not even arise, as for example in a fictional story. Suppose I write a science fiction story, and suppose this story involves various classes of people (human or otherwise!), among them being Gargatrons and Dacrons. Suppose I say the following about these two classes.

(1) all Dacrons are thieves
(2) no Gargatrons are thieves

(the latter is equivalent to: no thieves are Gargatrons).

What could the reader immediately conclude about the relation between Dacrons and Gargatrons?

(3) no Dacrons are Gargatrons (or: no Gargatrons are Dacrons)

I (the writer) would not have to say this explicitly for it to be true in my story; I would not have to say it for you (the reader) to know that it is true in my story; it follows from other things already stated. Furthermore, if I (the writer) were to introduce a character in a later chapter call it Persimion (unknown gender!), and if I were to say that Persimion is both a Dacron and a Gargatron, then I would be guilty of *logical inconsistency* in the story.

I would be guilty of inconsistency, because it is not possible for the first two statements above to be true without the third statement also being true. The third statement *follows from* the first two. There is no world (real or imaginary) in which the first two statements are true, but the third statement is false.

Thus, we can say that statement (3) follows from statements (1) and (2) without having any idea whether they are true or false. All we know is that in any world (real or imaginary), if (1) and (2) are true, then (3) must also be true.

Note that the argument from (1) and (2) to (3) has the form (F3) from the beginning of this section.

10. EXERCISES FOR CHAPTER 1

EXERCISE SET A

For each of the following say whether the statement is true (T) or false (F).

1. In any valid argument, the premises are all true.
2. In any valid argument, the conclusion is true.
3. In any valid argument, if the premises are all true, then the conclusion is also true.
4. In any factually correct argument, the premises are all true.
5. In any factually correct argument, the conclusion is true.
6. In any sound argument, the premises are all true.
7. In a sound argument the conclusion is true.
8. Every sound argument is factually correct.
9. Every sound argument is valid.
10. Every factually correct argument is valid.
11. Every factually correct argument is sound.
12. Every valid argument is factually correct.
13. Every valid argument is sound.
14. Every valid argument has a true conclusion.
15. Every factually correct argument has a true conclusion.
16. Every sound argument has a true conclusion.
17. If an argument is valid and has a false conclusion, then it must have at least one false premise.
18. If an argument is valid and has a true conclusion, then it must have all true premises.
19. If an argument is valid and has at least one false premise then its conclusion must be false.
20. If an argument is valid and has all true premises, then its conclusion must be true.

EXERCISE SET B

In each of the following, you are given an argument to analyze. In each case, answer the following questions.

(1) Is the argument **factually correct**?
(2) Is the argument **valid**?
(3) Is the argument **sound**?

Note that in many cases, the answer might legitimately be "can't tell". For example, in certain cases in which one does not know whether the premises are true or false, one cannot decide whether the argument is factually correct, and hence on cannot decide whether the argument is sound.

1. all dogs are reptiles
all reptiles are Martians
/ all dogs are Martians

2. some dogs are cats
all cats are felines
/ some dogs are felines

3. all dogs are Republicans
some dogs are flea-bags
/ some Republicans are flea-bags

4. all dogs are Republicans
some Republicans are flea-bags
/ some dogs are flea-bags

5. some cats are pets
some pets are dogs
/ some cats are dogs

6. all cats are mammals
all dogs are mammals
/ all cats are dogs

7. all lizards are reptiles
no reptiles are warm-blooded
/ no lizards are warm-blooded

8. all dogs are reptiles
no reptiles are warm-blooded
/ no dogs are warm-blooded

9. no cats are dogs
no dogs are cows
/ no cats are cows

10. no cats are dogs
some dogs are pets
/ some pets are not cats

11. only dogs are pets
some cats are pets
/ some cats are dogs

12. only bullfighters are macho
Max is macho
/ Max is a bullfighter

13. only bullfighters are macho
Max is a bullfighter
/ Max is macho

14. food containing DDT is dangerous
everything I cook is dangerous
/ everything I cook contains DDT

15. the only dogs I like are collies
Sean is a dog I like
/ Sean is a collie

16. the only people still working these exercises are masochists
I am still working on these exercises
/ I am a masochist

EXERCISE SET C

In the following, you are given several syllogistic arguments (some valid, some invalid). In each case, attempt to construct a counterexample. A valid argument does not admit a counterexample, so in some cases, you will not be able to construct a counterexample.

1. all dogs are reptiles
   all reptiles are Martians
   / all dogs are Martians

2. all dogs are mammals
   some mammals are pets
   / some dogs are pets

3. all ducks waddle
   nothing that waddles is graceful
   / no duck is graceful

4. all cows are eligible voters
   some cows are stupid
   / some eligible voters are stupid

5. all birds can fly
   some mammals can fly
   / some birds are mammals

6. all cats are vertebrates
   all mammals are vertebrates
   / all cats are mammals

7. all dogs are Republicans
   some Republicans are flea-bags
   / some dogs are flea-bags

8. all turtles are reptiles
   no turtles are warm-blooded
   / no reptiles are warm-blooded

9. no dogs are cats
   no cats are apes
   / no dogs are apes

10. no mammals are cold-blooded
    some lizards are cold-blooded
    / some mammals are not lizards

11. ANSWERS TO EXERCISES FOR CHAPTER 1

EXERCISE SET A

1. False
2. False
3. True
4. True
5. False
6. True
7. True
8. True
9. True
10. False
11. False
12. False
13. False
14. False
15. False
16. True
17. True
18. False
19. False
20. True

EXERCISE SET B

1. factually correct? NO
   valid? YES
   sound? NO

2. factually correct? NO
   valid? YES
   sound? NO

3. factually correct? NO
   valid? YES
   sound? NO

4. factually correct? NO
   valid? NO
   sound? NO

5. factually correct? YES
   valid? NO
   sound? NO

6. factually correct? YES
   valid? NO
   sound? NO

7. factually correct? YES
   valid? YES
   sound? YES

8. factually correct? NO
   valid? YES
   sound? NO

9. factually correct? YES
   valid? NO
   sound? NO

10. factually correct? YES
    valid? YES
    sound? YES

11. factually correct? NO
    valid? YES
    sound? NO

12. factually correct? NO
    valid? YES
    sound? NO

13. factually correct? NO
    valid? NO
    sound? NO

14. factually correct? can't tell
    valid? NO
    sound? NO

15. factually correct? can't tell
    valid? YES
    sound? can't tell

16. factually correct? can't tell
    valid? YES
    sound? can't tell

EXERCISE SET C

	Original Argument	Counterexample
1.	all dogs are reptiles all reptiles are Martians / all dogs are Martians	valid; admits no counterexample
2.	all dogs are mammals some mammals are pets / some dogs are pets	all dogs are mammals some mammals are cats / some dogs are cats
3.	all ducks waddle nothing that waddles is graceful / no duck is graceful	valid; admits no counterexample
4.	all cows are eligible voters some cows are stupid / some eligible voters are stupid	valid; admits no counterexample
5.	all birds can fly some mammals can fly / some birds are mammals	all birds lay eggs some mammals lay eggs (the platypus) / some birds are mammals
6.	all cats are vertebrates all mammals are vertebrates / all cats are mammals	all cats are vertebrates all reptiles are vertebrates / all cats are reptiles
7.	all dogs are Republicans some Republicans are flea-bags / some dogs are flea-bags	all dogs are mammals some mammals are cats / some dogs are cats
8.	all turtles are reptiles no turtles are warm-blooded / no reptiles are warm-blooded	all turtles are reptiles no turtles are lizards / no reptiles are lizards
9.	no dogs are cats no cats are apes / no dogs are apes	no dogs are cats no cats are poodles / no dogs are poodles
10.	no mammals are cold-blooded some lizards are cold-blooded / some mammals are not lizards	no mammals are cold-blooded some vertebrates are cold-blooded / some mammals are not vertebrates

2 TRUTH FUNCTIONAL CONNECTIVES

1. INTRODUCTION

As noted earlier, an argument is valid or invalid purely in virtue of its form. The form of an argument is a function of the arrangement of the terms in the argument, where the logical terms play a primary role. However, as noted earlier, what counts as a logical term, as opposed to a descriptive term, is not absolute. Rather, it depends upon the level of logical analysis we are pursuing.

In the previous chapter we briefly examined one level of logical analysis, the level of syllogistic logic. In syllogistic logic, the logical terms include 'all', 'some', 'no', 'are', and 'not', and the descriptive terms are all expressions that denote classes.

In the next few chapters, we examine a different branch of logic, which represents a different level of logical analysis; specifically, we examine sentential logic (also called propositional logic and statement logic). In sentential logic, the logical terms are truth-functional statement connectives, and nothing else.

2. STATEMENT CONNECTIVES

We begin by defining statement connective, or what we will simply call a connective.

> A **(statement)** ***connective*** is an expression with one or more blanks (places) such that, whenever the blanks are filled by statements the resulting expression is also a statement.

In other words, a (statement) connective takes one or more smaller statements and forms a larger statement. The following is a simple example of a connective.

__________ and __________

To say that this expression is a connective is to say that if we fill each blank with a statement then we obtain another statement. The following are examples of statements obtained in this manner.

(e1) snow is white **and** grass is green
(e2) all cats are felines **and** some felines are not cats
(e3) it is raining **and** it is sleeting

Notice that the blanks are filled with statements and the resulting expressions are also statements.

The following are further examples of connectives, which are followed by particular instances.

(c1) it is not true that __________________
(c2) the president believes that ___________
(c3) it is necessarily true that ____________

(c4) __________ or __________
(c5) if __________ then __________
(c6) __________ only if __________
(c7) __________ unless __________

(c8) __________ if __________; otherwise __________
(c9) __________ unless __________ in which case __________

(i1) it is not true that <u>all felines are cats</u>
(i2) the president believes that <u>snow is white</u>
(i3) it is necessarily true that <u>2+2=4</u>

(i4) <u>it is raining</u> or <u>it is sleeting</u>
(i5) if <u>it is raining</u> then <u>it is cloudy</u>
(i6) <u>I will pass</u> only if <u>I study</u>

(i7) <u>I will play tennis</u> unless <u>it rains</u>
(i8) <u>I will play tennis</u> if <u>it is warm</u>; otherwise <u>I will play racquetball</u>
(i9) <u>I will play tennis</u> unless <u>it rains</u> in which case <u>I will play squash</u>

Notice that the above examples are divided into three groups, according to how many blanks (places) are involved. This grouping corresponds to the following series of definitions.

> A ***one-place*** **connective** is a connective with one blank.

> A ***two-place*** **connective** is a connective with two blanks.

> A ***three-place*** **connective** is a connective with three blanks.

etc.

At this point, it is useful to introduce a further pair of definitions.

> A ***compound statement*** is a statement that is constructed from one or more smaller statements by the application of a statement connective.

> A ***simple statement*** is a statement that is *not* constructed out of smaller statements by the application of a statement connective.

We have already seen many examples of compound statements. The following are examples of *simple statements*.

(s1) snow is white
(s2) grass is green
(s3) I am hungry
(s4) it is raining
(s5) all cats are felines
(s6) some cats are pets

Note that, from the viewpoint of sentential logic, all statements in syllogistic logic are simple statements, which is to say that they are regarded by sentential logic as having no internal structure.

In all the examples we have considered so far, the constituent statements are all simple statements. A connective can also be applied to compound statements, as illustrated in the following example.

it is not true that all swans are white,
and
the president believes that all swans are white

In this example, the two-place connective '...and...' connects the following two statements,

it is not true that all swans are white

the president believes that all swans are white

which are themselves compound statements. Thus, in this example, there are three connectives involved:

it is not true that...

...and...

the president believes that...

The above statement can in turn be used to form an even larger compound statement. For example, we combine it with the following (simple) statement, using the two-place connective 'if...then...'.

the president is fallible

We accordingly obtain the following compound statement.

IF it is not true that all swans are white,
AND the president believes that all swans are white,
THEN the president is fallible

There is no *theoretical* limit on the complexity of compound statements constructed using statement connectives; *in principle*, we can form compound statements that are as long as we please (say a billion miles long!). However, there are *practical* limits to the complexity of compound statements, due to the limitation of space and time, and the limitation of human minds to comprehend excessively long and complex statements. For example, I doubt very seriously whether any human can understand a statement that is a billion miles long (or even one mile long!) However, this is a practical limit, not a theoretical limit.

By way of concluding this section, we introduce terminology that is often used in sentential logic.

> Simple statements are often referred to as ***atomic statements***, or simply **atoms**.
>
> Compound statements are often referred to as ***molecular statements***, or simply **molecules**.

The analogy, obviously, is with chemistry. Whereas *chemical* atoms (hydrogen, oxygen, etc.) are the smallest *chemical* units, *sentential* atoms are the smallest *sentential* units. The analogy continues. Although the word 'atom' literally means "that which is *indivisible*" or "that which has *no parts*", we know that the chemical atoms do have parts (neutrons, protons, etc.); however, these parts are not chemical in nature. Similarly, atomic sentences have parts, but these parts are not sentential in nature. These further (sub-atomic) parts are the topic of later chapters, on predicate logic.

3. TRUTH-FUNCTIONAL STATEMENT CONNECTIVES

In the previous section, we examined the *general* class of (statement) connectives. At the level we wish to pursue, sentential logic is not concerned with *all* connectives, but only special ones – namely, the *truth-functional* connectives.

Recall that a statement is a sentence that, when uttered, is either true or false. In logic it is customary to refer to *truth* and *falsity* as *truth values*, which are respectively abbreviated T and F. Furthermore, if a statement is true, then we say its *truth value* is T, and if a statement is false, then we say that its *truth value* is F. This is summarized as follows.

> The **truth value** of a true statement is **T**.

> The **truth value** of a false statement is **F**.

The truth value of a statement (say, 'it is raining') is analogous to the weight of a person. Just as we can say that the weight of John is 150 pounds, we can say that the truth value of 'it is raining' is T. Also, John's weight can *vary* from day to day; one day it might be 150 pounds; another day it might be 152 pounds. Similarly, for some statements at least, such as 'it is raining', the truth value can vary from occasion to occasion. On one occasion, the truth value of 'it is raining' might be T; on another occasion, it might be F. The difference between weight and truth-value is quantitative: whereas weight can take infinitely many values (the positive real numbers), truth value can only take two values, T and F.

The analogy continues. Just as we can apply *functions* to numbers (addition, subtraction, exponentiation, etc.), we can apply functions to truth values. Whereas the former are *numerical* functions, the latter are *truth-functions*.

In the case of a numerical function, like addition, the input are numbers, and so is the output. For example, if we input the numbers 2 and 3, then the output is 5. If we want to learn the addition function, we have to learn what the output number is for any two input numbers. Usually we learn a tiny fragment of this in elementary school when we learn the addition tables. The addition tables tabulate the output of the addition function for a few select inputs, and we learn it primarily by rote.

Truth-functions do not take numbers as input, nor do they produce numbers as output. Rather, truth-functions take truth values as input, and they produce truth values as output. Since there are only two truth values (compared with infinitely many numbers), learning a truth-function is considerably simpler than learning a numerical function.

Just as there are two ways to learn, and to remember, the addition tables, there are two ways to learn truth-function tables. On the one hand, you can simply memorize it (two plus two is four, two plus three is five, etc.) On the other hand, you can master the underlying concept (what are you doing when you add two numbers together?) The best way is probably a combination of these two techniques.

We will discuss several examples of truth functions in the following sections. For the moment, let's look at the definition of a truth-functional connective.

> A statement connective is ***truth-functional*** if and only if the truth value of *any* compound statement obtained by applying that connective is a function of (is completely determined by) the individual truth values of the constituent statements that form the compound.

This definition will be easier to comprehend after a few examples have been discussed. The basic idea is this: suppose we have a statement connective, call it +, and suppose we have any two statements, call them S_1 and S_2. Then we can form a compound, which is denoted S_1+S_2. Now, to say that the connective + is truth-functional is to say this: if we know the truth values of S_1 and S_2 individually, then we *automatically know*, or at least we can *compute*, the truth value of S_1+S_2. On the other hand, to say that the connective + is *not* truth-functional is to say this: merely knowing the truth values of S_1 and S_2 does not automatically tell us the truth value of S_1+S_2. An example of a connective that is *not* truth-functional is discussed later.

4. CONJUNCTION

The first truth-functional connective we discuss is conjunction, which corresponds to the English expression 'and'.

[Note: In traditional grammar, the word 'conjunction' is used to refer to *any* two-place statement connective. However, in logic, the word 'conjunction' refers exclusively to one connective – 'and'.]

Conjunction is a two-place connective. In other words, if we have two statements (simple or compound), we can form a compound statement by combining them with 'and'. Thus, for example, we can combine the following two statements

it is raining
it is sleeting

to form the compound statement

it is raining **and** it is sleeting.

In order to aid our analysis of logical form in sentential logic, we employ various symbolic devices. First, we abbreviate *simple* statements by upper case Roman letters. The letter we choose will usually be suggestive of the statement that is abbreviated; for example, we might use 'R' to abbreviate 'it is raining', and 'S' to abbreviate 'it is sleeting'.

Second, we use special symbols to abbreviate (truth-functional) connectives. For example, we abbreviate conjunction ('and') by the ampersand sign ('&'). Putting these abbreviations together, we abbreviate the above compound as follows.

R & S

Finally, we use parentheses to punctuate compound statements, in a manner similar to arithmetic. We discuss this later.

A word about terminology, R&S is called a *conjunction*. More specifically, R&S is called *the conjunction of* R and S, which individually are called *conjuncts*. By analogy, in arithmetic, x+y is called the *sum of* x and y, and x and y are individually called *summands*.

Conjunction is a truth-functional connective. This means that if we know the truth value of each conjunct, we can simply compute the truth value of the conjunction. Consider the simple statements R and S. Individually, these can be true or false, so in combination, there are four cases, given in the following table.

	R	S
case 1	T	T
case 2	T	F
case 3	F	T
case 4	F	F

In the first case, both statements are true; in the fourth case, both statements are false; in the second and third cases, one is true, the other is false.

Now consider the conjunction formed out of these two statements: R&S. What is the truth value of R&S in each of the above cases? Well, it seems plausible that the conjunction R&S is true *if* both the conjuncts are true individually, and R&S is false if either conjunct is false. This is summarized in the following table.

	R	S	R&S
case 1	T	T	T
case 2	T	F	F
case 3	F	T	F
case 4	F	F	F

The information contained in this table readily generalizes. We do not have to regard 'R' and 'S' as standing for *specific* statements. They can stand for any statements whatsoever, and this table still holds. No matter what R and S are specifically, if they are both true (case 1), then the conjunction R&S is also true, but if one or both are false (cases 2-4), then the conjunction R&S is false.

We can summarize this information in a number of ways. For example, each of the following statements summarizes the table in more or less ordinary English. Here, $\mathcal{A}$ and $\mathcal{B}$ stand for arbitrary statements.

> A conjunction $\mathcal{A}$&$\mathcal{B}$ is true
> if and only if
> both conjuncts are true.

> A conjunction $\mathcal{A}$&$\mathcal{B}$ is true if both conjuncts are true;
> otherwise, it is false.

We can also display the truth function for conjunction in a number of ways. The following three tables present the truth function for conjunction; they are followed by three corresponding tables for multiplication.

$\mathcal{A}$	$\mathcal{B}$	$\mathcal{A}$&$\mathcal{B}$
T	T	T
T	F	F
F	T	F
F	F	F

$\mathcal{A}$	&	$\mathcal{B}$
T	T	T
T	F	F
F	F	T
F	F	F

&	T	F
T	T	F
F	F	F

a	b	a×b
1	1	1
1	0	0
0	1	0
0	0	0

a	×	b
1	1	1
1	0	0
0	0	1
0	0	0

×	1	0
1	1	0
0	0	0

Note: The middle table is obtained from the first table simply by superimposing the three columns of the first table. Thus, in the middle table, the truth values of $\mathcal{A}$ are all under the $\mathcal{A}$, the truth values of $\mathcal{B}$ are under the $\mathcal{B}$, and the truth values of $\mathcal{A}\&\mathcal{B}$ are the &. Notice, also, that the final (output) column is also shaded, to help distinguish it from the input columns. This method saves much space, which is important later.

We can also express the content of these tables in a series of statements, just like we did in elementary school. The conjunction truth function may be conveyed by the following series of statements. Compare them with the corresponding statements concerning multiplication.

(1)	T & T = T	$1 \times 1 = 1$
(2)	T & F = F	$1 \times 0 = 0$
(3)	F & T = F	$0 \times 1 = 0$
(4)	F & F = F	$0 \times 0 = 0$

For example, the first statement may be read "T ampersand T is T" (analogously, "one times one is one"). These phrases may simply be memorized, but it is better to understand what they are about – namely, conjunctions.

5. DISJUNCTION

The second truth-functional connective we consider is called *disjunction*, which corresponds roughly to the English 'or'. Like conjunction, disjunction is a two-place connective: given any two statements S_1 and S_2, we can form the compound statement 'S_1 or S_2'. For example, beginning with the following simple statements,

(s1)	it is raining	R
(s2)	it is sleeting	S

we can form the following compound statement.

(c) it is raining **or** it is sleeting $R \vee S$

The symbol for disjunction is '∨' (wedge). Just as R&S is called the *conjunction* of R and S, R∨S is called the *disjunction* of R and S. Similarly, just as the constituents of a conjunction are called *conjuncts*, the constituents of a disjunction are called *disjuncts*.

In English, the word 'or' has at least two different meanings, or senses, which are respectively called the **exclusive sense** and the **inclusive sense**. The exclusive sense is typified by the following sentences.

(e1) would you like a baked potato, OR French fries
(e2) would you like squash, OR beans

In answering these questions, you cannot choose both disjuncts; choosing one disjunct *excludes* choosing the other disjunct.

On the other hand, the inclusive sense of disjunction is typified by the following sentences.

(i1) would you like coffee or dessert
(i2) would you like cream or sugar with your coffee

In answering these questions, you can choose both disjuncts; choosing one disjunct does not exclude choosing the other disjunct as well.

Latin has two different disjunctive words, 'vel' (inclusive) and 'aut' (exclusive). By contrast, English simply has one word 'or', which does double duty. This problem has led the legal profession to invent the expression 'and/or' to use when inclusive disjunction is intended. By using 'and/or' they are able to avoid ambiguity in legal contracts.

In logic, the *inclusive sense* of 'or' (the sense of 'vel' or 'and/or') is taken as basic; it is symbolized by wedge '$\vee$' (suggestive of 'v', the initial letter of 'vel'). The truth table for $\vee$ is given as follows.

$\mathcal{A}$	$\mathcal{B}$	$\mathcal{A}\vee\mathcal{B}$
T	T	T
T	F	T
F	T	T
F	F	F

$\mathcal{A}$	$\vee$	$\mathcal{B}$
T	T	T
T	T	F
F	T	T
F	F	F

$\vee$	T	F
T	T	T
F	T	F

The information conveyed in these tables can be conveyed in either of the following statements.

> A disjunction $\mathcal{A}\vee\mathcal{B}$ is false
> if and only if
> both disjuncts are false.

> A disjunction $\mathcal{A}\vee\mathcal{B}$ is false if both disjuncts are false;
> otherwise, it is true.

The following is an immediate consequence, which is worth remembering.

> If $\mathcal{A}$ is true, then so is $\mathcal{A}\vee\mathcal{B}$,
> regardless of the truth value of $\mathcal{B}$.

> If $\mathcal{B}$ is true, then so is $\mathcal{A}\vee\mathcal{B}$,
> regardless of the truth value of $\mathcal{A}$.

6. A STATEMENT CONNECTIVE THAT IS NOT TRUTH-FUNCTIONAL

Conjunction (&) and disjunction (∨) are both truth-functional connectives. In the present section, we discuss a connective that is not truth-functional – namely, the connective 'because'.

Like conjunction ('and') and disjunction ('or'), 'because' is a two-place connective; given any two statements S_1 and S_2, we can form the compound statement 'S_1 because S_2'. For example, given the following simple statements

(s1)	I am sad	S
(s2)	it is raining	R

we can form the following compound statements.

(c1)	I am sad **because** it is raining	S because R
(c2)	it is raining **because** I am sad	R because S

The simple statements (s1) and (s2) can be individually true or false, so there are four possible combinations of truth values. The question is, for each combination of truth values, what is the truth value of each resulting compound.

First of all, it seems fairly clear that if either of the simple statements is false, then the compound is false. On the other hand, if both statements are true, then it is not clear what the truth value of the compound is. This is summarized in the following *partial* truth table.

S	R	S because R	R because S
T	T	?	?
T	F	F	F
F	T	F	F
F	F	F	F

In the above table, the question mark (?) indicates that the truth value is unclear.

Suppose both S ('I am sad') and R ('it is raining') are true. What can we say about the truth value of 'S because R' and 'R because S'? Well, at least in the case of

it is raining because I am sad,

we can safely assume that it is false (unless the speaker in question is God, in which case all bets are off).

On the other hand, in the case of

I am sad because it is raining,

we cannot say whether it is true, or whether it is false. Merely knowing that the speaker is sad and that it is raining, we do not know whether the rain is *responsible* for the sadness. It might be, it might not. Merely knowing the individual truth values of S ('I am sad') and R ('it is raining'), we do not automatically know the truth value of the compound 'I am sad because it is raining'; additional information (of a complicated sort) is needed to decide whether the compound is true or false. In other words, 'because' is not a truth-functional connective.

Another way to see that 'because' is *not* truth-functional is to suppose to the contrary that it *is* truth-functional. If it is truth-functional, then we can replace the question mark in the above table. We have only two choices. If we replace '?' by 'T', then the truth table for 'R because S' is identical to the truth table for R&S. This would mean that for any statements $\mathcal{A}$ and $\mathcal{B}$, '$\mathcal{A}$ because $\mathcal{B}$' says no more than '$\mathcal{A}$ and $\mathcal{B}$'. This is absurd, for that would mean that both of the following statements are true.

grass is green *because* 2+2=4
2+2=4 *because* grass is green

Our other choice is to replace '?' by 'F'. This means that the output column consists entirely of F's, which means that '$\mathcal{A}$ because $\mathcal{B}$' is *always false*. This is also absurd, or at least implausible. For surely some statements of the form '$\mathcal{A}$ because $\mathcal{B}$' are true. The following might be considered an example.

grass is green *because* grass contains chlorophyll

7. NEGATION

So far, we have examined three two-place connectives. In the present section, we examine a one-place connective, negation, which corresponds to the word 'not'.

If we wish to deny a statement, for example,

it is raining,

the easiest way is to insert the word 'not' in a strategic location, thus yielding

it is **not** raining.

We can also deny the original statement by prefixing the whole sentence by the modifier

it is not true that

to obtain

it is not true that it is raining

The advantage of the first strategy is that it produces a colloquial sentence. The advantage of the second strategy is that it is simple to apply; one simply prefixes the statement in question by the modifier, and one obtains the denial. Furthermore, the second strategy employs a statement connective. In particular, the expression it is not true that ______________

meets our criterion to be a one-place connective; its single blank can be filled by any statement, and the result is also a statement.

This one-place connective is called *negation*, and is symbolized by '~' (tilde), which is a stylized form of 'n', short for negation. The following are variant negation expressions.

it is false that __________________
it is not the case that ____________

Next, we note that the negation connective (~) is truth-functional. In other words, if we know the truth value of a statement S, then we automatically know the truth value of the negation ~S; the truth value of ~S is simply the opposite of the truth value of S.

This is plausible. For ~S denies what S asserts; so if S is in fact false, then its denial (negation) is true, and if S is in fact true, then its denial is false. This is summarized in the following truth tables.

$\mathcal{A}$	$\sim\mathcal{A}$
T	F
F	T

~	$\mathcal{A}$
F	T
T	F

In the second table, the truth values of $\mathcal{A}$ are placed below the $\mathcal{A}$, and the resulting truth values for $\sim\mathcal{A}$ are placed below the tilde sign (~). The right table is simply a compact version of the left table. Both tables can be summarized in the following statement.

> $\sim\mathcal{A}$ has the opposite truth value of $\mathcal{A}$.

8. THE CONDITIONAL

In the present section, we introduce one of the two remaining truth-functional connectives that are customarily studied in sentential logic – the *conditional connective*, which corresponds to the expression

if __________, then __________.

The conditional connective is a two-place connective, which is to say that we can replace the two blanks in the above expression by any two statements, then the resulting expression is also a statement.

For example, we can take the following simple statements.

(1) I am relaxed
(2) I am happy

and we can form the following conditional statements, using if-then.

(c1) **if** I am relaxed, **then** I am happy
(c2) **if** I am happy, **then** I am relaxed

The symbol used to abbreviate if-then is the arrow (→), so the above compounds can be symbolized as follows.

(s1) R → H
(s2) H → R

Every conditional statement divides into two constituents, which do *not* play equivalent roles (in contrast to conjunction and disjunction). The constituents of a conditional $\mathcal{A} \rightarrow \mathcal{C}$ are respectively called the *antecedent* and the *consequent*. The word 'antecedent' means "that which leads", and the word 'consequent' means "that which follows". In a conditional, the first constituent is called the antecedent, and the second constituent is called the consequent. When a conditional is stated in standard form in English, it is easy to identify the antecedent and the consequent, according to the following rule.

'if' introduces the antecedent
'then' introduces the consequent

The fact that the antecedent and consequent do not play equivalent roles is related to the fact that $\mathcal{A} \rightarrow \mathcal{C}$ is not generally equivalent to $\mathcal{C} \rightarrow \mathcal{A}$. Consider the following two conditionals.

if my car runs out of gas, **then** my car stops R→S

if my car stops, **then** my car runs out of gas S→R

9. THE NON-TRUTH-FUNCTIONAL VERSION OF IF-THEN

In English, if-then is used in a variety of ways, many of which are *not* truth-functional. Consider the following conditional statements.

if I **lived** in L.A., then I **would** live in California

if I **lived** in N.Y.C., then I **would** live in California

The constituents of these two conditionals are given as follows; note that they are individually stated in the indicative mood, as required by English grammar.

L: I live in L.A. (Los Angeles)
N: I live in N.Y.C. (New York City)
C: I live in California

Now, for the author at least, all three simple statements are false. But what about the two conditionals? Well, it seems that the first one is *true*, since L.A. is entirely contained inside California (presently!). On the other hand, it seems that the second one is *false*, since N.Y.C. does not overlap California.

Thus, in the first case, two false constituents yield a *true* conditional, but in the second case, two false constituents yield a *false* conditional. It follows that the conditional connective employed in the above conditionals is *not* truth-functional.

The conditional connective employed above is customarily called the *subjunctive conditional* connective, since the constituent statements are usually stated in the subjunctive mood.

Since subjunctive conditionals are not truth-functional, they are not examined in sentential logic, at least at the introductory level. Rather, what is examined are the *truth functional conditional* connectives.

10. THE TRUTH-FUNCTIONAL VERSION OF IF-THEN

Insofar as we want to have a truth-functional conditional connective, we must construct its truth table. Of course, since not every use of 'if-then' in English is intended to be truth-functional, no truth functional connective is going to be completely plausible. Actually, the problem is to come up with a truth functional version of if-then that is even marginally plausible. Fortunately, there is such a connective.

By way of motivating the truth table for the truth-functional version of 'if-then', we consider *conditional promises* and *conditional requests*. Consider the following promise (made to the intro logic student by the intro logic instructor).

if you get a hundred on every exam, **then** I will give you an A

which may be symbolized

$H \rightarrow A$

Now suppose that the semester ends; under what circumstances has the instructor kept his/her promise. The relevant circumstances may be characterized as follows.

	H	A
case 1:	T	T
case 2:	T	F
case 3:	F	T
case 4:	F	F

The cases divide into two groups. In the first two cases, you get a hundred on every exam; the condition in question is activated; if the condition is activated, the question whether the promise is kept simply reduces to whether you do or don't get an A. In case 1, you get your A; the instructor has kept the promise. In case 2, you don't get your A, even though you got a hundred on every exam; the instructor has not kept the promise.

The remaining two cases are different. In these cases, you don't get a hundred on every exam, so the condition in question isn't activated. We have a choice now about evaluating the promise. We can say that no promise was made, so no obligation was incurred; or, we can say that a promise was made, and it was kept *by default.*

We follow the latter course, which produces the following truth table.

	H	A	H→A
case 1:	T	T	T
case 2:	T	F	F
case 3:	F	T	T
case 4:	F	F	T

Note carefully that in making the above promise, the instructor has not committed him(her)self about your grade when you don't get a hundred on every exam. It is a very simple promise, by itself, and may be combined with other promises. For example, the instructor has not promised *not* to give you an A if you do *not* get a hundred on every exam. Presumably, there are other ways to get an A; for example, a 99% average should also earn an A.

On the basis of these considerations, we propose the following truth table for the arrow connective, which represents the truth-functional version of 'if-then'.

$\mathcal{A}$	$\mathcal{C}$	$\mathcal{A}\to\mathcal{C}$
T	T	T
T	F	F
F	T	T
F	F	T

$\mathcal{A}$	$\to$	$\mathcal{C}$
T	T	T
T	F	F
F	T	T
F	T	F

The information conveyed in the above tables may be summarized by either of the following statements.

> A conditional $\mathcal{A}\to\mathcal{C}$ is false
> if and only if
> the antecedent $\mathcal{A}$ is true
> and the consequent $\mathcal{C}$ is false.

> A conditional $\mathcal{A}\to\mathcal{C}$ is false
> if the antecedent $\mathcal{A}$ is true
> and the consequent $\mathcal{C}$ is false;
> otherwise, it is true.

11. THE BICONDITIONAL

We have now examined four truth-functional connectives, three of which are two-place connectives (conjunction, disjunction, conditional), and one of which is a one-place connective (negation). There is one remaining connective that is generally studied in sentential logic, the biconditional, which corresponds to the English

______________ if and only if ______________

Like the conditional, the biconditional is a two-place connective; if we fill the two blanks with statements, the resulting expression is also a statement. For example, we can begin with the statements

I am happy
I am relaxed

and form the compound statement

I am happy **if and only if** I am relaxed

The symbol for the biconditional connective is '↔', which is called *double arrow*. The above compound can accordingly be symbolized thus.

H ↔ R

H↔R is called the *biconditional* of H and R, which are individually called *constituents*. The truth table for ↔ is quite simple. One can understand a biconditional $\mathcal{A}\leftrightarrow\mathcal{B}$ as saying that the two constituents are equal in truth value; accordingly, $\mathcal{A}\leftrightarrow\mathcal{B}$ is true if $\mathcal{A}$ and $\mathcal{B}$ have the same truth value, and is false if they don't have the same truth value. This is summarized in the following tables.

$\mathcal{A}$	$\mathcal{B}$	$\mathcal{A}\leftrightarrow\mathcal{B}$
T	T	T
T	F	F
F	T	F
F	F	T

$\mathcal{A}$	$\leftrightarrow$	$\mathcal{B}$
T	T	T
T	F	F
F	F	T
F	T	F

The information conveyed in the above tables may be summarized by any of the following statements.

A biconditional $\mathcal{A}\leftrightarrow\mathcal{B}$ is true
if and only if
the constituents $\mathcal{A}$, $\mathcal{B}$ have the same truth value.

A biconditional $\mathcal{A}\leftrightarrow\mathcal{B}$ is false
if and only if
the constituents $\mathcal{A}$, $\mathcal{B}$ have opposite truth values.

A biconditional $\mathcal{A} \leftrightarrow \mathcal{B}$ is true
if its constituents have the same truth value; otherwise,
it is false.

A biconditional $\mathcal{A} \leftrightarrow \mathcal{B}$ is false
if its constituents have opposite truth values; otherwise,
it is true.

12. COMPLEX FORMULAS

As noted in Section 2, a statement connective forms larger (compound) statements out of smaller statements. Now, these smaller statements may themselves be compound statements; that is, they may be constructed out of smaller statements by the application of one or more statement connectives. We have already seen examples of this in Section 2.

Associated with each statement (simple or compound) is a symbolic abbreviation, or translation. Each acceptable symbolic abbreviation is what is customarily called a ***formula***. Basically, a formula is simply a string of symbols that is grammatically acceptable. Any ungrammatical string of symbols is not a formula.

For example, the following strings of symbols are not formulas in sentential logic; they are ungrammatical.

(n1) &∨P(Q
(n2) P&∨Q
(n3) P(∨Q(
(n4))(P&Q

By contrast, the following strings count as formulas in sentential logic.

(f1) (P & Q)
(f2) (~(P & Q) ∨ R)
(f3) ~(P & Q)
(f4) (~(P & Q) ∨ (P & R))
(f5) ~((P & Q) ∨ (P & R))

In order to distinguish grammatical from ungrammatical strings, we provide the following formal definition of formula in sentential logic. In this definition, the script letters stand for strings of symbols. The definition tells us which strings of symbols are formulas of sentential logic, and which strings are not.

(1) any upper case Roman letter is a formula;
(2) if $\mathcal{A}$ is a formula, then so is $\sim\mathcal{A}$;
(3) if $\mathcal{A}$ and $\mathcal{B}$ are formulas, then so is $(\mathcal{A}\ \&\ \mathcal{B})$;
(4) if $\mathcal{A}$ and $\mathcal{B}$ are formulas, then so is $(\mathcal{A} \vee \mathcal{B})$;
(5) if $\mathcal{A}$ and $\mathcal{B}$ are formulas, then so is $(\mathcal{A} \rightarrow \mathcal{B})$;
(6) if $\mathcal{A}$ and $\mathcal{B}$ are formulas, then so is $(\mathcal{A} \leftrightarrow \mathcal{B})$;
(7) nothing else is a formula.

Let us do some examples of this definition. By clause 1, both P and Q are formulas, so by clause 2, the following are both formulas.

~P ~Q

So by clause 3, the following are all formulas.

(P & Q) (P & ~Q) (~P & Q) (~P & ~Q)

Similarly, by clause 4, the following expressions are all formulas.

(P ∨ Q) (P ∨ ~Q) (~P ∨ Q) (~P ∨ ~Q)

We can now apply clause 2 again, thus obtaining the following formulas.

~(P & Q) ~(P & ~Q) ~(~P & Q) ~(~P & ~Q)

~(P ∨ Q) ~(P ∨ ~Q) ~(~P ∨ Q) ~(~P ∨ ~Q)

We can now apply clause 3 to any pair of these formulas, thus obtaining the following *among others*.

((P ∨ Q) & (P ∨ ~Q)) ((P ∨ Q) & ~(P ∨ ~Q))

The process described here can go on *indefinitely*. There is no limit to how long a formula can be, although most formulas are too long for humans to write.

In addition to formulas, in the strict sense, given in the above definition, there are also formulas in a less strict sense. We call these strings *unofficial formulas*. Basically, an unofficial formula is a string of symbols that is obtained from an official formula by dropping the outermost parentheses. This applies only to official formulas that have outermost parenthesis; negations do not have outer parentheses. The following is the official definition of an unofficial formula.

An **unofficial formula** is any string of symbols that is obtained from an official formula by removing its outermost parentheses (if such exist).

We have already seen numerous examples of unofficial formulas in this chapter. For example, we symbolized the sentence

it is raining and it is sleeting

by the expression

R & S

Officially, the latter is not a formula; however, it is an unofficial formula.

The following represent the rough guidelines for dealing with unofficial formulas in sentential logic.

> When a formula stands by itself, one is permitted to drop its outermost parentheses (if such exist), thus obtaining an unofficial formula. However, an unofficial formula cannot be used to form a compound formula. In order to form a compound, one must restore the outermost parentheses, thereby converting the unofficial formula into an official formula.

Thus, the expression 'R & S', which is an unofficial formula, can be used to symbolize 'it is raining and it is sleeting'. On the other hand, if we wish to symbolize the denial of this statement, which is 'it is not both raining and sleeting', then we must first restore the outermost parentheses, and then prefix the resulting expression by '~'. This is summarized as follows.

it is raining and it is sleeting:	R & S
it is not both raining and sleeting:	~(R & S)

13. TRUTH TABLES FOR COMPLEX FORMULAS

There are infinitely many formulas in sentential logic. Nevertheless, no matter how complex a given formula $\mathcal{A}$ is, we can compute its truth value, provided we know the truth values of its constituent atomic formulas. This is because all the connectives used in constructing $\mathcal{A}$ are *truth-functional*. In order to ascertain the truth value of $\mathcal{A}$, we simply compute it starting with the truth values of the atoms, using the truth function tables.

In this respect, at least, sentential logic is exactly like arithmetic. In arithmetic, if we know the *numerical values* assigned to the variables x, y, z, we can routinely calculate the numerical value of any compound arithmetical expression involving these variables. For example, if we know the numerical values of x, y, z, then we can compute the numerical value of $((x+y)\times z)+((x+y)\times(x+z))$. This computation is particularly simple if we have a hand calculator (provided that we know how to enter the numbers in the correct order; some calculators even solve this problem for us).

The only significant difference between sentential logic and arithmetic is that, whereas arithmetic concerns *numerical values* (1,2,3...) and *numerical functions* (+,×, etc.), sentential logic concerns *truth values* (T, F) and truth functions (&, ∨, etc.). Otherwise, the computational process is completely analogous. In particular, one builds up a complex computation on the basis of simple computations, and each simple computation is based on a table (in the case of arithmetic, the tables are stored in calculators, which perform the simple computations).

Let us begin with a simple example of computing the truth value of a complex formula on the basis of the truth values of its atomic constituents. The example we consider is the negation of the conjunction of two simple formulas P and Q, which is the formula ~(P&Q). Now suppose that we substitute T for both P and Q; then we obtain the following expression: ~(T&T). But we know that T&T = T, so ~(T&T) = ~T, but we also know that ~T = F, so ~(T&T) = F; this ends our computation. We can also substitute T for P and F for Q, in which case we have ~(T&F). We know that T&F is F, so ~(T&F) is ~F, but ~F is T, so ~(T&F) is T. There are two other cases: substituting F for P and T for Q, and substituting F for both P and Q. They are computed just like the first two cases. We simply build up the larger computation on the basis of smaller computations.

These computations may be summarized in the following statements.

case 1: ~(T&T) = ~T = F
case 2: ~(T&F) = ~F = T
case 3: ~(F&T) = ~F = T
case 4: ~(F&F) = ~F = T

Another way to convey this information is in the following table.

Table 1

	P	Q	P&Q	~(P&Q)
case 1	T	T	T	F
case 2	T	F	F	T
case 3	F	T	F	T
case 4	F	F	F	T

This table shows the computations step by step. The first two columns are the initial input values for P and Q; the third column is the computation of the truth value of the conjunction (P&Q); the fourth column is the computation of the truth value of the negation ~(P&Q), which uses the third column as input.

Let us consider another simple example of computing the truth value of a complex formula. The formula we consider is a disjunction of (P&Q) and ~P, that is, it is the formula (P&Q)∨~P. As in the previous case, there are just two letters, so there are four combinations of truth values that can be substituted. The computations are compiled as follows, followed by the corresponding table.

case 1: (T&T) ∨ ~T =
T ∨ F = T

case 2: (T&F) ∨ ~T =
F ∨ F = F

case 3: (F&T) ∨ ~F =
F ∨ T = T

case 4: (F&F) ∨ ~F =
F ∨ T = T

By way of explanation, in case 1, the value of T&T is placed below the &, and the value of ~T is placed below the ~. These values in turn are combined by the ∨.

Table 2

	P	Q	P&Q	~P	(P&Q)∨~P
case 1	T	T	T	F	T
case 2	T	F	F	F	F
case 3	F	T	F	T	T
case 4	F	F	F	T	T

Let's now consider the formula that is obtained by conjoining the first formula (Table 1) with the second case formula (Table 2); the resulting formula is: ~(P&Q)&((P&Q)∨~P). Notice that the parentheses have been restored on the second formula before it was conjoined with the first formula. This formula has just two atomic formulas - P and Q - so there are just four cases to consider. The best way to compute the truth value of this large formula is simply to take the output columns of Tables 1 and 2 and combine them according to the conjunction truth table.

Table 3

	~(P&Q)	(P&Q)∨~P	~(P&Q)&((P&Q)∨~P)
case 1	F	T	F
case 2	T	F	F
case 3	T	T	T
case 4	T	T	T

In case 1, for example, the truth value of ~(P&Q) is F, and the truth value of (P&Q) ∨ ~P is T, so the value of their conjunction is F&T, which is F. If we were to construct the table for the complex formula from scratch, we would basically combine Tables 1 and 2. Table 3 represents the last three columns of such a table.

It might be helpful to see the computation of the truth value for ~(P&Q)&((P&Q)∨~P) done in complete detail for the first case. To begin with, we write down the formula, and we then substitute in the truth values for the first case. This yields the following.

~(P & Q) & ((P & Q) ∨ ~P)

case 1: ~(T & T) & ((T & T) ∨ ~T)

The first computation is to calculate T&T, which is T, so that yields

~T & (T ∨ ~T)

The next step is to calculate ~T, which is F, so this yields.

F & (T ∨ F)

Next, we calculate T ∨ F, which is T, which yields.

F & T

Finally, we calculate F&T, which is F, the final result in the computation.

This particular computation can be diagrammed as follows.

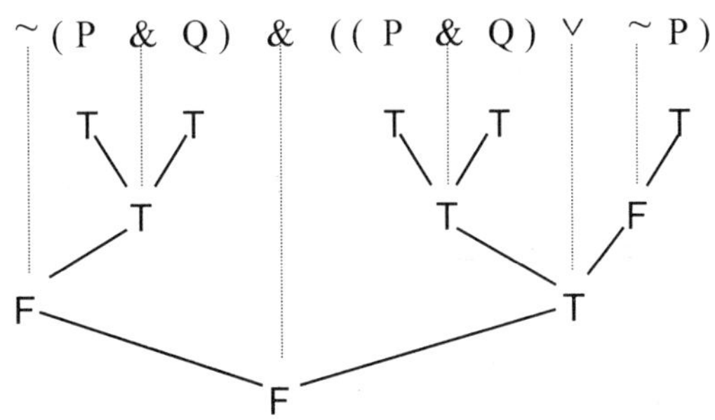

Case 2 can also be done in a similar manner, shown as follows.

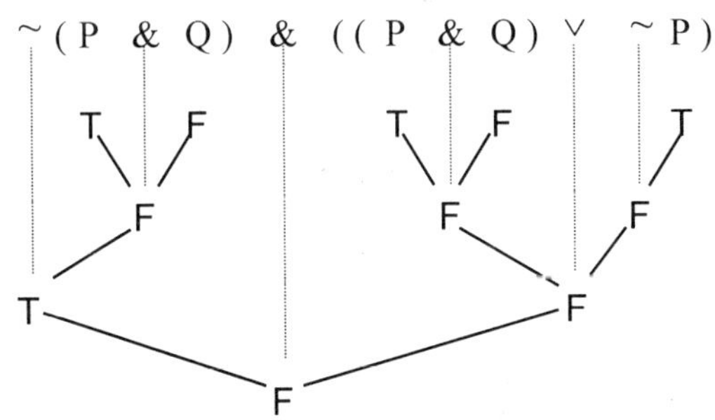

In the above diagrams, the broken lines indicate, in each simple computation, which truth function (connective) is employed, and the solid lines indicate the input values.

In principle, in each complex computation involving truth functions, one can construct a diagram like those above for each case. Unfortunately, however, this takes up a lot of space and time, so it is helpful to have a more compact method of presenting such computations. The method that I propose simply involves superimposing all the lines above into a single line, so that each case can be presented on a single line. This can be illustrated with reference to the formulas we have already discussed.

In the case of the first formula, presented in Table 1, we can present its truth table as follows.

Table 3

	~	(P	&	Q)
case 1	F	T	T	T
case 2	T	T	F	F
case 3	T	F	F	T
case 4	T	F	F	F

In this table, the truth values pertaining to each connective are placed beneath that connective. Thus, for example, in case 1, the first column is the truth value of ~(P&Q), and the third column is the truth value of (P&Q).

We can do the same with Table 2, which yields the following table.

Table 4

	(P	&	Q)	∨	~	P
case 1	T	T	T	T	F	T
case 2	T	F	F	F	F	T
case 3	F	F	T	T	T	F
case 4	F	F	F	T	T	F

In this table, the second column is the truth value of (P&Q), the fourth column is the truth value of the whole formula (P&Q)∨~P, and the fifth column is the truth value of ~P.

Finally, we can do the compact truth table for the conjunction of the formulas given in Tables 3 and 4.

Table 5

	~	(P	&	Q)	&	((P	&	Q)	∨	~	P)
case 1:	F	T	T	T	F	T	T	T	T	F	T
case 2:	T	T	F	F	F	T	F	F	F	F	T
case 3:	T	F	F	T	T	F	F	T	T	T	F
case 4:	T	F	F	F	T	F	F	F	T	T	F
	4		3		5		1		3	2	

The numbers at the bottom of the table indicate the order in which the columns are filled in. In the case of ties, this means that the order is irrelevant to the construction of the table.

In constructing compact truth tables, or in computing complex formulas, the following rules are useful to remember.

DO CONNECTIVES THAT ARE DEEPER BEFORE DOING CONNECTIVES THAT ARE LESS DEEP.

Here, the depth of a connective is determined by how many pairs of parentheses it is inside; a connective that is inside two pairs of parentheses is deeper than one that is inside of just one pair.

AT ANY PARTICULAR DEPTH, ALWAYS DO NEGATIONS FIRST.

These rules are applied in the above table, as indicated by the numbers at the bottom.

Before concluding this section, let us do an example of a formula that contains three atomic formulas P, Q, R. In this case, there are 8 combinations of truth values that can be assigned to the letters. These combinations are given in the following guide table.

Guide Table for any Formula Involving 3 Atomic Formulas

	P	Q	R
case 1	T	T	T
case 2	T	T	F
case 3	T	F	T
case 4	T	F	F
case 5	F	T	T
case 6	F	T	F
case 7	F	F	T
case 8	F	F	F

There are numerous ways of writing down all the combinations of truth values; this is just one particular one. The basic rule in constructing this guide table is that the rightmost column (R) is alternated T and F singly, the middle column (Q) is alternated T and F in doublets, and the leftmost column (P) is alternated T and F in quadruplets. It is simply a way of remembering all the cases.

Now let's consider a formula involving three letters P, Q, R, and its associated (compact) truth table.

Table 6

			1		2	3	4	5		6		7	8	9	10	
P	Q	R	~	[(	P	&	~	Q	)	∨	(	~	P	∨	R	)]
T	T	T	F		T	F	F	T		T		F	T	T	T	
T	T	F	T		T	F	F	T		F		F	T	F	F	
T	F	T	F		T	T	T	F		T		F	T	T	T	
T	F	F	F		T	T	T	F		T		F	T	F	F	
F	T	T	F		F	F	F	T		T		T	F	T	T	
F	T	F	F		F	F	F	T		T		T	F	T	F	
F	F	T	F		F	F	T	F		T		T	F	T	T	
F	F	F	F		F	F	T	F		T		T	F	T	F	
			5		1	3	2	1		4		2	1	3	1	

The guide table is not required, but is convenient, and is filled in first. The remaining columns, numbered 1-10 at the top, completed in the order indicated at the bottom. In the case of ties, the order doesn't matter.

In filling a truth table, it is best to understand the structure of the formula. In case of the above formula, it is a *negation*; in particular it is the negation of the formula (P&~Q)∨(~P∨R). This formula is a *disjunction*, where the individual disjuncts are P&~Q and P∨R respectively. The first disjunct P&~Q is a conjunction of P and the negation of Q; the second disjunct ~P∨R is a disjunction of ~P and R.

The structure of the formula is crucial, and is intimately related to the order in which the truth table is filled in. In particular, the order in which the table is filled in is exactly opposite from the order in which the formula is broken into its constituent parts, as we have just done.

In filling in the above table, the first thing we do is fill in three columns under the letters, which are the smallest parts; these are labeled 1 at the bottom. Next, we do the negations of letters, which corresponds to columns 4 and 7, but *not* column 1. Column 4 is constructed from column 5 on the basis of the tilde truth table, and column 7 is constructed from column 8 in a like manner. Next column 3 is constructed from columns 2 and 4 according to the ampersand truth table, and column 9 is constructed from columns 7 and 10 according to the wedge truth table. These two resulting columns, 3 and 9, in turn go into constructing column 6 according to the wedge truth table. Finally, column 6 is used to construct column 1 in accordance with the negation truth table.

The first two cases are diagrammed in greater detail below.

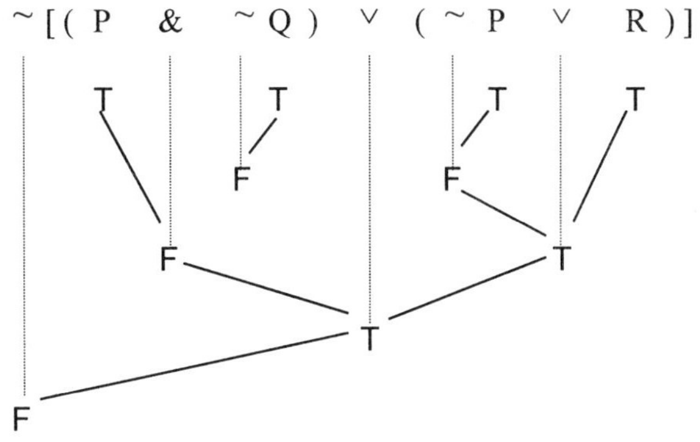

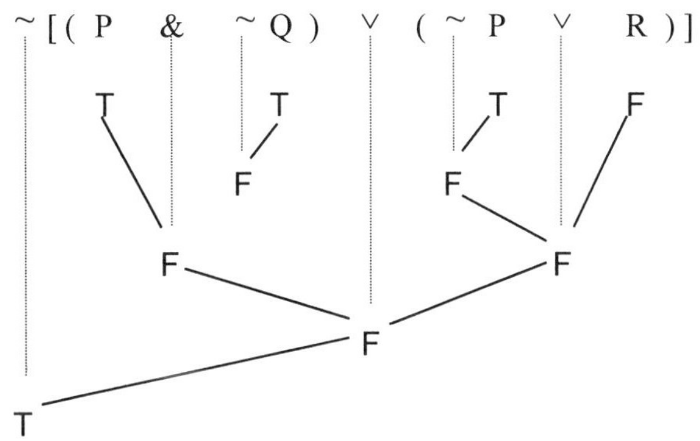

As in our previous example, the broken lines indicate which truth function is applied, and the solid lines indicate the particular input values, and output values.

14. EXERCISES FOR CHAPTER 2

EXERCISE SET A

Compute the truth values of the following symbolic statements, supposing that the truth value of A, B, C is T, and the truth value of X, Y, Z is F.

1. ~A ∨ B
2. ~B ∨ X
3. ~Y ∨ C
4. ~Z ∨ X
5. (A & X) ∨ (B & Y)
6. (B & C) ∨ (Y & Z)
7. ~(C & Y) ∨ (A & Z)
8. ~(A & B) ∨ (X & Y)
9. ~(X & Z) ∨ (B & C)
10. ~(X & ~Y) ∨ (B & ~C)
11. (A ∨ X) & (Y ∨ B)
12. (B ∨ C) & (Y ∨ Z)
13. (X ∨ Y) & (X ∨ Z)
14. ~(A ∨ Y) & (B ∨ X)
15. ~(X ∨ Z) & (~X ∨ Z)
16. ~(A ∨ C) ∨ ~(X & ~Y)
17. ~(B ∨ Z) & ~(X ∨ ~Y)
18. ~[(A ∨ ~C) ∨ (C ∨ ~A)]
19. ~[(B & C) & ~(C &B)]
20. ~[(A & B) ∨ ~(B & A)]
21. [A ∨ (B ∨ C)] & ~[(A ∨ B) ∨ C]
22. [X ∨ (Y & Z)] ∨ ~[(X ∨ Y) & (X ∨ Z)]
23. [A & (B ∨ C)] & ~[(A & B) ∨ (A & C)]
24. ~{[(~A & B) & (~X & Z)] & ~[(A & ~B) ∨ ~(~Y & ~Z)]}
25. ~{~[(B & ~C) ∨ (Y & ~Z)] & [(~B ∨ X) ∨ (B ∨ ~Y)]}

EXERCISE SET B

Compute the truth values of the following symbolic statements, supposing that the truth value of A, B, C is T, and the truth value of X, Y, Z is F.

1. $A \rightarrow B$
2. $A \rightarrow X$
3. $B \rightarrow Y$
4. $Y \rightarrow Z$
5. $(A \rightarrow B) \rightarrow Z$
6. $(X \rightarrow Y) \rightarrow Z$
7. $(A \rightarrow B) \rightarrow C$
8. $(X \rightarrow Y) \rightarrow C$
9. $A \rightarrow (B \rightarrow Z)$
10. $X \rightarrow (Y \rightarrow Z)$
11. $[(A \rightarrow B) \rightarrow C] \rightarrow Z$
12. $[(A \rightarrow X) \rightarrow Y] \rightarrow Z$
13. $[A \rightarrow (X \rightarrow Y)] \rightarrow C$
14. $[A \rightarrow (B \rightarrow Y)] \rightarrow X$
15. $[(X \rightarrow Z) \rightarrow C] \rightarrow Y$
16. $[(Y \rightarrow B) \rightarrow Y] \rightarrow Y$
17. $[(A \rightarrow Y) \rightarrow B] \rightarrow Z$
18. $[(A \& X) \rightarrow C] \rightarrow [(X \rightarrow C) \rightarrow X]$
19. $[(A \& X) \rightarrow C] \rightarrow [(A \rightarrow X) \rightarrow C]$
20. $[(A \& X) \rightarrow Y] \rightarrow [(X \rightarrow A) \rightarrow (A \rightarrow Y)]$
21. $[(A \& X) \vee ({\sim}A \& {\sim}X)] \rightarrow [(A \rightarrow X) \& (X \rightarrow A)]$
22. $\{[A \rightarrow (B \rightarrow C)] \rightarrow [(A \& B) \rightarrow C]\} \rightarrow [(Y \rightarrow B) \rightarrow (C \rightarrow Z)]$
23. $\{[(X \rightarrow Y) \rightarrow Z] \rightarrow [Z \rightarrow (X \rightarrow Y)]\} \rightarrow [(X \rightarrow Z) \rightarrow Y]$
24. $[(A \& X) \rightarrow Y] \rightarrow [(A \rightarrow X) \& (A \rightarrow Y)]$
25. $[A \rightarrow (X \& Y)] \rightarrow [(A \rightarrow X) \vee (A \rightarrow Y)]$

EXERCISE SET C

Construct the complete truth table for each of the following formulas.

1. (P & Q) ∨ (P & ~Q)
2. ~(P & ~P)
3. ~(P ∨ ~P)
4. ~(P&Q)∨(~P∨~Q)
5. ~(P ∨ Q) ∨ (~P & ~Q)
6. (P & Q) ∨ (~P & ~Q)
7. ~(P ∨ (P & Q))
8. ~(P ∨ (P & Q)) ∨ P
9. (P & (Q ∨ P)) & ~P
10. ((P → Q) → P) → P
11. ~(~(P → Q) → P)
12. (P → Q) ↔ ~P
13. P → (Q → (P & Q))
14. (P ∨ Q) ↔ (~P → Q)
15. ~(P ∨ (P → Q))
16. (P → Q) ↔ (Q → P)
17. (P → Q) ↔ (~Q → ~P)
18. (P ∨ Q) → (P & Q)
19. (P & Q) ∨ (P & R)
20. [P → (Q & R)] → [P → R]
21. [P → (Q ∨ R)] → [P → Q]
22. [(P ∨ Q) → R] → [P → R]
23. [(P & Q) → R] → [P → R]
24. [(P & Q) → R] → [(Q & ~R) → ~P]
25. [P ↔ (Q ↔ R)] ↔ [(P ↔ Q) ↔ R]

15. ANSWERS TO EXERCISES FOR CHAPTER 2

EXERCISE SET A

1. T
2. F
3. T
4. T
5. F
6. F
7. T
8. F
9. T
10. T
11. T
12. F
13. F
14. F
15. T
16. T
17. F
18. F
19. T
20. F
21. F
22. T
23. F
24. T
25. F

EXERCISE SET B

1. T
2. F
3. F
4. T
5. F
6. T
7. T
8. T
9. F
10. T
11. F
12. F
13. T
14. T
15. F
16. T
17. F
18. F
19. T
20. F
21. T
22. F
23. F
24. F
25. T

EXERCISE SET C

1.

(P	&	Q)	∨	(P	&	~	Q)
T	T	T	T	T	F	F	T
T	F	F	T	T	T	T	F
F	F	T	F	F	F	F	T
F	F	F	F	F	F	T	F

2.

~	(P	&	~	P)
T	T	F	F	T
T	F	F	T	F

3.

~	(P	∨	~	P)
F	T	T	F	T
F	F	T	T	F

4.

~	(P	&	Q)	∨	(~	P	∨	~	Q)
F	T	T	T	F	F	T	F	F	T
T	T	F	F	T	F	T	T	T	F
T	F	F	T	T	T	F	T	F	T
T	F	F	F	T	T	F	T	T	F

5.

~	(P	∨	Q)	∨	(~	P	&	~	Q)
F	T	T	T	F	F	T	F	F	T
F	T	T	F	F	F	T	F	T	F
F	F	T	T	F	T	F	F	F	T
T	F	F	F	T	T	F	T	T	F

6.

(P	&	Q)	∨	(~	P	&	~	Q)
T	T	T	T	F	T	F	F	T
T	F	F	F	F	T	F	T	F
F	F	T	F	T	F	F	F	T
F	F	F	T	T	F	T	T	F

7.

~	(P	∨	(P	&	Q))
F	T	T	T	T	T
F	T	T	T	F	F
T	F	F	F	F	T
T	F	F	F	F	F

8.

~	(P	∨	(P	&	Q))	∨	P
F	T	T	T	T	T	T	T
F	T	T	T	F	F	T	T
T	F	F	F	F	T	T	F
T	F	F	F	F	F	T	F

9.

(P	&	(Q	∨	P))	&	~	P
T	T	T	T	T	F	F	T
T	T	F	T	T	F	F	T
F	F	T	T	F	F	T	F
F	F	F	F	F	F	T	F

10.

((P	→	Q)	→	P)	→	P
T	T	T	T	T	T	T
T	F	F	T	T	T	T
F	T	T	F	F	T	F
F	T	F	F	F	T	F

11.

~	(~	(P	→	Q)	→	P)
F	F	T	T	T	T	T
F	T	T	F	F	T	T
F	F	F	T	T	T	F
F	F	F	T	F	T	F

12.

(P	→	Q)	↔	~	P
T	T	T	F	F	T
T	F	F	T	F	T
F	T	T	T	T	F
F	T	F	T	T	F

13.

P	→	(Q	→	(P	&	Q))
T	T	T	T	T	T	T
T	T	F	T	T	F	F
F	T	T	F	F	F	T
F	T	F	T	F	F	F

14.

(P	∨	Q)	↔	(~	P	→	Q)
T	T	T	T	F	T	T	T
T	T	F	T	F	T	T	F
F	T	T	T	T	F	T	T
F	F	F	T	T	F	F	F

15.

~	(P	∨	(P	→	Q))
F	T	T	T	T	T
F	T	T	T	F	F
F	F	T	F	T	T
F	F	T	F	T	F

16

(P	→	Q)	↔	(Q	→	P)
T	T	T	T	T	T	T
T	F	F	F	F	T	T
F	T	T	F	T	F	F
F	T	F	T	F	T	F

17.

(P	→	Q)	↔	(~	Q	→	~	P)
T	T	T	T	F	T	T	F	T
T	F	F	T	T	F	F	F	T
F	T	T	T	F	T	T	T	F
F	T	F	T	T	F	T	T	F

18.

(P	∨	Q)	→	(P	&	Q)
T	T	T	T	T	T	T
T	T	F	F	T	F	F
F	T	T	F	F	F	T
F	F	F	T	F	F	F

19.

(P	&	Q)	∨	(P	&	R)
T	T	T	T	T	T	T
T	T	T	T	T	F	F
T	F	F	T	T	T	T
T	F	F	F	T	F	F
F	F	T	F	F	F	T
F	F	T	F	F	F	F
F	F	F	F	F	F	T
F	F	F	F	F	F	F

20.

[P	→	(Q	&	R)]	→	[P	→	R]
T	T	T	T	T	T	T	T	T
T	F	T	F	F	T	T	F	F
T	F	F	F	T	T	T	T	T
T	F	F	F	F	T	T	F	F
F	T	T	T	T	T	F	T	T
F	T	T	F	F	T	F	T	F
F	T	F	F	T	T	F	T	T
F	T	F	F	F	T	F	T	F

21.

[P	→	(Q	∨	R)]	→	[P	→	Q]
T	T	T	T	T	T	T	T	T
T	T	T	T	F	T	T	T	T
T	T	F	T	T	F	T	F	F
T	F	F	F	F	T	T	F	F
F	T	T	T	T	T	F	T	T
F	T	T	T	F	T	F	T	T
F	T	F	T	T	T	F	T	F
F	T	F	F	F	T	F	T	F

22.

[(P ∨ Q) → R] → [P → R]

P	∨	Q	→	R	→	P	→	R
T	T	T	T	T	T	T	T	T
T	T	T	F	F	T	T	F	F
T	T	F	T	T	T	T	T	T
T	T	F	F	F	T	T	F	F
F	T	T	T	T	T	F	T	T
F	T	T	F	F	T	F	T	F
F	F	F	T	T	T	F	T	T
F	F	F	T	F	T	F	T	F

23.

[(P & Q) → R] → [P → R]

P	&	Q	→	R	→	P	→	R
T	T	T	T	T	T	T	T	T
T	T	T	F	F	T	T	F	F
T	F	F	T	T	T	T	T	T
T	F	F	T	F	F	T	F	F
F	F	T	T	T	T	F	T	T
F	F	T	T	F	T	F	T	F
F	F	F	T	T	T	F	T	T
F	F	F	T	F	T	F	T	F

24.

[(P & Q) → R] → [(Q & ~ R) → ~ P]

P	&	Q	→	R	→	Q	&	~	R	→	~	P
T	T	T	T	T	T	T	F	F	T	T	F	T
T	T	T	F	F	T	T	T	T	F	F	F	T
T	F	F	T	T	T	F	F	F	T	T	F	T
T	F	F	T	F	T	F	F	T	F	T	F	T
F	F	T	T	T	T	T	F	F	T	T	T	F
F	F	T	T	F	T	T	T	T	F	T	T	F
F	F	F	T	T	T	F	F	F	T	T	T	F
F	F	F	T	F	T	F	F	T	F	T	T	F

25.

[P ↔ (Q ↔ R)] ↔ [(P ↔ Q) ↔ R]

P	↔	Q	↔	R	↔	P	↔	Q	↔	R
T	T	T	T	T	T	T	T	T	T	T
T	F	T	F	F	T	T	T	T	F	F
T	F	F	F	T	T	T	F	F	F	T
T	T	F	T	F	T	T	F	F	T	F
F	F	T	T	T	T	F	F	T	F	T
F	T	T	F	F	T	F	F	T	T	F
F	T	F	F	T	T	F	T	F	T	T
F	F	F	T	F	T	F	T	F	F	F

3 VALIDITY IN SENTENTIAL LOGIC

1. TAUTOLOGIES, CONTRADICTIONS, AND CONTINGENT FORMULAS

In Chapter 2 we saw how to construct the truth table for any formula in sentential logic. In doing the exercises, you may have noticed that in some cases the final (output) column has all T's, in other cases the final column has all F's, and in still other cases the final column has a mixture of T's and F's. There are special names for formulas with these particular sorts of truth tables, which are summarized in the following definitions.

> A formula $\mathcal{A}$ is a ***tautology***
> if and only if
> the truth table of $\mathcal{A}$ is such that
> every entry in the final column is T.

> A formula $\mathcal{A}$ is a ***contradiction***
> if and only if
> the truth table of $\mathcal{A}$ is such that
> every entry in the final column is F.

> A formula $\mathcal{A}$ is a ***contingent*** formula
> if and only if
> $\mathcal{A}$ is neither a tautology nor a contradiction.

The following are examples of each of these types of formulas.

A Tautology:

P	∨	~	P
T	T	F	T
F	T	T	F

A Contradiction:

P	&	~	P
T	F	F	T
F	F	T	F

A Contingent Formula:

P	→	~	P
T	F	F	T
F	T	T	F

In each example, the final column is shaded. In the first example, the final column consists entirely of T's, so the formula is a tautology; in the second example, the final column consists entirely of F's, so the formula is a contradiction; in the third example, the final column consists of a mixture of T's and F's, so the formula is contingent.

Given the above definitions, and given the truth table for negation, we have the following theorems.

> If a formula $\mathcal{A}$ is a tautology, then its negation $\sim\mathcal{A}$ is a contradiction.

> If a formula $\mathcal{A}$ is a contradiction, then its negation $\sim\mathcal{A}$ is a tautology.

> If a formula $\mathcal{A}$ is contingent, then its negation $\sim\mathcal{A}$ is also contingent.

By way of illustrating these theorems, we consider the three formulas cited earlier. In particular, we write down the truth tables for their negations.

~	(	P	∨	~	P	)
F		T	T	F	T	
F		F	T	T	F	

~	(	P	&	~	P	)
T		T	F	F	T	
T		F	F	T	F	

~	(	P	→	~	P	)
T		T	F	F	T	
F		F	T	T	F	

Once again, the final column of each formula is shaded; the first formula is a contradiction, the second is a tautology, the third is contingent.

2. IMPLICATION AND EQUIVALENCE

We can use the notion of tautology to define two very important notions in sentential logic, the notion of *implication*, and the notion of *equivalence*, which are defined as follows.

> Formula $\mathcal{A}$ ***logically implies*** formula $\mathcal{B}$
> if and only if
> the conditional formula $\mathcal{A}\rightarrow\mathcal{B}$ is a tautology.

> Formulas $\mathcal{A}$ and $\mathcal{B}$ are ***logically equivalent***
> if and only if
> the biconditional formula $\mathcal{A}\leftrightarrow\mathcal{B}$ is a tautology.

[Note: The above definitions apply specifically to sentential logic. A more general definition is required for other branches of logic. Once we have a more general definition, it is customary to refer to the special cases as *tautological implication* and *tautological equivalence*.]

Let us illustrate these concepts with a few examples. To begin with, we note that whereas the formula ~P logically implies the formula ~(P&Q), the converse is not true; i.e., ~(P&Q) does not logically imply ~P). This can be shown by constructing truth tables for the associated pair of conditionals. In particular, the question whether ~P implies ~(P&Q) reduces to the question whether the formula ~P→~(P&Q) is a tautology. The following is the truth table for this formula.

~	P	→	~	(	P	&	Q	)
F	T	T	F		T	T	T	
F	T	T	T		T	F	F	
T	F	T	T		F	F	T	
T	F	T	T		F	F	F	

Notice that the conditional ~P→~(P&Q) is a tautology, so we conclude that its antecedent logically implies its consequent; that is, ~P logically implies ~(P&Q).

Considering the converse implication, the question whether ~(P&Q) logically implies ~P reduces to the question whether the conditional formula ~(P&Q)→~P is a tautology. The truth table follows.

~	(	P	&	Q	)	→	~	P
F		T	T	T		T	F	T
T		T	F	F		F	F	T
T		F	F	T		T	T	F
T		F	F	F		T	T	F

The formula is false in the second case, so it is not a tautology. We conclude that its antecedent does not imply its consequent; that is, ~(P&Q) does not imply ~P.

Next, we turn to logical equivalence. As our first example, we ask whether ~(P&Q) and ~P&~Q are logically equivalent. According to the definition of logical equivalence, this reduces to the question whether the biconditional formula ~(P&Q)↔(~P&~Q) is a tautology. Its truth table is given as follows.

~	(P	&	Q)	↔	(~	P	&	~	Q)
F	T	T	T	T	F	T	F	F	T
T	T	F	F	F	F	T	F	T	F
T	F	F	T	F	T	F	F	F	T
T	F	F	F	T	T	F	T	T	F
*							*		

In this table, the truth value of the biconditional is shaded, whereas the constituents are marked by '*'. Notice that the biconditional is false in cases 2 and 3, so it is not a tautology. We conclude that the two constituents – ~(P&Q) and ~P&~Q – are *not* logically equivalent.

As our second example, we ask whether ~(P&Q) and ~P∨~Q are logically equivalent. As before, this reduces to the question whether the biconditional formula ~(P&Q)↔(~P∨~Q) is a tautology. Its truth table is given as follows.

~	(P	&	Q)	↔	(~	P	∨	~	Q)
F	T	T	T	T	F	T	F	F	T
T	T	F	F	T	F	T	T	T	F
T	F	F	T	T	T	F	T	F	T
T	F	F	F	T	T	F	T	T	F
*							*		

Once again, the biconditional is shaded, and the constituents are marked by '*'. Comparing the two *-columns, we see they are the same in every case; accordingly, the shaded column is true in every case, which is to say that the biconditional formula is a tautology. We conclude that the two constituents – ~(P&Q) and ~P∨~Q – are logically equivalent.

We conclude this section by citing a theorem about the relation between implication and equivalence.

> Formulas $\mathcal{A}$ and $\mathcal{B}$ are logically equivalent
> if and only if
> $\mathcal{A}$ logically implies $\mathcal{B}$
> and
> $\mathcal{B}$ logically implies $\mathcal{A}$.

This follows from the fact that $\mathcal{A}\leftrightarrow\mathcal{B}$ is logically equivalent to $(\mathcal{A}\rightarrow\mathcal{B})\&(\mathcal{B}\rightarrow\mathcal{A})$, and the fact that two formulas $\mathcal{A}$ and $\mathcal{B}$ are tautologies if and only if the conjunction $\mathcal{A}\&\mathcal{B}$ is a tautology.

3. VALIDITY IN SENTENTIAL LOGIC

Recall that an argument is valid if and only if it is impossible for the premises to be true while the conclusion is false; equivalently, it is impossible for the premises to be true without the conclusion also being true. Possibility and impossibility are difficult to judge in general. However, in case of sentential logic, we may judge them by reference to truth tables. This is based on the following definition of 'impossible', relative to logic.

> To say that it is ***impossible*** that *S* is to say that there is **no case** in which *S*.

Here, ø is any statement. the sort of statement we are interested in is the following.

S: the premises of argument $\mathcal{A}$ are all true, and the conclusion is false.

Substituting this statement for *S* in the above definition, we obtain the following.

> To say that it is impossible that {the premises of argument A are all true, and the conclusion is false} is to say that there is no case in which {the premises of argument A are all true, and the conclusion is false}.

This is slightly complicated, but it is the basis for defining validity in sentential logic. The following is the resulting definition.

> An argument $\mathcal{A}$ is ***valid***
> if and only if
> there is **no case** in which
> the premises are true
> and the conclusion is false.

This definition is acceptable provided that we know what "cases" are. This term has already arisen in the previous chapter. In the following, we provide the official definition.

> The ***cases*** relevant to an argument $\mathcal{A}$ are precisely all the possible combinations of truth values that can be assigned to the atomic formulas (P, Q, R, etc.), as a group, that constitute the argument.

By way of illustration, consider the following sentential argument form.

Example 1

(a1) P → Q
~Q
/ ~P

In this argument form, there are two atomic formulas – P, Q – so the possible cases relevant to (a1) consist of all the possible combinations of truth values that can be assigned to P and Q. These are enumerated as follows.

	P	Q
case 1	T	T
case 2	T	F
case 3	F	T
case 4	F	F

As a further illustration, consider the following sentential argument form, which involves three atomic formulas – P, Q, R.

Example 2

(a2) $P \rightarrow Q$
$Q \rightarrow R$
$/ P \rightarrow R$

The possible combinations of truth values that can be assigned to P, Q, R are given as follows.

	P	Q	R
case 1	T	T	T
case 2	T	T	F
case 3	T	F	T
case 4	T	F	F
case 5	F	T	T
case 6	F	T	F
case 7	F	F	T
case 8	F	F	F

Notice that in constructing this table, the T's and F's are alternated in quadruples in the P column, in pairs in the Q column, and singly in the R column. Also notice that, in general, if there are n atomic formulas, then there are 2^n cases.

4. TESTING ARGUMENTS IN SENTENTIAL LOGIC

In the previous section, we noted that an argument is valid if and only if there is *no case* in which the premises are true and the conclusion is false. We also noted that the cases in sentential logic are the possible combinations of truth values that can be assigned to the atomic formulas (letters) in an argument.

In the present section, we use these ideas to test sentential argument forms for validity and invalidity.

The first thing we do is adopt a new method of displaying argument forms. Our present method is to display arguments in vertical lists, where the conclusion is at the bottom. In combination with truth tables, this is inconvenient, so we will henceforth write argument forms in horizontal lists. For example, the argument forms from earlier may be displayed as follows.

(a1) P → Q ; ~Q/~P
(a2) P → Q ; Q → R/P → R

In (a1) and (a2), the premises are separated by a semi-colon (;), and the conclusion is marked of by a forward slash (/). If there are three premises, then they are separated by two semi-colons; if there are four premises, then they are separated by three semi-colons, etc.

Using our new method of displaying argument forms, we can form *multiple truth tables*. Basically, a multiple truth table is a collection of truth tables that all use the same guide table. This may be illustrated in reference to argument form (a1).

GuideTable: **Argument:**

	P	Q	P	→	Q	;	~	Q	/	~	P
case 1	T	T	T	T	T		F	T		F	T
case 2	T	F	T	F	F		T	F		F	T
case 3	F	T	F	T	T		F	T		T	F
case 4	F	F	F	T	F		T	F		T	F

In the above table, the three formulas of the argument are written side by side, and their truth tables are placed beneath them. In each case, the final (output) column is shaded. Notice the following. If we were going to construct the truth table for ~Q *by itself*, then there would only be two cases to consider. But in relation to the whole collection of formulas, in which there are two atomic formulas – P and Q – there are four cases to consider in all. This is a property of multiple truth tables that makes them different from individual truth tables. Nevertheless, we can look at a multiple truth table simply as a set of several truth tables all put together. So in the above case, there are three truth tables, one for each formula, which all use the same guide table.

The above collection of formulas is not merely a collection; it is also an argument (form). So we can ask whether it is valid or invalid. According to our definition an argument is valid if and only if there is no case in which the premises are all true but the conclusion is false.

Let's examine the above (multiple) truth table to see whether there are any cases in which the premises are both true and the conclusion is false. The shaded columns are the only columns of interest at this point, so we simply extract them to form the following table.

	P	Q	P→Q	;	~Q	/	~P
case 1	T	T	T		F		F
case 2	T	F	F		T		F
case 3	F	T	T		F		T
case 4	F	F	T		T		T

In cases 1 through 3, one of the premises is false, so they won't do. In case 4, both the premises are true, but the conclusion is also true, so this case won't do either. Thus, there is no case in which the premises are all true and the conclusion is false. To state things equivalently, every case in which the premises are all true is also a case in which the conclusion is true. On the basis of this, we conclude that argument (a1) is *valid*.

Whereas argument (a1) is valid, the following similar looking argument (form) is *not* valid.

(a3) P → Q
~P
/ ~Q

The following is a concrete argument with this form.

(c3) if Bush is president, then the president is a U.S. citizen;
Bush is *not* president;
/ the president is *not* a U.S. citizen.

Observe that (c3) as the form (a3), that (c3) has all true premises, that (c3) has a false conclusion. In other words, (c3) is a counterexample to (a3); indeed, (c3) is a counterexample to any argument with the same form. It follows that (a3) is *not* valid; it is *invalid*.

This is one way to show that (a3) is invalid. We can also show that it is invalid using truth tables. To show that (a3) is invalid, we show that there is a case (line) in which the premises are both true but the conclusion is false. The following is the (multiple) truth table for argument (a3).

	P	Q	P	→	Q	;	~	P	/	~	Q
case 1	T	T	T	T	T		F	T		F	T
case 2	T	F	T	F	F		F	T		T	F
case 3	F	T	F	**T**	T		**T**	F		**F**	T
case 4	F	F	F	T	F		T	F		T	F

In deciding whether the argument form is valid or invalid, we look for a case in which the premises are all true and the conclusion is false. In the above truth table, cases 1 and 2 do not fill the bill, since the premises are not both true. In case 4, the premises are both true, but the conclusion is also true, so case 4 doesn't fill the bill either. On the other hand, in case 3 the premises are both true, and the conclusion is false. Thus, there is a case in which the premises are all true and the conclusion is false (namely, the 3rd case). On this basis, we conclude that argument (a3) is *invalid*.

Note carefully that case 3 in the above truth table demonstrates that argument (a3) is *invalid*; one case is all that is needed to show invalidity. But this is *not* to say that the argument is valid in the other three cases. This does not make any sense, for the notions of validity and invalidity do not apply to the individual cases, but to all the cases taken all together.

Having considered a couple of simple examples, let us now examine a couple of examples that are somewhat more complicated.

	P	Q	P → (~ P ∨ Q)						;	~ P → Q				;	Q → P			/	P & Q		
	P	Q	P	→	~	P	∨	Q		~	P	→	Q		Q	→	P		P	&	Q
1	T	T	T	T	F	T	T	T		F	T	T	T		T	T	T		T	T	T
2	T	F	T	F	F	T	F	F		F	T	T	F		F	T	T		T	F	F
3	F	T	F	T	T	F	T	T		T	F	T	T		T	F	F		F	F	T
4	F	F	F	T	T	F	T	F		T	F	F	F		F	T	F		F	F	F

In this example, the argument has three premises, but it only involves two atomic formulas (P, Q), so there are four cases to consider. What we are looking for is at least one case in which the premises are all true and the conclusion is false. As usual the final (output) columns are shaded, and these are the only columns that interest us. If we extract them from the above table, we obtain the following.

	P	Q	P→(~P∨Q) ;	~P→Q ;	Q→P /	P&Q
1	T	T	T	T	T	T
2	T	F	F	T	T	F
3	F	T	T	T	F	F
4	F	F	T	F	T	F

In case 1, the premises are all true, but so is the conclusion. In each of the remaining cases (2-4), the conclusion is false, but in each of these cases, at least one premise is also false. Thus, there is no case in which the premises are all true and the conclusion is false. From this we conclude that the argument is *valid*.

The final example we consider is an argument that involves three atomic formulas (letters). There are accordingly 8 cases to consider, not just four as in previous examples.

	P	Q	R	P ∨ (Q → R)					;	P → ~ R				/	~ (Q & ~ R)				
	P	Q	R	P	∨	Q	→	R		P	→	~	R		~	Q	&	~	R
1	T	T	T	T	T	T	T	T		T	F	F	T		T	T	F	F	T
2	T	T	F	T	T	T	F	F		T	T	T	F		F	T	T	T	F
3	T	F	T	T	T	F	T	T		T	F	F	T		T	F	F	F	T
4	T	F	F	T	T	F	T	F		T	T	T	F		T	F	F	T	F
5	F	T	T	F	T	T	T	T		F	T	F	T		T	T	F	F	T
6	F	T	F	F	F	T	F	F		F	T	T	F		F	T	T	T	F
7	F	F	T	F	T	F	T	T		F	T	F	T		T	F	F	F	T
8	F	F	F	F	T	F	T	F		F	T	T	F		T	F	F	T	F

As usual, the shaded columns are the ones that we are interested in as far as deciding the validity or invalidity of this argument. We are looking for a case in which the premises are all true and the conclusion is false. So in particular, we are looking for a case in which the conclusion is false. There are only two such cases – case 2 and case 6; the remaining question is whether the premises both true in either of these cases. In case 6, the first premise is false, but in case 2, the premises are both true. This is exactly what we are looking for – a case with all true premises and a false conclusion. Since such a case exists, as shown by the above truth table, we conclude that the argument is *invalid*.

5. THE RELATION BETWEEN VALIDITY AND IMPLICATION

Let us begin this section by recalling some earlier definitions. In Section 1, we noted that a formula $\mathcal{A}$ is a tautology if and only if it is *true in every case*. We can describe this by saying that a tautology is a formula that is true no matter what. By contrast, a *contradiction* is a formula that is false in every case, or false no matter what. Between these two extremes contingent formulas, which are true under some circumstances but false under others.

Next, in Section 2, we noted that a formula $\mathcal{A}$ logically *implies* (or simply implies) a formula $\mathcal{B}$ if and only if the conditional formula $\mathcal{A}\rightarrow\mathcal{B}$ is a tautology.

The notion of implication is intimately associated with the notion of validity. This may be illustrated first using the simplest example – an argument with just one premise. Consider the following argument form.

(a1) ~P /~(P&Q)

You might read this as saying that: it is not true that P; *so* it is not true that P&Q. On the other hand, consider the conditional formed by taking the premise as the antecedent, and the conclusion as the consequent.

(c1) ~P → ~(P&Q)

As far as the symbols are concerned, all we have done is to replace the '/' by '→'. The resulting conditional may be read as saying that: *if* it is not true that P, *then* it is not true that P&Q.

There seems to be a natural relation between (a1) and (c1), though it is clearly not the relation of identity. Whereas (a1) is a pair of formulas, (c1) is a single formula. Nevertheless they are intimately related, as can be seen by constructing the respective truth tables.

	P	Q		~	P	/	~	(P	&	Q)		~	P	→	~	(P	&	Q)
1	T	T		F	T		F	T	T	T		F	T	T	F	T	T	T
2	T	F		F	T		T	T	F	F		F	T	T	T	T	F	F
3	F	T		T	F		T	F	F	T		T	F	T	T	F	F	T
4	F	F		T	F		T	F	F	F		T	F	T	T	F	F	F

We now have two truth tables side by side, one for the argument ~P/~(P&Q), the other for the conditional ~P→~(P&Q).

Let's look at the conditional first. The third column is the final (output) column, and it has all T's, so we conclude that this formula is a tautology. In other words, no matter what, if it is not true that P, then it is not true that P&Q.

This is reflected in the corresponding argument to the left. In looking for a case that serves as a counterexample, we notice that every case in which the premise is true so is the conclusion. Thus, the argument is valid.

This can be stated as a general principle.

> Argument P/C is valid
> if and only if
> the conditional formula P→C is a tautology.

Since, by definition, a formula P implies a formula C if and only if the conditional P→C is a tautology, this principle can be restated as follows.

> Argument P/C is valid
> if and only if
> the premise P logically implies the conclusion C.

In order to demonstrate the truth of this principle, we can argue as follows. Suppose that the argument P/C is *not* valid. Then there is a case (call it case n) in which P is true but C is false. Consequently, in the corresponding truth table for the conditional P→C, there is a case (namely, case n) in which P is true and C is false. Accordingly, in case n, the truth value of P→C is T→F, i.e.,, F. It follows that P→C is *not* a tautology, so P does *not* imply C.

This demonstrates that if P/C is *not* valid, then P→C is *not* a tautology. We also have to show the converse conditional: if P→C is *not* a tautology, then P/C is *not* valid. Well, suppose that P→C isn't a tautology. Then there is a case in which P→C is false. But a conditional is false if and only if its antecedent is true and its consequent is false. So there is a case in which P is true but C is false. It immediately follows that P/C is not valid. This completes our argument.

[Note: What we have in fact demonstrated is this: the argument P/C is *not* valid *if and only if* the conditional P→C is *not* a tautology. This statement has the form: ~V↔~T. The student should convince him(her)self that ~V↔~T is *equivalent* to V↔T, which is to say that (~V↔~T)↔(V↔T) is a tautology.]

The above principle about validity and implication is not particularly useful because not many arguments have just one premise. It would be nice if there were a comparable principle that applied to arguments with two premises, arguments with three premises, in general to *all* arguments. There is such a principle.

What we have to do is to form a single formula out of an argument irrespective of how many premises it has. The particular formula we use begins with the premises, next forms a conjunction out of all these, next takes this conjunction and makes a conditional with it as the antecedent and the conclusion as the consequent. The following examples illustrate this technique.

	Argument	Associated conditional:
(1)	$P_1; P_2/C$	$(P_1 \& P_2) \rightarrow C$
(2)	$P_1; P_2; P_3/C$	$(P_1 \& P_2 \& P_3) \rightarrow C$
(3)	$P_1; P_2; P_3; P_4/C$	$(P_1 \& P_2 \& P_3 \& P_4) \rightarrow C$

In each case, we take the argument, first conjoin the premises, and then form the conditional with this conjunction as its antecedent and with the conclusion as its consequent. Notice that the above formulas are not strictly speaking formulas, since the parentheses are missing in connection with the ampersands. The removal of the extraneous parentheses is comparable to writing 'x+y+z+w' in place of the strictly correct '((x+y)+z)+z'.

Having described how to construct a conditional formula on the basis of an argument, we can now state the principle that relates these two notions.

> An argument $\mathcal{A}$ is **valid**
> if and only if
> the associated conditional is a **tautology**.

In virtue of the relation between implication and tautologies, this principle can be restated as follows.

> Argument $P_1;P_2;...P_n/C$ is **valid**
> if and only if
> the conjunction $P_1\&P_2\&...\&P_n$
> **logically implies**
> the conclusion C.

The interested reader should try to convince him(her)self that this principle is true, at least in the case of two premises. The argument proceeds like the earlier one, except that one has to take into account the truth table for conjunction (in particular, P&Q can be true only if both P and Q are true).

6. EXERCISES FOR CHAPTER 3

EXERCISE SET A

Go back to Exercise Set 2C in Chapter 2. For each formula, say whether it is a **tautology**, a **contradiction**, or a **contingent** formula.

EXERCISE SET B

In each of the following, you are given a pair generically denoted $\mathcal{A}$, $\mathcal{B}$. In each case, answer the following questions:

(1) Does $\mathcal{A}$ **logically imply** $\mathcal{B}$?
(2) Does $\mathcal{B}$ **logically imply** $\mathcal{A}$?
(3) Are $\mathcal{A}$ and $\mathcal{B}$ logically equivalent?

1. $\mathcal{A}$: ~(P&Q)
   $\mathcal{B}$: ~P&~Q
2. $\mathcal{A}$: ~(P&Q)
   $\mathcal{B}$: ~P∨~Q
3. $\mathcal{A}$: ~(P∨Q)
   $\mathcal{B}$: ~P∨~Q
4. $\mathcal{A}$: ~(P∨Q)
   $\mathcal{B}$: ~P&~Q
5. $\mathcal{A}$: ~(P→Q)
   $\mathcal{B}$: ~P→~Q
6. $\mathcal{A}$: ~(P→Q)
   $\mathcal{B}$: P&~Q
7. $\mathcal{A}$: ~(P↔Q)
   $\mathcal{B}$: ~P↔~Q
8. $\mathcal{A}$: ~(P↔Q)
   $\mathcal{B}$: P↔~Q
9. $\mathcal{A}$: ~(P↔Q)
   $\mathcal{B}$: ~P↔Q
10. $\mathcal{A}$: P↔Q
    $\mathcal{B}$: (P&Q) & (Q→P)
11. $\mathcal{A}$: P↔Q
    $\mathcal{B}$: (P→Q) & (Q→P)
12. $\mathcal{A}$: P→Q
    $\mathcal{B}$: Q→P
13. $\mathcal{A}$: P→Q
    $\mathcal{B}$: ~P→~Q
14. $\mathcal{A}$: P→Q
    $\mathcal{B}$: ~Q→~P
15. $\mathcal{A}$: P→Q
    $\mathcal{B}$: ~P∨Q
16. $\mathcal{A}$: P→Q
    $\mathcal{B}$: ~(P&~Q)
17. $\mathcal{A}$: ~P
    $\mathcal{B}$: ~(P&Q)
18. $\mathcal{A}$: ~P
    $\mathcal{B}$: ~(P∨Q)
19. $\mathcal{A}$: ~(P↔Q)
    $\mathcal{B}$: (P&Q) → R
20. $\mathcal{A}$: (P&Q) → R
    $\mathcal{B}$: P→R
21. $\mathcal{A}$: (P∨Q) → R
    $\mathcal{B}$: P→R
22. $\mathcal{A}$: (P&Q)→R
    $\mathcal{B}$: P → (Q→R)
23. $\mathcal{A}$: P → (Q&R)
    $\mathcal{B}$: P→Q
24. $\mathcal{A}$: P → (Q∨R)
    $\mathcal{B}$: P→Q

EXERCISE SET C

In each of the following, you are given an argument form from sentential logic, splayed horizontally. In each case, use the method of truth tables to decide whether the argument form is **valid** or **invalid**. Explain your answer.

1. P→Q; P/Q
2. P→Q; Q/P
3. P→Q; ~Q/~P
4. P→Q; ~P/~Q
5. P∨Q; ~P/Q
6. P∨Q; P/~Q
7. ~(P&Q); P/~Q
8. ~(P&Q); ~P/Q
9. P↔Q; ~P/~Q
10. P↔Q; Q/P
11. P∨Q; P→Q/Q
12. P∨Q; P→Q/P&Q
13. P→Q; P→~Q/~P
14. P→Q; ~P→Q/Q
15. P∨Q; ~P→~Q/P&Q
16. P→Q; ~P→~Q/P↔Q
17. ~P→~Q; ~Q→~P/P↔Q
18. ~P→~Q; ~Q→~P/P&Q
19. P∨~Q; P∨Q/P
20. P→Q; P∨Q/P↔Q
21. ~(P→Q); P→~P/~P&~Q
22. ~(P&Q); ~Q→P/P
23. P→Q; Q→R/P→R
24. P→Q; Q→R; ~P→R/R
25. P→Q; Q→R/P&R
26. P→Q; Q→R; R→P/P↔R
27. P→Q; Q→R/R
28. P→R; Q→R/(P∨Q)→R
29. P→Q; P→R/Q&R
30. P∨Q; P→R; Q→R/R
31. P→Q; Q→R; R→~P/~P
32. P→(Q∨R); Q&R/~P
33. P→(Q&R); Q→~R/~P
34. P&(Q∨R); P→~Q/R
35. P→(Q→R); P&~R/~Q
36. ~P∨Q; R→P; ~(Q&R)/~R

EXERCISE SET D

Go back to Exercise Set B. In each case, consider the argument $\mathcal{A}/\mathcal{B}$, as well as the converse argument $\mathcal{B}/\mathcal{A}$. Thus, there are a total of 48 arguments to consider. On the basis of your answers for Exercise Set B, decide which of these arguments are valid and which are invalid.

7. ANSWERS TO EXERCISES FOR CHAPTER 3

EXERCISE SET A

1. contingent
2. tautology
3. contradiction
4. contingent
5. contingent
6. contingent
7. contingent
8. tautology
9. contradiction
10. tautology
11. contradiction
12. contingent
13. tautology
14. tautology
15. contradiction
16. contingent
17. tautology
18. contingent
19. contingent
20. tautology
21. tautology
22. contingent
23. tautology
24. contingent
25. tautology

EXERCISE SET B

#1.

$\mathcal{A}$:

~	(P	&	Q)
F	T	T	T
T	T	F	F
T	F	F	T
T	F	F	F

$\mathcal{B}$:

~	P	&	~	Q
F	T	F	F	T
F	T	F	T	F
T	F	F	F	T
T	F	T	T	F

$\mathcal{A}$	→	$\mathcal{B}$
F	T	F
T	**F**	F
T	**F**	F
T	T	T

$\mathcal{B}$	→	$\mathcal{A}$
F	T	F
F	T	T
F	T	T
T	T	T

Does $\mathcal{A}$ logically imply $\mathcal{B}$? NO
Does $\mathcal{B}$ logically imply $\mathcal{A}$? YES
Are $\mathcal{A}$ and $\mathcal{B}$ logically equivalent? NO

#2.

$\mathcal{A}$:

~	(P	&	Q)
F	T	T	T
T	T	F	F
T	F	F	T
T	F	F	F

$\mathcal{B}$:

~	P	∨	~	Q
F	T	F	F	T
F	T	T	T	F
T	F	T	F	T
T	F	T	T	F

$\mathcal{A}$	→	$\mathcal{B}$
F	T	F
T	T	T
T	T	T
T	T	T

$\mathcal{B}$	→	$\mathcal{A}$
F	T	F
T	T	T
T	T	T
T	T	T

Does $\mathcal{A}$ logically imply $\mathcal{B}$? YES
Does $\mathcal{B}$ logically imply $\mathcal{A}$? YES
Are $\mathcal{A}$ and $\mathcal{B}$ logically equivalent? YES

#3.

$\mathcal{A}$: ~ (P ∨ Q)

~	P	∨	Q
F	T	T	T
F	T	T	F
F	F	T	T
T	F	F	F

$\mathcal{B}$: ~ P ∨ ~ Q

~	P	∨	~	Q
F	T	F	F	T
F	T	T	T	F
T	F	T	F	T
T	F	T	T	F

$\mathcal{A}$	→	$\mathcal{B}$
F	T	F
F	T	T
F	T	T
T	T	T

$\mathcal{B}$	→	$\mathcal{A}$
F	T	F
T	F	F
T	F	F
T	T	T

Does $\mathcal{A}$ logically imply $\mathcal{B}$? YES
Does $\mathcal{B}$ logically imply $\mathcal{A}$? NO
Are $\mathcal{A}$ and $\mathcal{B}$ logically equivalent? NO

#4.

$\mathcal{A}$: ~ (P ∨ Q)

~	P	∨	Q
F	T	T	T
F	T	T	F
F	F	T	T
T	F	F	F

$\mathcal{B}$: ~ P & ~ Q

~	P	&	~	Q
F	T	F	F	T
F	T	F	T	F
T	F	F	F	T
T	F	T	T	F

$\mathcal{A}$	→	$\mathcal{B}$
F	T	F
F	T	F
F	T	F
T	T	T

$\mathcal{B}$	→	$\mathcal{A}$
F	T	F
F	T	F
F	T	F
T	T	T

Does $\mathcal{A}$ logically imply $\mathcal{B}$? YES
Does $\mathcal{B}$ logically imply $\mathcal{A}$? YES
Are $\mathcal{A}$ and $\mathcal{B}$ logically equivalent? YES

#5.

$\mathcal{A}$: ~ (P → Q)

~	P	→	Q
F	T	T	T
T	T	F	F
F	F	T	T
F	F	T	F

$\mathcal{B}$: ~ P → ~ Q

~	P	→	~	Q
F	T	T	F	T
F	T	T	T	F
T	F	F	F	T
T	F	T	T	F

$\mathcal{A}$	→	$\mathcal{B}$
F	T	T
T	T	T
F	T	F
F	T	T

$\mathcal{B}$	→	$\mathcal{A}$
T	F	F
T	T	T
F	T	F
T	F	F

Does $\mathcal{A}$ logically imply $\mathcal{B}$? YES
Does $\mathcal{B}$ logically imply $\mathcal{A}$? NO
Are $\mathcal{A}$ and $\mathcal{B}$ logically equivalent? NO

#6.

$\mathcal{A}$: ~ (P → Q)

~	P	→	Q
F	T	T	T
T	T	F	F
F	F	T	T
F	F	T	F

$\mathcal{B}$: P & ~ Q

P	&	~	Q
T	F	F	T
T	T	T	F
F	F	F	T
F	F	T	F

$\mathcal{A}$	→	$\mathcal{B}$
F	T	F
T	T	T
F	T	F
F	T	F

$\mathcal{B}$	→	$\mathcal{A}$
F	T	F
T	T	T
F	T	F
F	T	F

Does $\mathcal{A}$ logically imply $\mathcal{B}$? YES
Does $\mathcal{B}$ logically imply $\mathcal{A}$? YES
Are $\mathcal{A}$ and $\mathcal{B}$ logically equivalent? YES

#7.

$\mathcal{A}$: ~ (P ↔ Q)

~	P	↔	Q
F	T	T	T
T	T	F	F
T	F	F	T
F	F	T	F

$\mathcal{B}$: ~ P ↔ ~ Q

~	P	↔	~	Q
F	T	T	F	T
F	T	F	T	F
T	F	F	F	T
T	F	T	T	F

$\mathcal{A} \rightarrow \mathcal{B}$

$\mathcal{A}$	→	$\mathcal{B}$
F	T	T
T	**F**	F
T	**F**	F
F	T	T

$\mathcal{B} \rightarrow \mathcal{A}$

$\mathcal{B}$	→	$\mathcal{A}$
T	**F**	F
F	T	T
F	T	F
T	**F**	F

Does $\mathcal{A}$ logically imply $\mathcal{B}$? NO
Does $\mathcal{B}$ logically imply $\mathcal{A}$? NO
Are $\mathcal{A}$ and $\mathcal{B}$ logically equivalent? NO

#8.

$\mathcal{A}$: ~ (P ↔ Q)

~	P	↔	Q
F	T	T	T
T	T	F	F
T	F	F	T
F	F	T	F

$\mathcal{B}$: P ↔ ~ Q

P	↔	~	Q
T	F	F	T
T	T	T	F
F	T	F	T
F	F	T	F

$\mathcal{A} \rightarrow \mathcal{B}$

$\mathcal{A}$	→	$\mathcal{B}$
F	T	F
T	T	T
T	T	T
F	T	F

$\mathcal{B} \rightarrow \mathcal{A}$

$\mathcal{B}$	→	$\mathcal{A}$
F	T	F
T	T	T
T	T	T
F	T	F

Does $\mathcal{A}$ logically imply $\mathcal{B}$? YES
Does $\mathcal{B}$ logically imply $\mathcal{A}$? YES
Are $\mathcal{A}$ and $\mathcal{B}$ logically equivalent? YES

#9.

$\mathcal{A}$: ~ (P ↔ Q)

~	P	↔	Q
F	T	T	T
T	T	F	F
T	F	F	T
F	F	T	F

$\mathcal{B}$: ~ P ↔ Q

~	P	↔	Q
F	T	F	T
F	T	T	F
T	F	T	T
T	F	F	F

$\mathcal{A} \rightarrow \mathcal{B}$

$\mathcal{A}$	→	$\mathcal{B}$
F	T	F
T	T	T
T	T	T
F	T	F

$\mathcal{B} \rightarrow \mathcal{A}$

$\mathcal{B}$	→	$\mathcal{A}$
F	T	F
T	T	T
T	T	T
F	T	F

Does $\mathcal{A}$ logically imply $\mathcal{B}$? YES
Does $\mathcal{B}$ logically imply $\mathcal{A}$? YES
Are $\mathcal{A}$ and $\mathcal{B}$ logically equivalent? YES

#10.

$\mathcal{A}$: P ↔ Q

P	↔	Q
T	T	T
T	F	F
F	F	T
F	T	F

$\mathcal{B}$: (P & Q) & (Q → P)

P	&	Q	&	Q	→	P
T	T	T	T	T	T	T
T	F	F	F	F	T	T
F	F	T	F	T	F	F
F	F	F	F	F	T	F

$\mathcal{A} \rightarrow \mathcal{B}$

$\mathcal{A}$	→	$\mathcal{B}$
T	T	T
F	T	F
F	T	F
T	**F**	F

$\mathcal{B} \rightarrow \mathcal{A}$

$\mathcal{B}$	→	$\mathcal{A}$
T	T	T
F	T	F
F	T	F
F	T	T

Does $\mathcal{A}$ logically imply $\mathcal{B}$? NO
Does $\mathcal{B}$ logically imply $\mathcal{A}$? YES
Are $\mathcal{A}$ and $\mathcal{B}$ logically equivalent? NO

#11.

$\mathcal{A}$:

P	↔	Q
T	T	T
T	F	F
F	F	T
F	T	F

$\mathcal{B}$:

(P	→	Q)	&	(Q	→	P)
T	T	T	T	T	T	T
T	F	F	F	F	T	T
F	T	T	F	T	F	F
F	T	F	T	F	T	F

$\mathcal{A}$	→	$\mathcal{B}$
T	T	T
F	T	F
F	T	F
T	T	T

$\mathcal{B}$	→	$\mathcal{A}$
T	T	T
F	T	F
F	T	F
T	T	T

Does $\mathcal{A}$ logically imply $\mathcal{B}$? YES
Does $\mathcal{B}$ logically imply $\mathcal{A}$? YES
Are $\mathcal{A}$ and $\mathcal{B}$ logically equivalent? YES

#12.

$\mathcal{A}$:

P	→	Q
T	T	T
T	F	F
F	T	T
F	T	F

$\mathcal{B}$:

Q	→	P
T	T	T
F	T	T
T	F	F
F	T	F

$\mathcal{A}$	→	$\mathcal{B}$
T	T	T
F	T	T
T	**F**	F
T	T	T

$\mathcal{B}$	→	$\mathcal{A}$
T	T	T
T	**F**	F
F	T	F
T	T	T

Does $\mathcal{A}$ logically imply $\mathcal{B}$? NO
Does $\mathcal{B}$ logically imply $\mathcal{A}$? NO
Are $\mathcal{A}$ and $\mathcal{B}$ logically equivalent? NO

#13.

$\mathcal{A}$:

P	→	Q
T	T	T
T	F	F
F	T	T
F	T	F

$\mathcal{B}$:

~	P	→	~	Q
F	T	T	F	T
F	T	T	T	F
T	F	F	F	T
T	F	T	T	F

$\mathcal{A}$	→	$\mathcal{B}$
T	T	T
F	T	T
T	**F**	F
T	T	T

$\mathcal{B}$	→	$\mathcal{A}$
T	T	T
T	**F**	F
F	T	T
T	T	T

Does $\mathcal{A}$ logically imply $\mathcal{B}$? NO
Does $\mathcal{B}$ logically imply $\mathcal{A}$? NO
Are $\mathcal{A}$ and $\mathcal{B}$ logically equivalent? NO

#14.

$\mathcal{A}$:

P	→	Q
T	T	T
T	F	F
F	T	T
F	T	F

$\mathcal{B}$:

~	Q	→	~	P
F	T	T	F	T
T	F	F	F	T
F	T	T	T	F
T	F	T	T	F

$\mathcal{A}$	→	$\mathcal{B}$
T	T	T
F	T	F
T	T	T
T	T	T

$\mathcal{B}$	→	$\mathcal{A}$
T	T	T
F	T	F
T	T	T
T	T	T

Does $\mathcal{A}$ logically imply $\mathcal{B}$? YES
Does $\mathcal{B}$ logically imply $\mathcal{A}$? YES
Are $\mathcal{A}$ and $\mathcal{B}$ logically equivalent? YES

#15.

$\mathcal{A}$: $\mathcal{B}$:

P	→	Q
T	T	T
T	F	F
F	T	T
F	T	F

~	P	∨	Q
F	T	T	T
F	T	F	F
T	F	T	T
T	F	T	F

$\mathcal{A}$	→	$\mathcal{B}$
T	T	T
F	T	F
T	T	T
T	T	T

$\mathcal{B}$	→	$\mathcal{A}$
T	T	T
F	T	F
T	T	T
T	T	T

Does $\mathcal{A}$ logically imply $\mathcal{B}$? YES
Does $\mathcal{B}$ logically imply $\mathcal{A}$? YES
Are $\mathcal{A}$ and $\mathcal{B}$ logically equivalent? YES

#16.

$\mathcal{A}$: $\mathcal{B}$:

P	→	Q
T	T	T
T	F	F
F	T	T
F	T	F

~	(	P	&	~	Q	)
T		T	F	F	T	
F		T	T	T	F	
T		F	F	F	T	
T		F	F	T	F	

$\mathcal{A}$	→	$\mathcal{B}$
T	T	T
F	T	F
T	T	T
T	T	T

$\mathcal{B}$	→	$\mathcal{A}$
T	T	T
F	T	F
T	T	T
T	T	T

Does $\mathcal{A}$ logically imply $\mathcal{B}$? YES
Does $\mathcal{B}$ logically imply $\mathcal{A}$? YES
Are $\mathcal{A}$ and $\mathcal{B}$ logically equivalent? YES

#17.

$\mathcal{A}$: $\mathcal{B}$:

~	P
F	T
F	T
T	F
T	F

~	(	P	&	Q	)
F		T	T	T	
T		T	F	F	
T		F	F	T	
T		F	F	F	

$\mathcal{A}$	→	$\mathcal{B}$
F	T	F
F	T	T
T	T	T
T	T	T

$\mathcal{B}$	→	$\mathcal{A}$
F	T	F
T	**F**	F
T	T	T
T	T	T

Does $\mathcal{A}$ logically imply $\mathcal{B}$? YES
Does $\mathcal{B}$ logically imply $\mathcal{A}$? NO
Are $\mathcal{A}$ and $\mathcal{B}$ logically equivalent? NO

#18.

$\mathcal{A}$: $\mathcal{B}$:

~	P
F	T
F	T
T	F
T	F

~	(	P	∨	Q	)
F		T	T	T	
F		T	T	F	
F		F	T	T	
T		F	F	F	

$\mathcal{A}$	→	$\mathcal{B}$
F	T	F
F	T	F
T	**F**	F
T	T	T

$\mathcal{B}$	→	$\mathcal{A}$
F	T	F
F	T	F
F	T	T
T	T	T

Does $\mathcal{A}$ logically imply $\mathcal{B}$? NO
Does $\mathcal{B}$ logically imply $\mathcal{A}$? YES
Are $\mathcal{A}$ and $\mathcal{B}$ logically equivalent? NO

#19.

$\mathcal{A}$: $\quad\quad$ $\mathcal{B}$:

~	(P	↔	Q)
F	T	T	T
F	T	T	T
T	T	F	F
T	T	F	F
T	F	F	T
T	F	F	T
F	F	T	F
F	F	T	F

(P	&	Q)	→	R
T	T	T	T	T
T	T	T	F	F
T	F	F	T	T
T	F	F	T	F
F	F	T	T	T
F	F	T	T	F
F	F	F	T	T
F	F	F	T	F

$\mathcal{A}$	→	$\mathcal{B}$
F	T	T
F	T	F
T	T	T
T	T	T
T	T	T
T	T	T
F	T	T
F	T	T

$\mathcal{B}$	→	$\mathcal{A}$
T	**F**	F
F	T	F
T	T	T
T	T	T
T	T	T
T	T	T
T	**F**	F
T	**F**	F

Does $\mathcal{A}$ logically imply $\mathcal{B}$? YES
Does $\mathcal{B}$ logically imply $\mathcal{A}$? NO
Are $\mathcal{A}$ and $\mathcal{B}$ logically equivalent? NO

#20.

$\mathcal{A}$: $\quad\quad$ $\mathcal{B}$:

(P	&	Q)	→	R
T	T	T	T	T
T	T	T	F	F
T	F	F	T	T
T	F	F	T	F
F	F	T	T	T
F	F	T	T	F
F	F	F	T	T
F	F	F	T	F

P	→	R
T	T	T
T	F	F
T	T	T
T	F	F
F	T	T
F	T	F
F	T	T
F	T	F

$\mathcal{A}$	→	$\mathcal{B}$
T	T	T
F	T	F
T	T	T
T	**F**	F
T	T	T
T	T	T
T	T	T
T	T	T

$\mathcal{B}$	→	$\mathcal{A}$
T	T	T
F	T	F
T	T	T
F	T	T
T	T	T
T	T	T
T	T	T
T	T	T

Does $\mathcal{A}$ logically imply $\mathcal{B}$? NO
Does $\mathcal{B}$ logically imply $\mathcal{A}$? YES
Are $\mathcal{A}$ and $\mathcal{B}$ logically equivalent? NO

#21.

$\mathcal{A}$: $\quad\quad$ $\mathcal{B}$:

(P	∨	Q)	→	R
T	T	T	T	T
T	T	T	F	F
T	T	F	T	T
T	T	F	F	F
F	T	T	T	T
F	T	T	F	F
F	F	F	T	T
F	F	F	T	F

P	→	R
T	T	T
T	F	F
T	T	T
T	F	F
F	T	T
F	T	F
F	T	T
F	T	F

$\mathcal{A}$	→	$\mathcal{B}$
T	T	T
F	T	F
T	T	T
F	T	F
T	T	T
F	T	T
T	T	T
T	T	T

$\mathcal{B}$	→	$\mathcal{A}$
T	T	T
F	T	F
T	T	T
F	T	F
T	T	T
T	**F**	F
T	T	T
T	T	T

Does $\mathcal{A}$ logically imply $\mathcal{B}$? YES
Does $\mathcal{B}$ logically imply $\mathcal{A}$? NO
Are $\mathcal{A}$ and $\mathcal{B}$ logically equivalent? NO

#22.

$\mathcal{A}$: $(P \,\&\, Q) \rightarrow R$

P	&	Q	→	R
T	T	T	T	T
T	T	T	F	F
T	F	F	T	T
T	F	F	T	F
F	F	T	T	T
F	F	T	T	F
F	F	F	T	T
F	F	F	T	F

$\mathcal{B}$: $P \rightarrow (Q \rightarrow R)$

P	→	Q	→	R
T	T	T	T	T
T	F	T	F	F
T	T	F	T	T
T	T	F	T	F
F	T	T	T	T
F	T	T	F	F
F	T	F	T	T
F	T	F	T	F

$\mathcal{A}$	→	$\mathcal{B}$
T	T	T
F	T	F
T	T	T
T	T	T
T	T	T
T	T	T
T	T	T
T	T	T

$\mathcal{B}$	→	$\mathcal{A}$
T	T	T
F	T	F
T	T	T
T	T	T
T	T	T
T	T	T
T	T	T
T	T	T

Does $\mathcal{A}$ logically imply $\mathcal{B}$? YES
Does $\mathcal{B}$ logically imply $\mathcal{A}$? YES
Are $\mathcal{A}$ and $\mathcal{B}$ logically equivalent? YES

#23.

$\mathcal{A}$: $P \rightarrow (Q \,\&\, R)$

P	→	Q	&	R
T	T	T	T	T
T	F	T	F	F
T	F	F	F	T
T	F	F	F	F
F	T	T	T	T
F	T	T	F	F
F	T	F	F	T
F	T	F	F	F

$\mathcal{B}$: $P \rightarrow Q$

P	→	Q
T	T	T
T	T	T
T	F	F
T	F	F
F	T	T
F	T	T
F	T	F
F	T	F

$\mathcal{A}$	→	$\mathcal{B}$
T	T	T
F	T	F
F	T	T
F	T	F
T	T	T
T	T	T
T	T	T
T	T	T

$\mathcal{B}$	→	$\mathcal{A}$
T	T	T
T	**F**	F
F	T	F
F	T	F
T	T	T
T	T	T
T	T	T
T	T	T

Does $\mathcal{A}$ logically imply $\mathcal{B}$? YES
Does $\mathcal{B}$ logically imply $\mathcal{A}$? NO
Are $\mathcal{A}$ and $\mathcal{B}$ logically equivalent? NO

#24.

$\mathcal{A}$: $P \rightarrow (Q \vee R)$

P	→	Q	∨	R
T	T	T	T	T
T	T	T	T	F
T	T	F	T	T
T	F	F	F	F
F	T	T	T	T
F	T	T	T	F
F	T	F	T	T
F	T	F	F	F

$\mathcal{B}$: $P \rightarrow Q$

P	→	Q
T	T	T
T	F	T
T	T	F
T	F	F
F	T	T
F	T	T
F	T	F
F	T	F

$\mathcal{A}$	→	$\mathcal{B}$
T	T	T
T	**F**	F
T	T	T
F	T	F
T	T	T
T	T	T
T	T	T
T	T	T

$\mathcal{B}$	→	$\mathcal{A}$
T	T	T
F	T	T
T	T	T
F	T	F
T	T	T
T	T	T
T	T	T
T	T	T

Does $\mathcal{A}$ logically imply $\mathcal{B}$? NO
Does $\mathcal{B}$ logically imply $\mathcal{A}$? YES
Are $\mathcal{A}$ and $\mathcal{B}$ logically equivalent? NO

EXERCISE SET C

1.

P	→	Q	;	P	/	Q
T	T	T		T		T
T	F	F		T		F
F	T	T		F		T
F	T	F		F		F

VALID

2.

P	→	Q	;	Q	/	P
T	T	T		T		T
T	F	F		F		T
F	**T**	T		**T**		**F**
F	T	F		F		F

INVALID

3.

P	→	Q	;	~	Q	/	~	P
T	T	T		F	T		F	T
T	F	F		T	F		F	T
F	T	T		F	T		T	F
F	T	F		T	F		T	F

VALID

4.

P	→	Q	;	~	P	/	~	Q
T	T	T		F	T		F	T
T	F	F		F	T		T	F
F	**T**	T		**T**	F		**F**	T
F	T	F		T	F		T	F

INVALID

5.

P	∨	Q	;	~	P	/	Q
T	T	T		F	T		T
T	T	F		F	T		F
F	T	T		T	F		T
F	F	F		T	F		F

VALID

6.

P	∨	Q	;	P	/	~	Q
T	**T**	T		**T**		**F**	T
T	T	F		T		T	F
F	T	T		F		F	T
F	F	F		F		T	F

INVALID

7.

~	(P	&	Q)	;	P	/	~	Q
F	T	T	T		T		F	T
T	T	F	F		T		T	F
T	F	F	T		F		F	T
T	F	F	F		F		T	F

VALID

8.

~	(P	&	Q)	;	~	P	/	Q
F	T	T	T		F	T		T
T	T	F	F		F	T		F
T	F	F	T		T	F		T
T	F	F	F		**T**	F		**F**

INVALID

9.

P	↔	Q	;	~	P	/	~	Q
T	T	T		F	T		F	T
T	F	F		F	T		T	F
F	F	T		T	F		F	T
F	T	F		T	F		T	F

VALID

10.

P	↔	Q	;	Q	/	P
T	T	T		T		T
T	F	F		F		T
F	F	T		T		F
F	T	F		F		F

VALID

11.

P	∨	Q	;	P	→	Q	/	Q
T	T	T		T	T	T		T
T	T	F		T	F	F		F
F	T	T		F	T	T		T
F	F	F		F	T	F		F

VALID

12.

P	∨	Q	;	P	→	Q	/	P	&	Q
T	T	T		T	T	T		T	T	T
T	T	F		T	F	F		T	F	F
F	**T**	T		F	**T**	T		F	**F**	T
F	F	F		F	T	F		F	F	F

INVALID

13.

P	→	Q	;	P	→	~	Q	/	~	P
T	T	T		T	F	F	T		F	T
T	F	F		T	T	T	F		F	T
F	T	T		F	T	F	T		T	F
F	T	F		F	T	T	F		T	F

VALID

14.

P	→	Q	;	~	P	→	Q	/	Q
T	T	T		F	T	T	T		T
T	F	F		F	T	T	F		F
F	T	T		T	F	T	T		T
F	T	F		T	F	F	F		F

VALID

15.

P	∨	Q	;	~	P	→	~	Q	/	P	&	Q
T	T	T		F	T	T	F	T		T	T	T
T	**T**	F		F	T	**T**	T	F		T	**F**	F
F	T	T		T	F	F	F	T		F	F	T
F	F	F		T	F	T	T	F		F	F	F

INVALID

16.

P	→	Q	;	~	P	→	~	Q	/	P	↔	Q
T	T	T		F	T	T	F	T		T	T	T
T	F	F		F	T	T	T	F		T	F	F
F	T	T		T	F	F	F	T		F	F	T
F	T	F		T	F	T	T	F		F	T	F

VALID

17.

~	P	→	~	Q	;	~	Q	→	~	P	/	P	↔	Q
F	T	T	F	T		F	T	T	F	T		T	T	T
F	T	T	T	F		T	F	F	F	T		T	F	F
T	F	F	F	T		F	T	T	T	F		F	F	T
T	F	T	T	F		T	F	T	T	F		F	T	F

VALID

18.

~ P → ~ Q ; ~ Q → ~ P / P & Q

~	P	→	~	Q
F	T	T	F	T
F	T	T	T	F
T	F	F	F	T
T	F	**T**	T	F

~	Q	→	~	P
F	T	T	F	T
T	F	F	F	T
F	T	T	T	F
T	F	**T**	T	F

P	&	Q
T	T	T
T	F	F
F	F	T
F	**F**	F

INVALID

19.

P ∨ ~ Q ; P ∨ Q / P

P	∨	~	Q
T	T	F	T
T	T	T	F
F	F	F	T
F	T	T	F

P	∨	Q
T	T	T
T	T	F
F	T	T
F	F	F

P
T
T
F
F

VALID

20.

P → Q ; P ∨ Q / P ↔ Q

P	→	Q
T	T	T
T	F	F
F	**T**	T
F	T	F

P	∨	Q
T	T	T
T	T	F
F	**T**	T
F	F	F

P	↔	Q
T	T	T
T	F	F
F	**F**	T
F	T	F

INVALID

21.

~ (P → Q) ; P → ~ P / ~ P & ~ Q

~	(	P	→	Q	)
F		T	T	T	
T		T	F	F	
F		F	T	T	
F		F	T	F	

P	→	~	P
T	F	F	T
T	F	F	T
F	T	T	F
F	T	T	F

~	P	&	~	Q
F	T	F	F	T
F	T	F	T	F
T	F	F	F	T
T	F	T	T	F

VALID

22.

~ (P & Q) ; ~ Q → P / P

~	(	P	&	Q	)
F		T	T	T	
T		T	F	F	
T		F	F	T	
T		F	F	F	

~	Q	→	P
F	T	T	T
T	F	T	T
F	T	**T**	F
T	F	F	F

P
T
T
F
F

INVALID

23.

P	→	Q	;	Q	→	R	/	P	→	R
T	T	T		T	T	T		T	T	T
T	T	T		T	F	F		T	F	F
T	F	F		F	T	T		T	T	T
T	F	F		F	T	F		T	F	F
F	T	T		T	T	T		F	T	T
F	T	T		T	F	F		F	T	F
F	T	F		F	T	T		F	T	T
F	T	F		F	T	F		F	T	F

VALID

24.

P	→	Q	;	Q	→	R	;	~	P	→	R	/	R
T	T	T		T	T	T		F	T	T	T		T
T	T	T		T	F	F		F	T	T	F		F
T	F	F		F	T	T		F	T	T	T		T
T	F	F		F	T	F		F	T	T	F		F
F	T	T		T	T	T		T	F	T	T		T
F	T	T		T	F	F		T	F	F	F		F
F	T	F		F	T	T		T	F	T	T		T
F	T	F		F	T	F		T	F	F	F		F

VALID

25.

P	→	Q	;	Q	→	R	/	P	&	R
T	T	T		T	T	T		T	T	T
T	T	T		T	F	F		T	F	F
T	F	F		F	T	T		T	T	T
T	F	F		F	T	F		T	F	F
F	**T**	T		T	**T**	T		F	**F**	T
F	T	T		T	F	F		F	F	F
F	**T**	F		F	**T**	T		F	**F**	T
F	T	F		F	T	F		F	F	F

INVALID

26.

P	→	Q	;	Q	→	R	;	R	→	P	/	P	↔	R
T	T	T		T	T	T		T	T	T		T	T	T
T	T	T		T	F	F		F	T	T		T	F	F
T	F	F		F	T	T		T	T	T		T	T	T
T	F	F		F	T	F		F	T	T		T	F	F
F	T	T		T	T	T		T	F	F		F	F	T
F	T	T		T	F	F		F	T	F		F	T	F
F	T	F		F	T	T		T	F	F		F	F	T
F	T	F		F	T	F		F	T	F		F	T	F

VALID

27.

P	→	Q	;	Q	→	R	/	R
T	T	T		T	T	T		T
T	T	T		T	F	F		F
T	F	F		F	T	T		T
T	F	F		F	T	F		F
F	T	T		T	T	T		T
F	T	T		T	F	F		F
F	T	F		F	T	T		T
F	T	F		F	T	F		F

INVALID

28.

P	→	R	;	Q	→	R	/	(P	∨	Q)	→	R
T	T	T		T	T	T		T	T	T	T	T
T	F	F		T	F	F		T	T	T	F	F
T	T	T		F	T	T		T	T	F	T	T
T	F	F		F	T	F		T	T	F	F	F
F	T	T		T	T	T		F	T	T	T	T
F	T	F		T	F	F		F	T	T	F	F
F	T	T		F	T	T		F	F	F	T	T
F	T	F		F	T	F		F	F	F	T	F

VALID

29.

P	→	Q	;	P	→	R	/	Q	&	R
T	T	T		T	T	T		T	T	T
T	T	T		T	F	F		T	F	F
T	F	F		T	T	T		F	F	T
T	F	F		T	F	F		F	F	F
F	T	T		F	T	T		T	T	T
F	T	T		F	T	F		T	F	F
F	T	F		F	T	T		F	F	T
F	T	F		F	T	F		F	F	F

INVALID

30.

P	∨	Q	;	P	→	R	;	Q	→	R	/	R
T	T	T		T	T	T		T	T	T		T
T	T	T		T	F	F		T	F	F		F
T	T	F		T	T	T		F	T	T		T
T	T	F		T	F	F		F	T	F		F
F	T	T		F	T	T		T	T	T		T
F	T	T		F	T	F		T	F	F		F
F	F	F		F	T	T		F	T	T		T
F	F	F		F	T	F		F	T	F		F

VALID

31.

P	→	Q	;	Q	→	R	;	R	→	~	P	/	~	P
T	T	T		T	T	T		T	F	F	T		F	T
T	T	T		T	F	F		F	T	F	T		F	T
T	F	F		F	T	T		T	F	F	T		F	T
T	F	F		F	T	F		F	T	F	T		F	T
F	T	T		T	T	T		T	T	T	F		T	F
F	T	T		T	F	F		F	T	T	F		T	F
F	T	F		F	T	T		T	T	T	F		T	F
F	T	F		F	T	F		F	T	T	F		T	F

VALID

32.

P	→	(Q	∨	R)	;	Q	&	R	/	~	P
T	T	T	T	T		T	T	T		F	T
T	T	T	T	F		T	F	F		F	T
T	T	F	T	T		F	F	T		F	T
T	F	F	F	F		F	F	F		F	T
F	T	T	T	T		T	T	T		T	F
F	T	T	T	F		T	F	F		T	F
F	T	F	T	T		F	F	T		T	F
F	T	F	F	F		F	F	F		T	F

INVALID

33.

P	→	(Q	&	R)	;	Q	→	~	R	/	~	P
T	T	T	T	T		T	F	F	T		F	T
T	F	T	F	F		T	T	T	F		F	T
T	F	F	F	T		F	T	F	T		F	T
T	F	F	F	F		F	T	T	F		F	T
F	T	T	T	T		T	F	F	T		T	F
F	T	T	F	F		T	T	T	F		T	F
F	T	F	F	T		F	T	F	T		T	F
F	T	F	F	F		F	T	T	F		T	F

VALID

34.

P	&	(Q	∨	R)	;	P	→	~	Q	/	R
T	T	T	T	T		T	F	F	T		T
T	T	T	T	F		T	F	F	T		F
T	T	F	T	T		T	T	T	F		T
T	F	F	F	F		T	T	T	F		F
F	F	T	T	T		F	T	F	T		T
F	F	T	T	F		F	T	F	T		F
F	F	F	T	T		F	T	T	F		T
F	F	F	F	F		F	T	T	F		F

VALID

35.

P	→	(Q	→	R)	;	P	&	~	R	/	~	Q
T	T	T	T	T		T	F	F	T		F	T
T	F	T	F	F		T	T	T	F		F	T
T	T	F	T	T		T	F	F	T		T	F
T	T	F	T	F		T	T	T	F		T	F
F	T	T	T	T		F	F	F	T		F	T
F	T	T	F	F		F	F	T	F		F	T
F	T	F	T	T		F	F	F	T		T	F
F	T	F	T	F		F	F	T	F		T	F

VALID

36.

~	P	∨	Q	;	R	→	P	;	~	(Q	&	R)	/	~	R
F	T	T	T		T	T	T		F	T	T	T		F	T
F	T	T	T		F	T	T		T	T	F	F		T	F
F	T	F	F		T	T	T		T	F	F	T		F	T
F	T	F	F		F	T	T		T	F	F	F		T	F
T	F	T	T		T	F	F		F	T	T	T		F	T
T	F	T	T		F	T	F		T	T	F	F		T	F
T	F	T	F		T	F	F		T	F	F	T		F	T
T	F	T	F		F	T	F		T	F	F	F		T	F

VALID

EXERCISE SET D

1. $\mathcal{A}$: ~(P&Q) $\mathcal{B}$: ~P&~Q
   (1) $\mathcal{A}/\mathcal{B}$ INVALID
   (2) $\mathcal{B}/\mathcal{A}$ VALID

2. $\mathcal{A}$:~(P&Q) $\mathcal{B}$: ~P∨~Q
   (1) $\mathcal{A}/\mathcal{B}$ VALID
   (2) $\mathcal{B}/\mathcal{A}$ VALID

3. $\mathcal{A}$: ~(P∨Q) $\mathcal{B}$: ~P∨~Q
   (1) $\mathcal{A}/\mathcal{B}$ VALID
   (2) $\mathcal{B}/\mathcal{A}$ INVALID

4. $\mathcal{A}$: ~(P∨Q) $\mathcal{B}$: ~P&~Q
   (1) $\mathcal{A}/\mathcal{B}$ VALID
   (2) $\mathcal{B}/\mathcal{A}$ VALID

5. $\mathcal{A}$: ~(P→Q) $\mathcal{B}$: ~P→~Q
   (1) $\mathcal{A}/\mathcal{B}$ VALID
   (2) $\mathcal{B}/\mathcal{A}$ INVALID

6. $\mathcal{A}$: ~(P→Q) $\mathcal{B}$: P&~Q
   (1) $\mathcal{A}/\mathcal{B}$ VALID
   (2) $\mathcal{B}/\mathcal{A}$ VALID

7. $\mathcal{A}$: ~(P↔Q) $\mathcal{B}$: ~P↔~Q
   (1) $\mathcal{A}/\mathcal{B}$ INVALID
   (2) $\mathcal{B}/\mathcal{A}$ INVALID

8. $\mathcal{A}$: ~(P↔Q) $\mathcal{B}$: P↔~Q
   (1) $\mathcal{A}/\mathcal{B}$ VALID
   (2) $\mathcal{B}/\mathcal{A}$ VALID

9 $\mathcal{A}$: ~(P↔Q) $\mathcal{B}$: ~P↔Q
   (1) $\mathcal{A}/\mathcal{B}$ VALID
   (2) $\mathcal{B}/\mathcal{A}$ VALID

10. $\mathcal{A}$: P↔Q $\mathcal{B}$: (P&Q) & (Q→P)
   (1) $\mathcal{A}/\mathcal{B}$ INVALID
   (2) $\mathcal{B}/\mathcal{A}$ VALID

11. $\mathcal{A}$: P↔Q $\mathcal{B}$: (P→Q) & (Q→P)
   (1) $\mathcal{A}/\mathcal{B}$ VALID
   (2) $\mathcal{B}/\mathcal{A}$ VALID

12. $\mathcal{A}$: P→Q $\mathcal{B}$: Q→P
   (1) $\mathcal{A}/\mathcal{B}$ INVALID
   (2) $\mathcal{B}/\mathcal{A}$ INVALID

13. $\mathcal{A}$: P→Q $\mathcal{B}$: ~P→~Q
   (1) $\mathcal{A}/\mathcal{B}$ INVALID
   (2) $\mathcal{B}/\mathcal{A}$ INVALID

14. $\mathcal{A}$: P→Q $\mathcal{B}$: ~Q→~P
   (1) $\mathcal{A}/\mathcal{B}$ VALID
   (2) $\mathcal{B}/\mathcal{A}$ VALID

15. $\mathcal{A}$: P→Q $\mathcal{B}$: ~P∨Q
   (1) $\mathcal{A}/\mathcal{B}$ VALID
   (2) $\mathcal{B}/\mathcal{A}$ VALID

16. $\mathcal{A}$: P→Q $\mathcal{B}$: ~(P&~Q)
   (1) $\mathcal{A}/\mathcal{B}$ VALID
   (2) $\mathcal{B}/\mathcal{A}$ VALID

17. $\mathcal{A}$: ~P $\mathcal{B}$ ~(P&Q)
   (1) $\mathcal{A}/\mathcal{B}$ VALID
   (2) $\mathcal{B}/\mathcal{A}$ INVALID

18. $\mathcal{A}$: ~P $\mathcal{B}$ ~(P∨Q)
   (1) $\mathcal{A}/\mathcal{B}$ INVALID
   (2) $\mathcal{B}/\mathcal{A}$ VALID

19. $\mathcal{A}$: ~(P↔Q) $\mathcal{B}$: (P&Q) → R
   (1) $\mathcal{A}/\mathcal{B}$ VALID
   (2) $\mathcal{B}/\mathcal{A}$ INVALID

20. $\mathcal{A}$: (P&Q) → R $\mathcal{B}$: P→R
   (1) $\mathcal{A}/\mathcal{B}$ INVALID
   (2) $\mathcal{B}/\mathcal{A}$ VALID

21. $\mathcal{A}$: (P∨Q) → R $\mathcal{B}$: P→R
   (1) $\mathcal{A}/\mathcal{B}$ VALID
   (2) $\mathcal{B}/\mathcal{A}$ INVALID

22. $\mathcal{A}$: (P&Q)→R $\mathcal{B}$: P → (Q→R)
   (1) $\mathcal{A}/\mathcal{B}$ VALID
   (2) $\mathcal{B}/\mathcal{A}$ VALID

23. $\mathcal{A}$: P → (Q&R) $\mathcal{B}$: P→Q
   (1) $\mathcal{A}/\mathcal{B}$ VALID
   (2) $\mathcal{B}/\mathcal{A}$ INVALID

24. $\mathcal{A}$: P → (Q∨R) $\mathcal{B}$: P→Q
   (1) $\mathcal{A}/\mathcal{B}$ INVALID
   (2) $\mathcal{B}/\mathcal{A}$ VALID

4 TRANSLATIONS IN SENTENTIAL LOGIC

1. INTRODUCTION

In the present chapter, we discuss how to translate a variety of English statements into the language of sentential logic.

From the viewpoint of sentential logic, there are five *standard connectives* – 'and', 'or', 'if...then', 'if and only if', and 'not'. In addition to these standard connectives, there are in English numerous non-standard connectives, including 'unless', 'only if', 'neither...nor', among others. There is nothing linguistically special about the five "standard" connectives; rather, they are the connectives that logicians have found most useful in doing symbolic logic.

The translation process is primarily a process of *paraphrase* – saying the same thing using different words, or expressing the same proposition using different sentences. Paraphrase is translation from English into English, which is presumably easier than translating English into, say, Japanese.

In the present chapter, we are interested chiefly in two aspects of paraphrase. The first aspect is paraphrasing statements involving various non-standard connectives into equivalent statements involving only standard connectives.

The second aspect is paraphrasing simple statements into straightforwardly equivalent compound statements. For example, the statement 'it is *not* raining' is straightforwardly equivalent to the more verbose '*it is not true that* it is raining'. Similarly, 'Jay *and* Kay are Sophomores' is straightforwardly equivalent to the more verbose 'Jay is a Sophomore, *and* Kay is a Sophomore'.

An English statement is said to be in *standard form*, or to be *standard*, if all its connectives are standard and it contains no simple statement that is straightforwardly equivalent to a compound statement; otherwise, it is said to be *non-standard*.

Once a statement is paraphrased into standard form, the only remaining task is to symbolize it, which consists of symbolizing the simple (atomic) statements and symbolizing the connectives. Simple statements are symbolized by upper case Roman letters, and the standard connectives are symbolized by the already familiar symbols – ampersand, wedge, tilde, arrow, and double-arrow.

In translating simple statements, the particular letter one chooses is not terribly important, although it is usually helpful to choose a letter that is suggestive of the English statement. For example, 'R' can symbolize either 'it is raining' or 'I am running'; however, if both of these statements appear together, then they must be symbolized by different letters. In general, in any particular context, different letters must be used to symbolize non-equivalent statements, and the same letter must be used to symbolize equivalent statements.

2. THE GRAMMAR OF SENTENTIAL LOGIC; A REVIEW

Before proceeding, let us review the grammar of sentential logic. First, recall that statements may be divided into *simple* statements and *compound* statements. Whereas the latter are constructed from smaller statements using statement connectives, the former are not so constructed.

The grammar of sentential logic reflects this grammatical aspect of English. In particular, formulas of sentential logic are divided into *atomic* formulas and *molecular* formulas. Whereas molecular formulas are constructed from other formulas using connectives, atomic formulas are structureless, they are simply upper case letters (of the Roman alphabet).

Formulas are strings of symbols. In sentential logic, the symbols include all the upper case letters, the five connective symbols, as well as left and right parentheses. Certain strings of symbols count as formulas of sentential logic, and others do not, as determined by the following definition.

Definition of Formula in Sentential Logic:

(1) every upper case letter is a formula;
(2) if $\mathcal{A}$ is a formula, then so is $\sim\mathcal{A}$;
(3) if $\mathcal{A}$ and $\mathcal{B}$ are formulas, then so is $(\mathcal{A} \,\&\, \mathcal{B})$;
(4) if $\mathcal{A}$ and $\mathcal{B}$ are formulas, then so is $(\mathcal{A} \vee \mathcal{B})$;
(5) if $\mathcal{A}$ and $\mathcal{B}$ are formulas, then so is $(\mathcal{A} \rightarrow \mathcal{B})$;
(6) if $\mathcal{A}$ and $\mathcal{B}$ are formulas, then so is $(\mathcal{A} \leftrightarrow \mathcal{B})$;
(7) nothing else is a formula.

In the above definition, the script letters stand for arbitrary strings of symbols. So for example, clause (2) says that if you have a string $\mathcal{A}$ of symbols, then provided $\mathcal{A}$ is a formula, the result of prefixing a tilde sign in front of $\mathcal{A}$ is also a formula. Also, clause (3) says that if you have a pair of strings, $\mathcal{A}$ and $\mathcal{B}$, then provided both strings are formulas, the result of infixing an ampersand and surrounding the resulting expression by parentheses is also a formula.

As noted earlier, in addition to formulas in the strict sense, which are specified by the above definition, we also have formulas in a less strict sense. These are called unofficial formulas, which are defined as follows.

An **unofficial formula** is any string of symbols obtained from an official formula by removing its outermost parentheses, if such exist.

The basic idea is that, although the outermost parentheses of a formula are crucial when it is used to form a larger formula, the outermost parentheses are optional when the formula stands alone. For example, the answers to the exercises, at the back of the chapter, are mostly unofficial formulas.

3. CONJUNCTIONS

The standard English expression for conjunction is 'and', but there are numerous other conjunction-like expressions, including the following.

(c1) but
(c2) yet
(c3) although
(c4) though
(c5) even though
(c6) moreover
(c7) furthermore
(c8) however
(c9) whereas

Although these expressions have different connotations, they are all truth-functionally equivalent to one another. For example, consider the following statements.

(s1) it is raining, *but* I am happy
(s2) *although* it is raining, I am happy
(s3) it is raining, *yet* I am happy
(s4) it is raining *and* I am happy

For example, under what conditions is (s1) true? Answer: (s1) is true precisely when 'it is raining' and 'I am happy' are both true, which is to say precisely when (s4) is true. In other words, (s1) and (s4) are true under precisely the same circumstances, which is to say that they are truth-functionally equivalent.

When we utter (s1)-(s3), we intend to emphasize a *contrast* that is not emphasized in the standard conjunction (s4), or we intend to convey (a certain degree of) *surprise*. The difference, however, pertains to appropriate usage rather than semantic content.

Although they connote differently, (s1)-(s4) have the same truth conditions, and are accordingly symbolized the same:

R & H

4. DISGUISED CONJUNCTIONS

As noted earlier, certain simple statements are straightforwardly equivalent to compound statements. For example,

(e1) Jay and Kay are Sophomores

is equivalent to

(p1) Jay is a Sophomore, *and* Kay is a Sophomore

which is symbolized:

(s1) J & K

Other examples of disguised conjunctions involve relative pronouns ('who', 'which', 'that'). For example,

(e2) Jones is a former player *who* coaches basketball

is equivalent to

(p2) Jones is a former (basketball) player, *and* Jones coaches basketball,

which may be symbolized:

(s2) F & C

Further examples do not use relative pronouns, but are easily paraphrased using relative pronouns. For example,

(e3) Pele is a Brazilian soccer player

may be paraphrased as

(p3) Pele is a Brazilian who is a soccer player

which is equivalent to

(p3') Pele is a Brazilian, and Pele is a soccer player,

which may be symbolized:

(s3) B & S

Notice, of course, that

(e4) Jones is a *former* basketball player

is *not* a conjunction, such as the following absurdity.

(??) Jones is a former, and Jones is a basketball player

Sentence (e4) is rather symbolized as a simple (atomic) formula.

5. THE RELATIONAL USE OF 'AND'

As noted in the previous section, the statement,

(c) Jay and Kay are Sophomores,

is equivalent to the conjunction,

Jay is a Sophomore, *and* Kay is a Sophomore,

and is accordingly symbolized:

J & K

Other statements look very much like (c), but are not equivalent to conjunctions. Consider the following statements.

(r1) Jay and Kay are cousins
(r2) Jay and Kay are siblings
(r3) Jay and Kay are neighbors
(r4) Jay and Kay are roommates
(r5) Jay and Kay are lovers

These are definitely not symbolized as conjunctions. The following is an incorrect translation.

(?) J & K WRONG!!!

For example, consider (r1), the standard reading of which is

(r1') Jay and Kay are cousins *of each other*.

In proposing J&K as the analysis of (r1'), we must specify which particular atomic statement each letter stands for. The following is the only plausible choice.

J: Jay is a cousin

K: Kay is a cousin

Accordingly, the formula J&K is read

Jay is a cousin, *and* Kay is a cousin.

But to say that Jay is a cousin is to say that he is a cousin of someone, but not necessarily Kay. Similarly, to say that Kay is a cousin is to say that she a cousin of someone, but not necessarily Jay. In other words, J&K does *not* say that Jay and Kay are cousins *of each other*.

The resemblance between statements like (r1)-(r5) and statements like

(c1) Jay and Kay are Sophomores
(c2) Jay and Kay are Republicans
(c3) Jay and Kay are basketball players

is grammatically superficial. Each of (c1)-(c3) states something about Jay *independently* of Kay, and something about Kay *independently* of Jay.

By contrast, each of (r1)-(r5) states that a particular *relationship* holds between Jay and Kay. The relational quality of (r1)-(r5) may be emphasized by restating them in either of the following ways.

(r1') Jay is a cousin of Kay
(r2') Jay is a sibling of Kay
(r3') Jay is a neighbor of Kay
(r4') Jay is a roommate of Kay
(r5') Jay is a lover of Kay

(r1) Jay and Kay are cousins of each other
(r2) Jay and Kay are siblings of each other
(r3) Jay and Kay are neighbors of each other
(r4) Jay and Kay are roommates of each other
(r5) Jay and Kay are lovers of each other

On the other hand, notice that one cannot paraphrase (c1) as

(??) Jay is a Sophomore of Kay
(??) Jay and Kay are Sophomores of each other

Relational statements like (r1)-(r5) are *not* correctly paraphrased as conjunctions. In fact, they are not correctly paraphrased by any compound statement. From the viewpoint of sentential logic, these statements are *simple*; they have no internal structure, and are accordingly symbolized by atomic formulas.

[NOTE: Later, in predicate logic, we will see how to uncover the internal structure of relational statements such as (r1)-(r5), internal structure that is inaccessible to sentential logic.]

We have seen so far that 'and' is used both *conjunctively*, as in

Jay and Kay are Sophomores,

and *relationally*, as in

Jay and Kay are cousins (of each other).

In other cases, it is not obvious whether 'and' is used conjunctively or relationally. Consider the following.

(s2) Jay and Kay are married

There are two plausible interpretations of this statement. On the one hand, we can interpret it as

(i1) Jay and Kay are married *to each other*,

in which case it expresses a relation, and is symbolized as an atomic formula, say: M. On the other hand, we can interpret it as

(i2) Jay is married, and Kay is married,
(perhaps, but *not necessarily*, to each other),

in which case it is symbolized by a conjunction, say: J&K. The latter simply reports the marital status of Jay, independently of Kay, and the marital status of Kay, independently of Jay.

We can also say things like the following.

(s3) Jay and Kay are married, but not to each other.

This is equivalent to

(p3) Jay is married, and Kay is married,
but Jay and Kay are not married to each other,

which is symbolized:

(J & K) & ~M

[Note: This latter formula does not uncover all the logical structure of the English sentence; it only uncovers its connective structure, but that is all sentential logic is concerned with.]

6. CONNECTIVE-USES OF 'AND' DIFFERENT FROM AMPERSAND

As seen in the previous section, 'and' is used both as a connective and as a separator in relation-statements.

In the present section, we consider how 'and' is occasionally used as a connective different in meaning from the ampersand connective (&). There are two cases of this use.

First, sentences that have the form 'P and Q' sometimes mean 'P *and then* Q'. For example, consider the following statements.

(s1) I went home and went to bed
(s2) I went to bed and went home

As they are colloquially understood at least, these two statements do not express the same proposition, since 'and' here means 'and then'.

Note, in particular, that the above use of 'and' to mean 'and then' is *not truth-functional*. Merely knowing that P is true, and merely knowing that Q is true, one does not automatically know the *order* of the two events, and hence one does not know the truth-value of the compound 'P and then Q'.

Sometimes 'and' does not have exactly the same meaning as the ampersand connective. Other times, 'and' has a quite different meaning from ampersand.

(e1) keep trying, and you will succeed

(e2) keep it up buster, and I will clobber you

(e3) give him an inch, and he will take a mile

(e4) give me a place to stand, and I will move the world (Archimedes, in reference to the power of levers)

(e5) give us the tools of war, and we will finish the job (Churchill, in reference to WW2)

Consider (e1) paraphrased as a conjunction, for example:

(?) K & S

In proposing (?) as an analysis of (e1), we must specify what particular statements K and S abbreviate. The only plausible answer is:

K: you will keep trying

S: you will succeed

Accordingly, the conjunction K&S reads:

you will keep trying, *and* you will succeed

But the original,

keep trying, and you will succeed,

does not say this at all. It does not say the addressee will keep trying, nor does it say that the addressee will succeed. Rather, it merely says (promises, predicts) that the addressee will succeed *if* he/she keeps trying.

Similarly, in the last example, it should be obvious that Churchill was not predicting that the addressee (i.e., Roosevelt) would in fact give him military aid *and* Churchill would in fact finish the job (of course, that was what Churchill was hoping!). Rather, Churchill was saying that he would finish the job *if* Roosevelt were to give him military aid. (As it turned out, of course, Roosevelt eventually gave substantial direct military aid.)

Thus, under very special circumstances, involving requests, promises, threats, warnings, etc., the word 'and' can be used to state conditionals. The appropriate paraphrases are given as follows.

(p1) *if* you keep trying, *then* you will succeed
(p2) *if* you keep it up buster, *then* I will clobber you
(p3) *if* you give him an inch, *then* he will take a mile
(p4) *if* you give me a place to stand, *then* I will move the world
(p5) *if* you give us the tools of war, *then* we will finish the job

The treatment of conditionals is discussed in a later section.

7. NEGATIONS, STANDARD AND IDIOMATIC

The standard form of the negation connective is

it is not true that _____

The following expressions are standard variants.

it is not the case that _____

it is false that _____

Given any statement, we can form its *standard negation* by placing 'it is not the case that' (or a variant) in front of it.

As noted earlier, standard negations seldom appear in colloquial-idiomatic English. Rather, the usual colloquial-idiomatic way to negate a statement is to place the modifier 'not' in a strategic place within the statement, usually immediately after the verb. The following is a simple example.

statement: it is raining
idiomatic negation: it is *not* raining
standard negation: *it is not true that* it is raining

Idiomatic negations are symbolized in sentential logic exactly like standard negations, according to the following simple principle.

> If sentence S is symbolized by the formula $\mathcal{A}$, then the negation of S (standard or idiomatic) is symbolized by the formula $\sim\mathcal{A}$.

Note carefully that this principle applies whether S is simple or compound. As an example of a compound statement, consider the following statement.

(e1) Jay is a Freshman basketball player.

As noted in Section 2, this may be paraphrased as a conjunction:

(p1) Jay is a Freshman, *and* Jay is a basketball player.

Now, there is no simple idiomatic negation of the latter, although there is a standard negation, namely

(n1) it is not true that (Jay is a Freshman and Jay is a basketball player)

The parentheses indicate the scope of the negation modifier.

However, there is a simple idiomatic negation of the former, namely,

(n1′) Jay is *not* a Freshman basketball player.

We consider (n1) and (n1′) further in the next section.

8. NEGATIONS OF CONJUNCTIONS

As noted earlier, the sentence

(s1) Jay is a Freshman basketball player,

may be paraphrased as a conjunction,

(p1) Jay is a Freshman, *and* Jay is a basketball player,

which is symbolized:

(f1) F & B

Also, as noted earlier, the idiomatic negation of (p1) is

(n1) Jay is *not* a Freshman basketball player.

Although there is no simple idiomatic negation of (p1), its standard negation is:

(n2) *it is not true that* (Jay is a Freshman, *and* Jay is a Basketball player),

which is symbolized:

~(F & B)

Notice carefully that, when the conjunction stands by itself, the outer parentheses may be dropped, as in (f2), but when the formula is negated, the outer parentheses *must* be restored before prefixing the negation sign. Otherwise, we obtain:

~F & B,

which is reads:

Jay is *not* a Freshman, *and* Jay is a Basketball player,

which is *not* equivalent to ~(F&B), as may be shown using truth tables.

How do we read the negation

~(F & B)?

Many students suggest the following *erroneous* paraphrase,

Jay is *not* a Freshman,
and
Jay is *not* a basketball player, WRONG!!!

which is symbolized:

~J & ~B.

But this is clearly not equivalent to (n1). To say that Jay isn't a Freshman basketball player is to say that one of the following states of affairs obtains.

(1) Jay is a Freshman who does not play Basketball;

(2) Jay is a Basketball player who is not a Freshman;

(3) Jay is neither a Freshman nor a Basketball player.

On the other hand, to say that Jay is not a Freshman and not a Basketball player is to say precisely that the last state of affairs (3) obtains.

We have already seen the following, in a previous chapter (voodoo logic notwithstanding!)

$\sim(\mathcal{A}\ \&\ \mathcal{B})$ is ***NOT*** logically equivalent to $(\sim\mathcal{A}\ \&\ \sim\mathcal{B})$

This is easily demonstrated using truth-tables. Whereas the latter entails the former, the former does not entail the latter.

The correct logical equivalence is rather:

$\sim(\mathcal{A}\ \&\ \mathcal{B})$ is logically equivalent to $(\sim\mathcal{A} \vee \sim\mathcal{B})$

The disjunction may be read as follows.

Jay is *not* a Freshman *and/or* Jay is *not* a Basketball player.

One more example might be useful. The colloquial negation of the sentence

Jay and Kay are *both* Republicans $\qquad$ J & K

is

Jay and Kay are *not both* Republicans $\qquad$ ~(J & K)

This is definitely not the same as

Jay and Kay are both non-Republicans,

which is symbolized:

~J & ~K.

The latter says that *neither* of them is a Republican (see later section concerning 'neither'), whereas the former says less – that at least one of them isn't a Republican, *perhaps* neither of them is a Republican.

9. DISJUNCTIONS

The standard English expression for disjunction is 'or', a variant of which is 'either...or'. As noted in a previous chapter, 'or' has two senses – an inclusive sense and an exclusive sense.

The legal profession has invented an expression to circumvent this ambiguity – 'and/or'. Similarly, Latin uses two different words: one, 'vel', expresses the inclusive sense of 'or'; the other, 'aut', expresses the exclusive sense.

The standard connective of sentential logic for disjunction is the wedge '∨', which is suggestive of the first letter of 'vel'. In particular, the wedge connective of sentential logic corresponds to the inclusive sense of 'or', which is the sense of 'and/or' and 'vel'.

Consider the following statements, where the inclusive sense is distinguished (parenthetically) from the exclusive sense.

(is) Jones will win *or* Smith will win (possibly both)

(es) Jones will win *or* Smith will win (but not both)

We can imagine a scenario for each. In the first scenario, Jones and Smith, and a third person, Adams, are the only people running in an election in which two people are elected. So Jones or Smith will win, maybe both. In the second scenario, Jones and Smith are the two finalists in an election in which only one person is elected. In this case, one will win, the other will lose.

These two statements may be symbolized as follows.

(f1) J ∨ S

(f2) (J ∨ S) & ~(J & S)

We can read (f1) as saying that Jones will win and/or Smith will win, and we can read (f2) as saying that Jones will win or Smith will win *but* they won't both win (recall previous section on negations of conjunctions).

As with conjunctions, certain simple statements are straightforwardly equivalent to disjunctions, and are accordingly symbolized as such. The following are examples.

(s1) it is raining or sleeting
(d1) it raining, *or* it is sleeting R ∨ S

(s2) Jones is a fool or a liar
(d2) Jones is a fool, *or* Jones is a liar F ∨ L

10. 'NEITHER...NOR'

Having considered disjunctions, we next look at negations of disjunctions. For example, consider the following statement.

(e1) Kay *isn't either* a Freshman *or* a Sophomore

This may be paraphrased in the following, non-idiomatic, way.

(p1) it is not true that (Kay is either a Freshman or a Sophomore)

This is a negation of a disjunction, and is accordingly symbolized as follows.

(s1) $\sim(F \vee S)$

Now, an alternative, idiomatic, paraphrase of (e1) uses the expression 'neither...nor', as follows.

(p1') Kay is *neither* a Freshman *nor* a Sophomore

Comparing (p1') with the original statement (e1), we can discern the following principle.

> 'neither...nor'
> is the negation of
> 'either...or'

This suggests introducing a non-standard connective, neither-nor with the following defining property.

> neither $\mathcal{A}$ nor $\mathcal{B}$
> is logically equivalent to
> $\sim(\mathcal{A} \vee \mathcal{B})$

Note carefully that neither-nor in its connective guise is highly non-idiomatic. In particular, in order to obtain a grammatically *general* reading of it, we must read it as follows.

> neither $\mathcal{A}$ nor $\mathcal{B}$
> is officially read:
> *neither is it true that* $\mathcal{A}$
> *nor is it true that* $\mathcal{B}$

This is completely analogous to the standard (grammatically general) reading of 'not P' as 'it is not the case that P'.

For example, if R stands for 'it is raining' and S stands for 'it is sleeting', then 'neither R nor S' is read

neither is it true that it is raining
nor is it true that it is sleeting

This awkward reading of neither-nor is required in order to insure that 'neither P nor Q' is grammatical irrespective of the actual sentences P and Q. Of course, as with simple negation, one can usually transform the sentence into a more colloquial form. For example, the above sentence is more naturally read

neither is it raining nor is it sleeting,

or more naturally still,

it is neither raining nor sleeting.

We have suggested that neither-nor is the negation of either-or. Other uses of the word 'neither' suggest another, equally natural, paraphrase of neither-nor. Consider the following sentences.

neither Jay nor Kay is a Sophomore

Jay is not a Sophomore, and neither is Kay

A bit of linguistic reflection reveals that these two sentences are equivalent to one another. Further reflection reveals that the latter sentence is simply a stylistic variant of the more monotonous sentence

Jay is *not* a Sophomore, *and* Kay is *not* a Sophomore

The latter is a conjunction of two negations, and is accordingly symbolized:

$\sim J \,\&\, \sim K$

Thus, we see that a neither-nor sentence can be symbolized as a conjunction of two negations. This is entirely consistent with the truth-functional behavior of 'and', 'or', and 'not', since the following pair are logically equivalent, as is easily demonstrated using truth-tables.

$\sim(\mathcal{A} \vee \mathcal{B})$ is logically equivalent to $(\sim\mathcal{A} \,\&\, \sim\mathcal{B})$

We accordingly have two equally natural paraphrases of sentences involving neither-nor, given by the following principle.

neither $\mathcal{A}$ nor $\mathcal{B}$ may be paraphrased $\sim(\mathcal{A} \vee \mathcal{B})$ or equivalently $\sim\mathcal{A} \,\&\, \sim\mathcal{B}$

11. CONDITIONALS

The standard English expression for the conditional connective is 'if...then'. A standard conditional (statement) is a statement of the form

if $\mathcal{A}$, then $\mathcal{C}$,

where $\mathcal{A}$ and $\mathcal{B}$ are any statements (simple or compound), and is symbolized:

$$\mathcal{A} \rightarrow \mathcal{C}$$

Whereas $\mathcal{A}$ is called the *antecedent* of the conditional, $\mathcal{C}$ is called the *consequent* of the conditional. Note that, unlike conjunction and disjunction, the constituents of a conditional do not play symmetric roles.

There are a number of idiomatic variants of 'if...then'. In particular, all of the following statement forms are equivalent ($\mathcal{A}$ and $\mathcal{C}$ being any statements whatsoever).

(c1) if $\mathcal{A}$, then $\mathcal{C}$

(c2) if $\mathcal{A}$, $\mathcal{C}$
(c2') $\mathcal{C}$ if $\mathcal{A}$

(c3) provided (that) $\mathcal{A}$, $\mathcal{C}$
(c3') $\mathcal{C}$ provided (that) $\mathcal{A}$

(c4) in case $\mathcal{A}$, $\mathcal{C}$
(c4') $\mathcal{C}$ in case $\mathcal{A}$

(c5) on the condition that $\mathcal{A}$, $\mathcal{C}$
(c5') $\mathcal{C}$ on the condition that $\mathcal{A}$

In particular, all of the above statement forms are symbolized in the same manner:

$$\mathcal{A} \rightarrow \mathcal{C}$$

As the reader will observe, the order of antecedent and consequent is not fixed: in idiomatic English usage, sometimes the antecedent goes first, sometimes the consequent goes first. The following principles, however, should enable one systematically to identify the antecedent and consequent.

'if' always introduces the antecedent

'then' always introduces the consequent

'provided (that)', 'in case', and 'on the condition that' are variants of 'if'

12. 'EVEN IF'

The word 'if' frequently appears in combination with other words, the most common being 'even' and 'only', which give rise to the expressions 'even if', 'only if'.

In the present section, we deal very briefly with 'even if', leaving 'only if' to the next section.

The expression 'even if' is actually quite tricky. Consider the following examples.

(e1) the Allies would have won *even if* the U.S. had *not* entered the war (in reference to WW2)

(i1) the Allies would have won *if* the U.S. had *not* entered the war

These two statements *suggest* quite different things. Whereas (e1) suggests that the Allies did win, (i1) suggests that the Allies didn't win. A more apt use of 'if' would be:

(i2) the Axis powers would have won *if* the U.S. had not entered the war.

Notwithstanding the pragmatic matters of appropriate, sincere usage, it seems that the *pure semantic content* of 'even if' is the same as the pure semantic content of 'if'. The difference is not one of meaning but of *presupposition*, on the part of the speaker. In such examples, we tend to use 'even if' when we presuppose that the consequent is true, and we tend to use 'if' when we presuppose that the consequent is false. This is summarized as follows.

it would have been the case that $\mathcal{B}$ **if** it had been the case that $\mathcal{A}$ pragmatically presupposes $\sim\mathcal{B}$

it would have been the case that $\mathcal{B}$ **even if** it had been the case that $\mathcal{A}$ pragmatically presupposes $\mathcal{B}$

To say that one statement $\mathcal{A}$ *pragmatically presupposes* another statement $\mathcal{B}$ is to say that when one (sincerely) asserts $\mathcal{A}$, one takes for granted the truth of $\mathcal{B}$.

Given the subtleties of content versus presupposition, we will not consider 'even if' any further in this text.

13. 'ONLY IF'

The word 'if' frequently appears in combination with other words, the most common being 'even' and 'only', which give rise to the expressions 'even if', 'only if'.

The expression 'even if' is very complex, and somewhat beyond the scope of intro logic, so we do not consider it any further. So, let us turn to the other expression, 'only if', which involves its own subtleties, but subtleties that can be dealt with in intro logic.

First, we note that 'only if' is definitely *not* equivalent to 'if'. Consider the following statements involving 'only if'.

(o1) I will get an A in logic *only if* I take all the exams
(o2) I will get into law school *only if* I take the LSAT

Now consider the corresponding statements obtained by replacing 'only if' by 'if'.

(i1) I will get an A in logic *if* I take all the exams
(i2) I will get into law school *if* I take the LSAT

Whereas the 'only if' statements are true, the corresponding 'if' statements are false. It follows that 'only if' is not equivalent to 'if'.

The above considerations show that an 'only if' statement does not imply the corresponding 'if' statement. One can also produce examples of 'if' statements that do not imply the corresponding 'only if' statements. Consider the following examples.

(i3) I will pass logic *if* I score 100 on every exam
(i4) I am guilty of a felony *if* I murder someone

(o3) I will pass logic *only if* I score 100 on every exam
(o4) I am guilty of a felony *only if* I murder someone

Whereas both 'if' statements are true, both 'only if' statements are false. Thus, 'A if B' does not imply 'A only if B', and 'A only if B' does not imply 'A if B'.

So how do we paraphrase 'only if' statements using the standard connectives? The answer is fairly straightforward, being related to the general way in which the word 'only' operates in English – as a special dual-negative modifier.

As an example of 'only' in ordinary discourse, a sign that reads 'employees only' means to *exclude* anyone who is *not* an employee. Also, if I say 'Jay loves only Kay', I mean that he does *not* love anyone *except* Kay.

In the case of the connective 'only if', 'only' modifies 'if' by introducing two negations; in particular, the statement

$\mathcal{A}$ only if $\mathcal{B}$ is paraphrased not $\mathcal{A}$ if not $\mathcal{B}$

In other words, the 'if' stays put, and in particular continues to introduce the antecedent, but the 'only' becomes two negations, one in front of the antecedent (introduced by 'if'), the other in front of the consequent.

With this in mind, let us go back to original examples, and paraphrase them in accordance with this principle. In each case, we use a colloquial form of negation.

(p1) I will *not* get an A in logic *if* I do *not* take all the exams
(p2) I will *not* get into law school *if* I do *not* take the LSAT

Now, (p1) and (p2) are not in standard form, the problem being the relative position of antecedent and consequent. Recalling that '$\mathcal{A}$ if $\mathcal{B}$' is an idiomatic variant of 'if $\mathcal{B}$, then $\mathcal{A}$', we further paraphrase (p1) and (p2) as follows.

(p1') *if* I do *not* take all the exams, *then* I will *not* get an A in logic
(p2') *if* I do *not* take the LSAT, *then* I will *not* get into law school

These are symbolized, respectively, as follows.

(s1) $\sim T \rightarrow \sim A$
(s2) $\sim T \rightarrow \sim L$

Combining the paraphrases of 'only if' and 'if', we obtain the following principle.

$\mathcal{A}$ only if $\mathcal{B}$ is paraphrased not $\mathcal{A}$ if not $\mathcal{B}$ which is further paraphrased if not $\mathcal{B}$, then not $\mathcal{A}$ which is symbolized $\sim\mathcal{B} \rightarrow \sim\mathcal{A}$

14. A PROBLEM WITH THE TRUTH-FUNCTIONAL IF-THEN

The reader will recall that the truth-functional version of 'if...then' is characterized by the truth-function that makes '$\mathcal{A}\rightarrow\mathcal{B}$' false precisely when $\mathcal{A}$ is true and $\mathcal{B}$ is false. As noted already, this is not a wholly satisfactory analysis of English 'if...then'; rather, it is simply the best we can do by way of a truth-functional version of 'if...then'. Whereas the truth-functional analysis of 'if...then' is well suited to the timeless, causeless, eventless realm of mathematics, it is not so well suited to the realm of ordinary objects and events.

In the present section, we examine one of the problems resulting from the truth-functional analysis of 'if...then', a problem specifically having to do with the expression 'only if'.

We have paraphrased '$\mathcal{A}$ only if $\mathcal{B}$' as 'not $\mathcal{A}$ if not $\mathcal{B}$', which is paraphrased 'if not $\mathcal{B}$, then not $\mathcal{A}$', which is symbolized '$\sim\mathcal{B}\rightarrow\sim\mathcal{A}$'. The reader may recall that, using truth tables, one can show the following.

$$\sim\mathcal{B} \rightarrow \sim\mathcal{A}$$

is equivalent to

$$\mathcal{A} \rightarrow \mathcal{B}$$

Now, $\sim\mathcal{B}\rightarrow\sim\mathcal{A}$ is the translation of '$\mathcal{A}$ only if $\mathcal{B}$', whereas $\mathcal{A}\rightarrow\mathcal{B}$ is the translation of 'if $\mathcal{A}$, then $\mathcal{B}$'. Therefore, since $\sim\mathcal{B}\rightarrow\sim\mathcal{A}$ is truth-functionally equivalent to $\mathcal{A}\rightarrow\mathcal{B}$, we are led to conclude that '$\mathcal{A}$ only if $\mathcal{B}$' is truth-functionally equivalent to 'if $\mathcal{A}$, then $\mathcal{B}$'.

This means, in particular that our original examples,

(o1) I will get an A in logic *only if* I take the exams
(o2) I will get into law school *only if* I take the LSAT

are truth-functionally equivalent to the following, respectively:

(e1) *if* I get an A in logic, *then* I will take the exams
(e2) *if* I get into law school, *then* I will take the LSAT

Compared with the original statements, these sound odd indeed. Consider the last one. My response is that, if you get into law school, why bother taking the LSAT!

The oddity we have just discovered further underscores the shortcomings of the truth-functional if-then connective. The particular difficulty is summarized as follows.

$\mathcal{A}$ only if $\mathcal{B}$
is equivalent (in English) to
not $\mathcal{A}$ if not $\mathcal{B}$
which is equivalent (in English) to
if not $\mathcal{B}$, then not $\mathcal{A}$
which is symbolized

$$\sim\mathcal{B} \rightarrow \sim\mathcal{A}$$

which is equivalent (by truth tables) to

$$\mathcal{A} \rightarrow \mathcal{B}$$

which is the symbolization of
if $\mathcal{A}$ then $\mathcal{B}$.

To paraphrase '$\mathcal{A}$ only if $\mathcal{B}$' as 'if $\mathcal{A}$ then $\mathcal{B}$' is at the very least misleading in cases involving temporal or causal factors. Consider the following example.

(o3) my tree will grow *only if* it receives adequate light

is best paraphrased

(p3) my tree will *not* grow *if* it does *not* receive adequate light

which is quite different from

(e3) *if* my tree grows, *then* it will receive adequate light.

The latter statement may indeed be true, but it suggests that the growing leads to, and precedes, getting adequate light (as often happens with trees competing with one another for available light). By contrast, the former suggests that getting adequate light is required, and hence precedes, growing (as happens with all photosynthetic organisms).

A major problem with (e1)-(e3) is with the tense in the consequents. The word 'then' makes it natural to use future tense, probably because 'then' is used both in a *logical* sense and in a *temporal* sense (for example, recall 'and then').

If we insist on translating 'only if' statements into 'if... then' statements, following the method above, then we must adjust the tenses appropriately. So, for example, getting adequate light precedes growing, so the appropriate tense is not simple future but future perfect. Adjusting the tenses in this manner, we obtain the following re-paraphrases of (e1)-(e3).

(p1') *if* I get an A in logic, *then* I *will have* taken the exams
(p2') *if* I get into law school, *then* I *will have* taken the LSAT
(p3') *if* my tree grows, *then* it *will have* received adequate light

Unlike the corresponding statements using simple future, these statements, which use future perfect tense, are more plausible paraphrases of the original 'only if' statements.

Nonetheless, 'not $\mathcal{A}$ if not $\mathcal{B}$' remains the generally most accurate paraphrase of '$\mathcal{A}$ only if B'.

15. 'IF AND ONLY IF'

Having examined 'if', and having examined 'only if', we next consider their natural conjunction, which is 'if and only if'. Consider the following sentence.

(e) you will pass *if and only if* you average at least fifty

This is naturally thought of as dividing into two halves, a promise-half and a threat-half. The promise is

(p) you will pass *if* you average at least fifty,

and the threat is

(t) you will pass *only if* you average at least fifty,

which we saw in the previous section may be paraphrased:

(t') you will *not* pass *if* you do *not* average at least fifty.

So (e) may be paraphrased as a conjunction:

(t'') you will pass *if* you average at least fifty,
and
you will *not* pass *if* you do *not* average at least fifty.

The first conjunct is symbolized:

$$A \rightarrow P$$

and the second conjunct is symbolized:

$$\sim A \rightarrow \sim P$$

so the conjunction is symbolized:

$$(A \rightarrow P) \ \& \ (\sim A \rightarrow \sim P)$$

The reader may recall that our analysis of the biconditional connective $\leftrightarrow$ is such that the above formula is truth-functionally equivalent to

$$P \leftrightarrow A$$

So $P \leftrightarrow A$ also counts as an acceptable symbolization of 'P if and only if A', although it does not do full justice to the internal logical structure of 'if and only if' statements, which are more naturally thought of as conjunctions of 'if' statements and 'only if' statements.

16. 'UNLESS'

There are numerous ways to express conditionals in English. We have already seen several conditional-forming expressions, including 'if', 'provided', 'only if'. In the present section, we consider a further conditional-forming expression – 'unless'.

'Unless' is very similar to 'only if', in the sense that it has a built-in negation. The difference is that, whereas 'only if' incorporates two negations, 'unless' incorporates only one. This means, in particular, that in order to paraphrase 'only if' statements using 'unless', one must add one explicit negation to the sentence. The following are examples of 'only if' statements, followed by their respective paraphrases using 'unless'.

(o1) I will graduate *only if* I pass logic
(u1) I will *not* graduate *unless* I pass logic
(u1') *unless* I pass logic, I will *not* graduate

(o2) I will pass logic *only if* I study
(u2) I will *not* pass logic *unless* I study
(u2') *unless* I study, I will *not* pass logic

Let us concentrate on the first one. We already know how to paraphrase and symbolize (o1), as follows.

(p1) I will *not* graduate *if* I do *not* pass logic
(p1') *if* I do *not* pass logic, *then* I will *not* graduate
(s1) $\sim P \rightarrow \sim G$

Now, comparing (u1) and (u1') with the last three items, we discern the following principle concerning 'unless'.

'unless' is equivalent to 'if not'

Here, 'if not' is short for 'if it is not true that'. Notice that this principle applies when 'unless' appears at the beginning of the statement, as well as when it appears in the middle of the statement.

The above principle may be restated as follows.

$\mathcal{A}$ unless $\mathcal{B}$ is equivalent to $\mathcal{A}$ if not $\mathcal{B}$ which is symbolized $\sim\mathcal{B} \rightarrow \mathcal{A}$	unless $\mathcal{A}$, $\mathcal{B}$ is equivalent to if not $\mathcal{A}$, then $\mathcal{B}$ which is symbolized $\sim\mathcal{A} \rightarrow \mathcal{B}$

17. THE STRONG SENSE OF 'UNLESS'

As with many words in English, the word 'unless' is occasionally used in a way different from its "official" meaning. As with the word 'or', which has both a weak (inclusive) sense and a strong (exclusive) sense, the word 'unless' also has both a weak and strong sense.

Just as we opt for the weak (inclusive) sense of 'or' in logic, we also opt for the weak sense of 'unless', which is summarized in the following principle.

the weak sense of 'unless' is equivalent to 'if not'

Unfortunately, 'unless' is not always intended in the weak sense. In addition to the meaning 'if not', various Webster Dictionaries give 'except when' and 'except on the condition that' as further meanings.

First, let us consider the meaning of 'except'; for example, consider the following fairly ordinary 'except' statement, which is taken from a grocery store sign.

(e1) open 24 hours a day *except* Sundays

It is plausible to suppose that (e1) means that the store is open 24 hours Monday-Saturday, and is *not* open 24 hours on Sunday (on Sunday, it may not be open at all, or it may only be open 8 hours). Thus, there are two implicit conditionals, as follows, where we let 'open' abbreviate 'open 24 hours'.

(c1) *if* it is *not* Sunday, *then* the store is open
(c2) *if* it is Sunday, *then* the store is *not* open

These two can be combined into the following biconditional.

(b) the store is open *if and only if* it is *not* Sunday

which is symbolized:

(s) $O \leftrightarrow {\sim}S$

Now, similar statements can be made using 'unless'. Consider the following statement from a sign on a swimming pool.

(u1) the pool may *not* be used *unless* a lifeguard is on duty

Following the dictionary definition, this is equivalent to:

(u1') the pool may *not* be used *except when* a lifeguard is on duty which amounts to the conjunction,

(c) the pool may *not* be used *if* a lifeguard is *not* on duty, *and* the pool may be used *if* a lifeguard is on duty.

which, as noted earlier, is equivalent to the following biconditional,

(b) the pool may be used *if and only if* a lifeguard is on duty

By comparing (b) with the original statement (u1), we can discern the following principle about the strong sense of 'unless'.

> the strong sense of
> 'unless'
> is equivalent to
> 'if and only if not'

Or stating it using our symbols, we may state the principle as follows.

> $\mathcal{A}$ unless $\mathcal{B}$
> (in the strong sense of unless)
> is equivalent to
> $\mathcal{A} \leftrightarrow \sim\mathcal{B}$

It is not always clear whether 'unless' is intended in the strong or in the weak sense. Most often, the overall context is important for determining this. The following rules of thumb may be of some use.

> Usually, if it is intended in the strong sense, 'unless' is placed in the middle of a sentence; (the converse, however, is not true).

> Usually, if 'unless' is at the beginning of a statement, then it is intended in the weak sense.

> If it is not obvious that 'unless' is intended in the strong sense, you should assume that it is intended in the weak sense.

Note carefully: Although 'unless' is occasionally used in the strong sense, you may assume that every exercise uses 'unless' in the weak sense.

Exercise (an interesting coincidence): show that, whereas the weak sense of 'unless' is truth-functionally equivalent to the weak (inclusive) sense of 'or', the strong sense of 'unless' is truth-functionally equivalent to the strong (exclusive) sense of 'or'.

18. NECESSARY CONDITIONS

There are still other words used in English to express conditionals, most importantly the words 'necessary' and 'sufficient'. In the present section, we examine conditional statements that involve 'necessary', and in the next section, we do the same thing with 'sufficient'.

The following expressions are some of the common ways in which 'necessary' is used.

(n1) in order that...it is necessary that...
(n2) in order for...it is necessary for...
(n3) in order to...it is necessary to...
(n4) ...is a necessary condition for...
(n5) ...is necessary for...

The following are examples of mutually equivalent statements using 'necessary'.

(N1) *in order that* I get an A, *it is necessary that* I take all the exams
(N2) *in order for* me to get an A, *it is necessary for* me to take all the exams
(N3) *in order to* get an A, *it is necessary to* take all the exams
(N4) taking all the exams *is a necessary condition for* getting an A
(N5) taking all the exams *is necessary for* getting an A

Statements involving 'necessary' can all be paraphrased using 'only if'. A more direct approach, however, is first to paraphrase the sentence into the simplest form, which is:

(f) $\mathcal{A}$ is necessary for $\mathcal{B}$

Now, to say that one state of affairs (event) $\mathcal{A}$ is necessary for another state of affairs (event) $\mathcal{B}$ is just to say that *if* the first thing does *not* obtain (happen), *then neither* does the second. Thus, for example, to say

taking all the exams *is necessary for* getting an A

is just to say that if E (i.e., taking-the-exams) doesn't obtain then neither does A (i.e., getting-an-A). The sentence is accordingly paraphrased and symbolized as follows.

if not E, then not A [$\sim E \rightarrow \sim A$]

The general paraphrase principle is as follows.

$\mathcal{A}$ is necessary for $\mathcal{B}$ is paraphrased if not $\mathcal{A}$, then not $\mathcal{B}$

19. SUFFICIENT CONDITIONS

The natural logical counterpart of 'necessary' is 'sufficient', which is used in the following ways, completely analogous to 'necessary'.

(s1) in order that...it is sufficient that...
(s2) in order for....it is sufficient for...
(s3) in order to....it is sufficient to....
(s4) ...is a sufficient condition for...
(s5) ...is sufficient for...

The following are examples of mutually equivalent statements using these different forms.

(S1) *in order that* I get an A *it is sufficient that* I get a 100 on every exam

(S2) *in order for* me to get an A *it is sufficient for* me to get a 100 on every exam

(S3) *in order to* get an A *it is sufficient* to get a 100 on every exam

(S4) getting a 100 on every exam *is a sufficient condition* for getting an A

(S5) getting a 100 on every exam *is sufficient for* getting an A

Just as necessity statements can be paraphrased like 'only if' statements, sufficiency statements can be paraphrased like 'if' statements. The direct approach is first to paraphrase the sufficiency statement in the following form.

(f) $\mathcal{A}$ is sufficient for $\mathcal{B}$

Now, to say that one state of affairs (event) $\mathcal{A}$ is sufficient for another state of affairs (event) $\mathcal{B}$ is just to say that $\mathcal{B}$ obtains (happens) *provided* (*if*) $\mathcal{A}$ obtains (happens). So for example, to say that

getting a 100 on every exam *is sufficient for* getting an A

is to say that

getting-an-A happens *provided* (*if*) getting-a-100 happens

which may be symbolized quite simply as:

$H \rightarrow A$

The general principle is as follows.

$\mathcal{A}$ is sufficient for $\mathcal{B}$ is paraphrased if $\mathcal{A}$, then $\mathcal{B}$

20. NEGATIONS OF NECESSITY AND SUFFICIENCY

First, note carefully that necessary conditions are quite different from sufficient conditions. For example,

> taking all the exams *is necessary for* getting an A,
> but
> taking all the exams *is not sufficient for* getting an A.

Similarly,

> getting a 100 *is sufficient for* getting an A,
> but
> getting a 100 *is not necessary for* getting an A.

This suggests that we can combine necessity and sufficiency in a number of ways to obtain various statements about the relation between two events (states of affairs). For example, we can say all the following, with respect to $\mathcal{A}$ and $\mathcal{B}$.

(c1) $\mathcal{A}$ is necessary for $\mathcal{B}$
(c2) $\mathcal{A}$ is sufficient for $\mathcal{B}$
(c3) $\mathcal{A}$ is not necessary for $\mathcal{B}$
(c4) $\mathcal{A}$ is not sufficient for $\mathcal{B}$

(c5) $\mathcal{A}$ is both necessary and sufficient for $\mathcal{B}$
(c6) $\mathcal{A}$ is necessary but not sufficient for $\mathcal{B}$
(c7) $\mathcal{A}$ is sufficient but not necessary for $\mathcal{B}$
(c8) $\mathcal{A}$ is neither necessary nor sufficient for $\mathcal{B}$

We have already discussed how to paraphrase (c1)-(c2). In the present section, we consider how to paraphrase (c3)-(c4), leaving (c5)-(c8) to a later section.

We start with the following example involving 'not necessary'.

(1) attendance *is not necessary* for passing logic

This may be regarded as the negation of

(2) attendance *is necessary* for passing logic

As seen earlier, the latter may be paraphrased and symbolized as follows.

(p2) if I do *not* attend class, *then* I will *not* pass logic

(s2) $\sim A \rightarrow \sim P$

So the negation of (2), which is (1), may be paraphrased and symbolized as follows.

(p1) *it is not true that if* I do *not* attend class, *then* I will *not* pass logic;

(s1) $\sim(\sim A \rightarrow \sim P)$

Notice, once again, that voodoo does not prevail in logic; there is no obvious simplification of the three negations in the formula. The negations do not simply cancel each other out. In particular, the latter is *not* equivalent to the following.

(voodoo) $A \rightarrow P$

The latter says (roughly) that attendance will ensure passing; this is, of course, not true. Your dog can attend every class, if you like, but it won't pass the course. The former says that attendance is not necessary for passing; this is true, in the sense that attendance is not an official requirement.

Next, consider the following example involving 'not sufficient'.

(3) taking all the exams is *not sufficient* for passing logic

This may be regarded as the negation of

(4) taking all the exams *is sufficient* for passing logic.

The latter is paraphrased and symbolized as follows.

(p4) *if* I take all the exams, *then* I will pass logic

(s4) $E \rightarrow P$

So the negation of (4), which is (3), may be paraphrased and symbolized as follows.

(p3) *it is not true that if* I take all the exams, *then* I will pass logic

(s4) $\sim(E \rightarrow P)$

As usual, there is no simple-minded (voodoo) transformation of the negation. The negation of an English conditional does not have a straightforward simplification. In particular, it is *not* equivalent to the following

(voodoo) $\sim E \rightarrow \sim P$

The former says (roughly) that taking all the exams does *not* ensure passing; this is true; after all, you can fail all the exams. On the other hand, the latter says that if you don't take all the exams, then you won't pass. This is not true, a mere 70 on each of the first three exams will guarantee a pass, in which case you don't have to take all the exams in order to pass.

21. YET ANOTHER PROBLEM WITH THE TRUTH-FUNCTIONAL IF-THEN

According to our analysis, to say that one state of affairs (event) $\mathcal{A}$ is *not sufficient* for another state of affairs (event) $\mathcal{B}$ is to say that it is not true that if the first obtains (happens), then so will the second. In other words,

$\mathcal{A}$ is not sufficient for $\mathcal{B}$ is paraphrased: it is not true that if $\mathcal{A}$ then $\mathcal{B}$, which is symbolized: $\sim(\mathcal{A} \rightarrow \mathcal{B})$

As noted in the previous section, there is no obvious simple transformation of the latter formula. On the other hand, the latter formula can be simplified in accordance with the following truth-functional equivalence, which can be verified using truth tables.

$\sim(\mathcal{A} \rightarrow \mathcal{B})$ is truth-functionally equivalent to $\mathcal{A}\ \&\ \sim\mathcal{B}$

Consider our earlier example,

(1) taking all the exams is *not sufficient* for passing logic

Our proposed paraphrase and symbolization is:

(p1) *it is not true that if* I take all the exams *then* I will pass logic
(s1) $\sim(E \rightarrow P)$

But this is truth-functionally equivalent to:

(s2) $E\ \&\ \sim P$
(p2) I will take all the exams, *and* I will *not* pass

However, to say that taking the exams is *not sufficient* for passing logic is not to say you *will* take all the exams yet you won't pass; rather, it says that it is *possible* (in some sense) for you to take the exams and yet not pass.

However, possibility is not a truth-functional concept; some falsehoods are possible; some falsehoods are impossible. Thus, possibility cannot be analyzed in truth-functional logic.

We have dealt with negations of conditionals, which lead to difficulties with the truth-functional analysis of necessity and sufficiency. Nevertheless, our paraphrase technique involving 'if...then' is not impugned, only the truth-functional analysis of 'if...then'.

22. COMBINATIONS OF NECESSITY AND SUFFICIENCY

Recall that the possible combinations of statements about necessity and sufficiency are as follows.

(c1) $\mathcal{A}$ is necessary for $\mathcal{B}$
(c2) $\mathcal{A}$ is sufficient for $\mathcal{B}$
(c3) $\mathcal{A}$ is not necessary for $\mathcal{B}$
(c4) $\mathcal{A}$ is not sufficient for $\mathcal{B}$

(c5) $\mathcal{A}$ is both necessary and sufficient for $\mathcal{B}$
(c6) $\mathcal{A}$ is necessary, but not sufficient, for $\mathcal{B}$
(c7) $\mathcal{A}$ is sufficient, but not necessary, for $\mathcal{B}$
(c8) $\mathcal{A}$ is neither necessary nor sufficient for $\mathcal{B}$

We have already dealt with (c1)-(c4). We now turn to (c5)-(c8).

First, notice carefully that (c1)-(c4) are less informative than (c5)-(c8). For example, if I say $\mathcal{A}$ is necessary for $\mathcal{B}$, and leave it at that, I am not saying whether $\mathcal{A}$ is sufficient for $\mathcal{B}$, one way or the other. Similarly, if I say that Jay is a Sophomore, and leave it at that, I have said nothing concerning whether Kay is a Sophomore, one way or the other.

Consider the following example of combination (c5).

(e5) averaging at least 50 *is both necessary and sufficient for* passing

This is quite clearly the conjunction of a necessity statement and a sufficiency statement, as follows.

averaging at least fifty *is necessary for* passing,
and
averaging at least fifty *is sufficient for* passing

The latter is symbolized:

$$(\sim F \rightarrow \sim P) \;\&\; (F \rightarrow P)$$

Reading this back into English, we obtain

if I do *not* average at least fifty, *then* I will *not* pass,
and
if do average at least fifty, *then* I will pass

Next, consider the following example of combination (c6).

(e6) taking all the exams *is necessary, but not sufficient, for* getting an A

This is a somewhat more complex conjunction:

taking all the exams *is necessary for* getting an A,
but
taking all the exams *is not sufficient for* getting an A

which is symbolized:

$$(\sim T \rightarrow \sim A) \;\&\; \sim(T \rightarrow A)$$

Reading this back into English, we obtain

if I do *not* take all the exams, *then* I will *not* get an A, *but it is not true that if* I do take all the exams *then* I will get an A

Next, consider the following example of combination (c7).

(e7) getting 100 on every exam *is sufficient, but not necessary, for* getting an A

This too is a conjunction:

getting 100 on every exam *is sufficient for* getting an A,
but
getting 100 on every exam *is not necessary for* getting an A

which is symbolized:

$$(H \rightarrow A) \;\&\; \sim(\sim H \rightarrow \sim A)$$

Reading this back into English, we obtain

if I get a 100 on every exam, *then* I will get an A,
but it is not true that
if I do *not* get a 100 on every exam *then* I will *not* get an A

Finally, consider the following example of combination (c8).

(e8) attending class is neither necessary nor sufficient for passing

which may be paraphrased as a complex conjunction:

attending class *is not necessary for* passing,
and
attending class *is not sufficient for* passing

which is symbolized:

$$\sim(\sim A \rightarrow \sim P) \;\&\; \sim(A \rightarrow P)$$

Reading this back into English, we obtain

it is not true that
    if I do *not* attend class
        then I will *not* pass,
nor is it true that
    if I do attend class
        then I will pass

23. 'OTHERWISE'

In the present section, we consider two three-place connective expressions that are used to express conditionals in English. The key words are 'otherwise' and 'in which case'.

First, the general forms for 'otherwise' statements are the following:

(o1) if $\mathcal{A}$, then $\mathcal{B}$; otherwise C
(o2) if $\mathcal{A}$, $\mathcal{B}$; otherwise C
(o3) $\mathcal{B}$ if $\mathcal{A}$; otherwise C

The following is a typical example.

(e1) *if* it is sunny, I'll play tennis
otherwise, I'll play racquetball

This statement asserts what the speaker will do *if* it is sunny, and it further asserts what the speaker will do *otherwise*, i.e., *if* it is *not* sunny. In other words, (e1) can be paraphrased as a conjunction, as follows.

(p1) *if* it is sunny, *then* I'll play tennis,
and
if it is *not* sunny, *then* I'll play racquetball

The latter statement is symbolized:

(s1) $(S \to T) \& (\sim S \to R)$

The general principle governing the paraphrase of 'otherwise' statements is as follows.

if $\mathcal{A}$, then $\mathcal{B}$; otherwise C is paraphrased if $\mathcal{A}$, then $\mathcal{B}$, and if not $\mathcal{A}$, then C, which is symbolized $(\mathcal{A} \to \mathcal{B}) \& (\sim\mathcal{A} \to C)$

A simple variant of 'otherwise' is 'else', which is largely interchangeable with 'otherwise'. In a number of high level programming languages, including BASIC and PASCAL, 'else' is used in conjunction with 'if...then' to issue commands. For example, the following is a typical BASIC command.

```
(c)    if X<=100 then goto 300 else goto 400
```

This is equivalent to two commands in succession:

```
if X<=100 then goto 300
if not(X<=100) then goto 400
```

In a computer language, such as BASIC, there is always a "default" 'else' command, namely to go to the next line and follow that command. So, for example, the command line

```
if X<=100 then goto 400
```

standing alone means

```
if X<=100 then goto 400 else goto next line
```

Unlike 'if...then' statements in computer languages, English 'if...then' statements do not incorporate default 'else' clauses. For example, the statement

(e2) I'll go to the doctor *if* I break my arm

says nothing about what the speaker will or won't do if he/she does *not* break an arm. Similarly, if I say I won't play tennis if it is raining, and leave it at that, I am not committing myself to anything in case it is not raining; I leave that case open, or undetermined.

That brings us to an expression that is very similar to 'otherwise' – namely, 'in which case'. Consider the following example.

(e2) I'll play tennis *unless* it is raining, *in which case* I'll play squash

Recall that 'unless' is equivalent to 'if not'. So, as with 'otherwise' statements, there are two cases considered – it rains; it doesn't rain. Statement (e2) asserts what the speaker will do in each case – *in case* it is *not* raining, and *in case* it is raining. Recall 'in case' is a variant of 'if'.

The paraphrase of (e2) is similar to that of (e1).

(p) *if* it is *not* raining, *then* I'll play tennis,
and
if it is raining, *then* I'll play squash

The latter is symbolized:

(s) $(\sim R \rightarrow T) \& (R \rightarrow S)$

The overall paraphrase pattern is given by the following principle.

$\mathcal{A}$ unless $\mathcal{B}$, in which case $\mathcal{C}$ is paraphrased if not $\mathcal{B}$, then $\mathcal{A}$, and if $\mathcal{B}$ then $\mathcal{C}$ which is symbolized $(\sim\mathcal{B} \rightarrow \mathcal{A}) \& (\mathcal{B} \rightarrow \mathcal{C})$

24. PARAPHRASING COMPLEX STATEMENTS

As noted earlier, compound statements may be built up from statements which are themselves compound statements. There are no theoretical limits to the complexity of compound statements, although there are practical limits, based on human linguistic capabilities.

We have already dealt with a number of complex statements in connection with the various non-standard connectives. We now systematically consider complex statements that involve various combinations of non-standard connectives. For example, we are interested in what happens when both 'unless' and 'only if' appear in the same sentence.

In paraphrasing and symbolizing complex statements, it is best to proceed systematically, in small steps. As one gets better, many intermediate steps can be done in one's head. On the easy ones, perhaps all the intermediate steps can be done in one's head. Still, it is a good idea to reason through the easy ones systematically, in order to provide practice in advance of doing the hard ones.

The first step in paraphrasing statements is:

Step 1:	Identify the simple (atomic) statements, and abbreviate them by upper case letters.

In most of the exercises, certain words are entirely capitalized in order to suggest to the student what the atomic statements are. For example, in the statement 'JAY and KAY are Sophomores' the atomic formulas are 'J' and 'K'.

At this stage of analysis, it is important to be clear concerning what each atomic formula stands for; it is especially important to be clear that each letter abbreviates a *complete sentence*. For example, in the above statement, 'J' does not stand for 'Jay', since this is not a sentence. Rather, it stands for 'Jay is a Sophomore'. Similarly, 'K' does not stand for 'Kay', but rather 'Kay is a Sophomore'.

Having identified the simple statements, and having established their abbreviations, the next step is:

Step 2:	Identify all the connectives, noting which ones are standard, and which ones are not standard.

Having identified the atomic statements and the connectives, the next step is:

Step 3:	Write down the first hybrid formula, making sure to retain internal punctuation.

The first hybrid formula is obtained from the original statement by replacing the simple statements by their abbreviations. A hybrid formula is so called because it contains both English words and symbols from sentential logic. Punctuation provides important clues about the logical structure of the sentence.

The first three steps may be better understood by illustration. Consider the following example.

Example 1

(e1) if neither Jay nor Kay is working, then we will go on vacation.

In this example, the simple statements are:

J: Jay is working
K: Kay is working
V: we go on vacation

and the connectives are:

if...then	(standard)
neither...nor	(non-standard)

Thus, our first hybrid formula is:

(h1) if neither J nor K, then V

Having obtained the first hybrid formula, the next step is to

> **Step 4:** Identify the major connective.

Here, the commas are important clues. In (h1), the placement of the comma indicates that the major connective is 'if...then', the structure being:

if neither J nor K,
then V

Having identified the major connective, we go on to the next step.

> **Step 5:** Symbolize the major connective if it is standard; otherwise, paraphrase it into standard form, and go back to step 4, and work on the resulting (hybrid) formula.

In (h1), the major connective is 'if...then', which is standard, so we symbolize it, which yields the following hybrid formula.

(h2) (neither J nor K) $\rightarrow$ V

Notice that, as we symbolize the connectives, we must provide the necessary logical punctuation (i.e., parentheses).

At this point, the next step is:

Step 6: Work on the constituent formulas separately.

In (h2), the constituent formulas are:

(c1) neither J nor K
(c2) V

The latter formula is fully symbolic, so we are through with it. The former is not fully symbolic, so we must work on it further. It has only one connective, 'neither...nor', which is therefore the major connective. It is not standard, so we must paraphrase it, which is done as follows.

(c1) neither J nor K
(p1) not J and not K

The latter formula is in standard form, so we symbolize it as follows.

(s1) $\sim J \ \& \sim K$

Having dealt with the constituent formulas, the next step is:

Step 7: Substitute symbolizations of constituents back into (original) hybrid formula.

In our first example, this yields:

(s2) $(\sim J \ \& \sim K) \rightarrow V$

Once you have a purely symbolic formula, the final step is:

Step 8: Translate the formula back into English and compare with the original statement.

This is to make sure the final formula says the same thing as the original statement. In our example, translating yields the following.

(t1) *if* Jay is *not* working and Kay is *not* working, *then* we will go on vacation.

Comparing this with the original,

(e1) if neither Jay nor Kay is working, then we will go on vacation

we see they are equivalent, so we are through.

Our first example is simple insofar as the major connective is standard. In many statements, all the connectives are non-standard, and so they have to be paraphrased in accordance with the principles discussed in previous sections. Consider the following example.

Example 2

(e2) you will pass unless you goof off, provided that you are intelligent.

In this statement, the simple statements are:

I: you are intelligent
P: you pass
G: you goof off

and the connectives are:

unless	(non-standard)
provided that	(non-standard)

Thus, the first stage of the symbolization yields the following hybrid formula.

(h1) P unless G, provided that I

Next, we identity the major connective. Once again, the placement of the comma tells us that 'provided that' is the major connective, the overall structure being:

P unless G,
provided that I

We cannot directly symbolize 'provided that', since it is non-standard. We must first paraphrase it. At this point, we recall that 'provided that' is equivalent to 'if', which is a simple variant of 'if...then'. This yields the following successive paraphrases.

(h2) P unless G, if I
(h3) if I, then P unless G

In (h3), the major connective is 'if...then', which is standard, so we symbolize it, which yields:

(h4) I $\rightarrow$ (P unless G)

We next work on the parts. The antecedent is finished, so we more to the consequent.

(c) P unless G

This has one connective, 'unless', which is non-standard, so we paraphrase and symbolize it as follows.

(c) P unless G
(p) P if not G,

(p') if not G, then P,
(s) $\sim$G $\rightarrow$ P

Substituting the parts back into the whole, we obtain the final formula.

(f) $I \rightarrow (\sim G \rightarrow P)$

Finally, we translate (f) back into English, which yields:

(t) *if* you are intelligent, *then* if you do *not* goof off *then* you will pass

Although this is not the exact same sentence as the original, it should be clear that they are equivalent in meaning.

Let us consider an example similar to Example 2.

Example 3

(e3) unless the exam is very easy, I will make a hundred only if I study

In this example, the simple statements are:

E: the exam is very easy
H: I make a hundred
S: I study

and the connectives are:

unless	(non-standard)
only if	(non-standard)

Having identified the logical parts, we write down the first hybrid formula.

(h1) unless E, H only if S

Next, we observe that 'unless' is the principal connective. Since it is non-standard, we cannot symbolize it directly, so we paraphrase it, as follows.

(h2) if not E, then H only if S

We now work on the new hybrid formula (h2). We first observe that the major connective is 'if...then'; since it is standard, we symbolize it, which yields:

(h3) not E $\rightarrow$ (H only if S)

Next, we work on the separate parts. The antecedent is simple, and is standard form, being symbolized:

(a) $\sim E$

The consequent has just one connective 'only if', which is non-standard, so we paraphrase and symbolize it as follows.

(c) H only if S
(p) not H if not S
(p') if not S, then not H
(s) $\sim S \rightarrow \sim H$

Next, we substitute the parts back into (h3), which yields:

(f) $\sim E \rightarrow (\sim S \rightarrow \sim H)$

Finally, we translate (f) back into English, which yields:

(t) *if* the exam is *not* very easy,
    then if I do *not* study
        then I will *not* get a hundred

Comparing this statement with the original statement, we see that they say the same thing.

The next example is slightly more complicated, being a conditional in which both constituents are conditionals.

Example 4

(e4) *if* Jones will work *only if* Smith is fired, *then* we should fire Smith *if* we want the job finished

In (e4), the simple statements are:

J: Jones works
F: we do fire Smith
S: we should fire Smith
W: we want the job finished

and the connectives are:

if...then	(standard)
only if	(non-standard)
if	(non-standard)

Next, we write down the first hybrid formula, which is:

(h1) if J only if F, then S if W

The comma placement indicates that the principal connective is 'if...then'. It is standard, so we symbolize it, which yields:

(h2) (J only if F) $\rightarrow$ (S if W)

Next, we work on the constituents separately. The antecedent is paraphrased and symbolized as follows.

(a) J only if F
(p) not J if not F
(p') if not F, then not J
(s) $\sim F \rightarrow \sim J$

The consequent is paraphrased and symbolized as follows.

(c) S if W
(p) if W, then S
(s) $W \rightarrow S$

Substituting the constituent formulas back into (h2) yields:

(f) $(\sim F \rightarrow \sim J) \rightarrow (W \rightarrow S)$

The direct translation of (f) into English reads as follows.

(t) *if if* we do *not* fire Smith *then* Jones does *not* work,
then if we want the job finished
then we should fire Smith

The complexity of the conditional structure of this sentence renders a direct translation difficult to understand. The major problem is the "stuttering" at the beginning of the sentence. The best way to avoid this problem is to opt for a more idiomatic translation (just as we do with negations); specifically, we replace some if-then's by simple variant forms. The following is an example of a more natural, idiomatic translation.

(t') *if* Jones will *not* work *if* Smith is *not* fired,
then if we want the job finished we should fire Smith

Comparing this paraphrase, in more idiomatic English, with the original statement, we see that they are equivalent in meaning.

Our last example involves the notion of necessary condition.

Example 5

(e5) in order to put on the show it will be necessary to find a substitute, if neither the leading lady nor her understudy recovers from the flu

In (e5), the simple statements are:

P: we put on the show
S: we find a substitute
L: the leading lady recovers from the flu
U: the understudy recovers from the flu

and the connectives are:

in order to... it is necessary to	(non-standard)
if	(non-standard)
neither...nor	(non-standard)

The first hybrid formula is:

(h1) in order that P it is necessary that S, if neither L nor U

Next, the principal connective is 'if', which is not in standard form; converting it into standard form yields:

(h2) if neither L nor U, then in order that P it is necessary that S

Here, the principal connective is 'if...then', which is standard, so we symbolize it as follows.

(h3) (neither L nor U) $\rightarrow$ (in order that P it is necessary that S)

We next attack the constituents. The antecedent is paraphrased as follows.

(a) neither L nor U
(p) not L and not U
(s) ~L & ~U

The consequent is paraphrased as follows.

(c) in order that P it is necessary that S
(p) S is necessary for P
(p') if not S, then not P
(s) ~S $\rightarrow$ ~P

Substituting the parts back into (h3), we obtain:

(f) (~L & ~U) $\rightarrow$ (~S $\rightarrow$ ~P)

Translating (f) back into English, we obtain:

(t) *if* the leading lady does *not* recover from the flu *and* her understudy does *not* recover from the flu, *then if* we do *not* find a substitute *then* we do *not* put on the show

Comparing (t) with the original statement, we see that they are equivalent in meaning.

By way of concluding this chapter, let us review the basic steps involved in symbolizing complex statements.

25. GUIDELINES FOR TRANSLATING COMPLEX STATEMENTS

Step 1:	Identify the simple (atomic) statements, and abbreviate them by upper case letters. What complete sentence does each letter stand for?

Step 2:	Identify all the connectives, noting which ones are standard, and which ones are non-standard.

Step 3:	Write down the first hybrid formula, making sure to retain internal punctuation.

Step 4:	Identify the major connective.

Step 5:	Symbolize the major connective if it is standard, introducing parentheses as necessary; otherwise, paraphrase it into standard form, and go back to step 4, and work on the resulting (hybrid) formula.

Step 6:	Work on the constituent formulas separately, which means applying steps 4-5 to each constituent formula.

Step 7:	Substitute symbolizations of constituents back into (original) hybrid formula.

Step 8:	Translate the formula back into English and compare with the original statement.

26. EXERCISES FOR CHAPTER 4

Directions: Translate each of the following statements into the language of sentential logic. Use the suggested abbreviations (capitalized words), if provided; otherwise, devise an abbreviation scheme of your own. In each case, write down what atomic statement each letter stands for, making sure it is a *complete sentence.* Letters should stand for positively stated sentences, not negatively stated ones; for example, the negative sentence 'I am not hungry' should be symbolized as '~H' using 'H' to stand for 'I am hungry'.

EXERCISE SET A

1. Although it is RAINING, I plan to go JOGGING this afternoon.
2. It is not RAINING, but it is still too WET to play.
3. JAY and KAY are Sophomores.
4. It is DINNER time, but I am not HUNGRY.
5. Although I am TIRED, I am not QUITTING.
6. Jay and Kay are roommates, but they hate one another.
7. Jay and Kay are Republicans, but they both hate Nixon.
8. KEEP trying, and the answer will APPEAR.
9. GIVE him an inch, and he will TAKE a mile.
10. Either I am CRAZY or I just SAW a flying saucer.
11. Either Jones is a FOOL or he is DISHONEST.
12. JAY and KAY won't both be present at graduation.
13. JAY will win, or KAY will win, but not both.
14. Either it is RAINING, or it is SUNNY and COLD.
15. It is RAINING or OVERCAST, but in any case it is not SUNNY.
16. If JONES is honest, then so is SMITH.
17. If JONES isn't a crook, then neither is SMITH.
18. Provided that I CONCENTRATE, I will not FAIL.
19. I will GRADUATE, provided I pass both LOGIC and HISTORY.
20. I will not GRADUATE if I don't pass both LOGIC and HISTORY.

EXERCISE SET B

21. Neither JAY nor KAY is able to attend the meeting.
22. Although I have been here a LONG time, I am neither TIRED nor BORED.
23. I will GRADUATE this semester only if I PASS intro logic.
24. KAY will attend the party only if JAY does not.
25. I will SUCCEED only if I WORK hard and take RISKS.
26. I will go to the BEACH this weekend, unless I am SICK.
27. Unless I GOOF off, I will not FAIL intro logic.
28. I won't GRADUATE unless I pass LOGIC and HISTORY.
29. In order to ACE intro logic, it is sufficient to get a HUNDRED on every exam.
30. In order to PASS, it is necessary to average at least FIFTY.
31. In order to become a PHYSICIAN, it is necessary to RECEIVE an M.D. and do an INTERNSHIP.
32. In order to PASS, it is both necessary and sufficient to average at least FIFTY.
33. Getting a HUNDRED on every exam is sufficient, but not necessary, for ACING intro logic.
34. TAKING all the exams is necessary, but not sufficient, for ACING intro logic.
35. In order to get into MEDICAL school, it is necessary but not sufficient to have GOOD grades and take the ADMISSIONS exam.
36. In order to be a BACHELOR it is both necessary and sufficient to be ELIGIBLE but not MARRIED.
37. In order to be ARRESTED, it is sufficient but not necessary to COMMIT a crime and GET caught.
38. If it is RAINING, I will play BASKETBALL; otherwise, I will go JOGGING.
39. If both JAY and KAY are home this weekend, we will go to thereat; otherwise, we will STAY home.
40. JONES will win the championship unless he gets INJURED, in which case SMITH will win.

EXERCISE SET C

41. We will have DINNER and attend the CONCERT, provided that JAY and KAY are home this weekend.
42. If neither JAY nor KAY can make it, we should either POSTPONE or CANCEL the trip.
43. Both Jay and Kay will go to the beach this weekend, provided that neither of them is sick.
44. I'm damned if I do, and I'm damned if I don't.
45. If I STUDY too hard I will not ENJOY college, but at the same time I will not ENJOY college if I FLUNK out.
46. If you NEED a thing, you will have THROWN it away, and if you THROW a thing away, you will NEED it.
47. If you WORK hard only if you are THREATENED, then you will not SUCCEED.
48. If I do not STUDY, then I will not PASS unless the prof ACCEPTS bribes.
49. Provided that the prof doesn't HATE me, I will PASS if I STUDY.
50. Unless logic is very DIFFICULT, I will PASS provided I CONCENTRATE.
51. Unless logic is EASY, I will PASS only if I STUDY.
52. Provided that you are INTELLIGENT, you will FAIL only if you GOOF off.
53. If you do not PAY, Jones will KILL you unless you ESCAPE.
54. If he CATCHES you, Jones will KILL you unless you PAY.
55. Provided that he has made a BET, Jones is HAPPY if and only if his horse WINS.
56. If neither JAY nor KAY comes home this weekend, we shall not stay HOME unless we are SICK.
57. If you MAKE an appointment and do not KEEP it, then I shall be ANGRY unless you have a good EXCUSE.
58. If I am not FEELING well this weekend, I will not GO out unless it is WARM and SUNNY.
59. If JAY will go only if KAY goes, then we will CANCEL the trip unless KAY goes.

EXERCISE SET D

60. If KAY will come to the party only if JAY does not come, then provided we WANT Kay to come we should DISSUADE Jay from coming.

61. If KAY will go only if JAY does not go, then either we will CANCEL the trip or we will not INVITE Jay.

62. If JAY will go only if KAY goes, then we will CANCEL the trip unless KAY goes.

63. If you CONCENTRATE only if you are INSPIRED, then you will not SUCCEED unless you are INSPIRED.

64. If you are HAPPY only if you are DRUNK, then unless you are DRUNK you are not HAPPY.

65. In order to be ADMITTED to law school, it is necessary to have GOOD grades, unless your family makes a large CONTRIBUTION to the law school.

66. I am HAPPY only if my assistant is COMPETENT, but if my assistant is COMPETENT, then he/she is TRANSFERRED to a better job and I am not HAPPY.

67. If you do not CONCENTRATE well unless you are ALERT, then you will FLY an airplane only if you are SOBER; provided that you are not a MANIAC.

68. If you do not CONCENTRATE well unless you are ALERT, then provided that you are not a MANIAC you will FLY an airplane only if you are SOBER.

69. If you CONCENTRATE well only if you are ALERT, then provided that you are WISE you will not FLY an airplane unless you are SOBER.

70. If you CONCENTRATE only if you are THREATENED, then you will not PASS unless you are THREATENED – provided that CONCENTRATING is a necessary condition for PASSING.

71. If neither JAY nor KAY is home this weekend, we will go to the BEACH; otherwise, we will STAY home.

27. ANSWERS TO EXERCISES FOR CHAPTER 4

1. R & J
2. ~R & W
3. J & K
4. D & ~H
5. T & ~Q
6. R & (J & K)
   R: Jay and Kay are roommates
   J: Jay hates Kay
   K: Kay hates Jay
7. (J & K) & (H & N)
   J: Jay is a Republican; K: Kay is a Republican
   H: Jay hates Nixon; N: Kay hates Nixon
8. K → A
9. G → T
10. C ∨ S
11. F ∨ D
12. ~(J & K)
13. (J ∨ K) & ~(J & K)
14. R ∨ (S & C)
15. (R ∨ O) & ~S
16. J → S
17. ~J → ~S
18. C → ~F
19. (L & H) → G
20. ~(L & H) → ~G
21. ~J & ~K [or: ~(J ∨ K)]
22. L & (~T & ~B) [or: L & ~(T ∨ B)]
23. ~P → ~G
24. ~~J → ~K [J → ~K]
25. ~(W & R) → ~S
26. ~S → B
27. ~G → ~F
28. ~(L & H) → ~G
29. H → A
30. ~F → ~P
31. ~(R & I) → ~P
32. (~F → ~P) & (F → P)
33. (H → A) & ~(~H → ~A)
34. (~T → ~A) & ~(T → A)
35. [~(G & A) → ~M] & ~[(G & A) → M]
36. [~(E & ~M) → ~B] & [(E & ~M) → B]
37. [(C & G) → A] & ~[~(C & G) → ~A]
38. (R → B) & (~R → J)
39. [(J & K) → B] & [~(J & K) → S]

40. (~I → J) & (I → S)
41. (J & K) → (D & C)
42. (~J & ~K) → (P ∨ C)
43. (~S & ~T) → (J & K)
S: Jay is sick; T: Kay is sick;
J: Jay will go to the beach; K: Kay will go to the beach.
44. (A → D) & (~A → D)
A: I do (what ever action is being discussed);
D: I am damned.
45. (S → ~E) & (F → ~E)
46. (N → T) & (T → N)
47. (~T → ~W) → ~S
48. ~S → (~A → ~P)
49. ~H → (S → P)
50. ~D → (C → P)
51. ~E → (~S → ~P)
52. I → (~G → ~F)
53. ~P → (~E → K)
54. C → (~P → K)
55. B → [(W → H) & (~W → ~H)]
56. (~J & ~K) → (~S → ~H)
57. (M & ~K) → (~E → A)
58. ~F → [~(W & S) → ~G]
59. (~K → ~J) → (~K → C)
60. (J → ~K) → (W → D)
61. (~~J → ~K) → (C ∨ ~I)
62. (~K → ~J) → (~K → C)
63. (~I → ~C) → (~I → ~S)
64. (~D → ~H) → (~D → ~H)
65. ~C → (~G → ~A)
66. (~C → ~H) & (C → [T & ~H])
67. ~M → [(~A → ~C) → (~S → ~F)]
68. (~A → ~C) → [~M → (~S → ~F)]
69. (~A → ~C) → [W → (~S → ~F)]
70. (~C → ~P) → [(~T → ~C) → (~T → ~P)]
71. [(~J & ~K) → B] & [~(~J & ~K) → S]

5 DERIVATIONS IN SENTENTIAL LOGIC

1. INTRODUCTION

In an earlier chapter, we studied a method of deciding whether an argument form of sentential logic is valid or invalid – the method of truth-tables. Although this method is infallible (when applied correctly), in many instances it can be tedious.

For example, if an argument form involves five distinct atomic formulas (say, P, Q, R, S, T), then the associated truth table contains 32 rows. Indeed, every additional atomic formula doubles the size of the associated truth-table. This makes the truth-table method impractical in many cases, unless one has access to a computer. Even then, due to the "doubling" phenomenon, there are argument forms that even a very fast main-frame computer cannot solve, at least in a reasonable amount of time (say, less than 100 years!)

Another shortcoming of the truth-table method is that it does not require much in the way of reasoning. It is simply a matter of mechanically following a simple set of directions. Accordingly, this method does not afford much practice in reasoning, either formal or informal.

For these two reasons, we now examine a second technique for demonstrating the validity of arguments – the method of *formal derivation*, or simply *derivation*. Not only is this method less tedious and mechanical than the method of truth tables, it also provides practice in symbolic reasoning.

Skill in symbolic reasoning can in turn be transferred to skill in practical reasoning, although the transfer is not direct. By analogy, skill in any game of strategy (say, chess) can be transferred indirectly to skill in general strategy (such as war, political or corporate). Of course, chess does not apply directly to any real strategic situation.

Constructing a derivation requires more thinking than filling out truth-tables. Indeed, in some instances, constructing a derivation demands considerable ingenuity, just like a good combination in chess.

Unfortunately, the method of formal derivation has its own shortcoming: unlike truth-tables, which can show both validity and invalidity, derivations can only show validity. If one succeeds in constructing a derivation, then one knows that the corresponding argument is valid. However, if one fails to construct a derivation, it does not mean that the argument is invalid. In the past, humans repeatedly failed to fly; this did not mean that flight was impossible. On the other hand, humans have repeatedly tried to construct perpetual motion machines, and they have failed. Sometimes failure is due to lack of cleverness; sometimes failure is due to the impossibility of the task!

2. THE BASIC IDEA

Underlying the method of formal derivations is the following fundamental idea.

> Granting the validity of a few selected argument forms, we can demonstrate the validity of other argument forms.

A simple illustration of this procedure might be useful. In an earlier chapter, we used the method of truth-tables to demonstrate the validity of numerous arguments. Among these, a few stand out for special mention. The first, and simplest one perhaps, is the following.

(MP)
$$\begin{array}{l} P \rightarrow Q \\ P \\ \hline Q \end{array}$$

This argument form is traditionally called *modus ponens*, which is short for *modus ponendo ponens*, which is a Latin expression meaning the mode of affirming by affirming. It is so called because, in this mode of reasoning, one goes from an affirmative premise to an affirmative conclusion.

It is easy to show that (MP) is a valid argument, using truth-tables. But we can use it to show other argument forms are also valid. Let us consider a simple example.

(a1)
$$\begin{array}{l} P \\ P \rightarrow Q \\ Q \rightarrow R \\ \hline R \end{array}$$

We can, of course, use truth-tables to show that (a1) is valid. Since there are three atomic formulas, 8 cases must be considered. However, we can also convince ourselves that (a1) is valid by reasoning as follows.

> Proof: Suppose the premises are all true. Then, in particular, the first two premises are both true. But if P and P→Q are both true, then Q must be true. Why? Because Q follows from P and P→Q by *modus ponens*. So now we know that the following formulas are all true: P, P→Q, Q, Q→R. This means that, in particular, both Q and Q→R are true. But R follows from Q and Q→R, by *modus ponens*, so R (the conclusion) must also be true. Thus, if the premises are all true, then so is the conclusion. In other words, the argument form is valid.

What we have done is show that (a1) is valid assuming that (MP) is valid.

Another important classical argument form is the following.

(MT)

$$\begin{array}{l} P \to Q \\ \sim Q \\ \hline \sim P \end{array}$$

This argument form is traditionally called *modus tollens*, which is short for *modus tollendo tollens*, which is a Latin expression meaning the mode of denying by denying. It is so called because, in this mode of reasoning, one goes from a negative premise to a negative conclusion.

Granting (MT), we can show that the following argument form is also valid.

(a2)

$$\begin{array}{l} P \to Q \\ Q \to R \\ \sim R \\ \hline \sim P \end{array}$$

Once again, we can construct a truth-table for (a2), which involves 8 lines. But we can also demonstrate its validity by the following reasoning.

> Proof: Suppose that the premises are all true. Then, in particular, the last two premises are both true. But if Q→R and ~R are both true, then ~Q is also true. For ~Q follows from Q→R and ~R, in virtue of *modus tollens*. So, if the premises are all true, then so is ~Q. That means that all the following formulas are true – P→Q, Q→R, ~R, ~Q. So, in particular, P→Q and ~Q are both true. But if these are true, then so is ~P (the conclusion), because ~P follows from P→Q and ~Q, in virtue of *modus tollens*. Thus, if the premises are all true, then so is the conclusion. In other words, the argument form is valid.

Finally, let us consider an example of reasoning that appeals to both *modus ponens* and *modus tollens*.

(a3)

$$\begin{array}{l} \sim P \\ \sim P \to \sim R \\ Q \to R \\ \hline \sim Q \end{array}$$

Proof: Suppose that the premises are all true. Then, in particular, the first two premises are both true. But if ~P and ~P→~R are both true, then so is ~R, in virtue of *modus ponens*. Then ~R and Q→R are both true, but then ~Q is true, in virtue of *modus tollens*. Thus, if the premises are all true, then the conclusion is also true, which is to say the argument is valid.

3. ARGUMENT FORMS AND SUBSTITUTION INSTANCES

In the previous section, the alert reader probably noticed a slight discrepancy between the official argument forms (MP) and (MT), on the one hand, and the actual argument forms appearing in the proofs of the validity of (a1)-(a3).

For example, in the proof of (a3), I said that ~R follows from ~P and ~P→~R, in virtue of *modus ponens*. Yet the argument forms are quite different.

(MP) $\quad$ P → Q
$\qquad\quad$ P
$\qquad\quad$ ―――
$\qquad\quad$ Q

(MP*) $\quad$ ~P → ~R
$\qquad\quad$ ~P
$\qquad\quad$ ―――
$\qquad\quad$ ~R

(MP*) looks somewhat like (MP); if we squinted hard enough, we might say they looked the same. But, clearly, (MP*) is not exactly the same as (MP). In particular, (MP) has no occurrences of negation, whereas (MP*) has 4 occurrences. So, in what sense can I say that (MP*) is valid *in virtue of* (MP)?

The intuitive idea is that "the overall form" of (MP*) is the same as (MP). (MP*) is an argument form with the following overall form.

conditional formula	() → []
antecedent	()
―――――――	―――
consequent	[]

The fairly imprecise notion of overall form can be made more precise by appealing to the notion of a substitution instance. We have already discussed this notion earlier. The slight complication here is that, rather than substituting a concrete argument for an argument form, we substitute one argument form for another argument form,

The following is the official definition.

> **Definition:**
> If ***A*** is an argument form of sentential logic, then a *substitution instance* of ***A*** is any argument form ***A***** that is obtained from ***A*** by substituting formulas for letters in ***A***.

There is an affiliated definition for formulas.

> **Definition:**
> If ***F*** is a formula of sentential logic, then a *substitution instance* of ***F*** is any formula ***F****** obtained from ***F*** by substituting formulas for letters in ***F***.

Note carefully: it is understood here that if a formula replaces a given letter in one place, then the formula replaces the letter in every place. One cannot substitute different formulas for the same letter. However, one is permitted to replace two different letters by the same formula. This gives rise to the notion of uniform substitution instance.

> **Definition:**
> A substitution instance is a *uniform substitution instance* if and only if distinct letters are replaced by distinct formulas.

These definitions are best understood in terms of specific examples. First, (MP*) is a (uniform) substitution of (MP), obtained by substituting ~P for P, and ~R for Q. The following are examples of substitution instances of (MP)

~P → ~Q	(P & Q) → ~R	(P → Q) → (P → R)
~P	P & Q	P → Q
―――――	―――――	―――――
~Q	~R	P → R

Whereas (MP*) is a substitution instance of (MP), the converse is not true: (MP) is not a substitution instance of (MP*). There is no way to substitute formulas for letters in (MP*) in such a way that (MP) is the result. (MP*) has four negations, and (MP) has none. A substitution instance F* always has at least as many occurrences of a connective as the original form F.

The following are substitution instances of (MP*).

~(P & Q) → ~(P → Q)	~~P → ~(Q ∨ R)
~(P & Q)	~~P
―――――	―――――
~(P → Q)	~(Q ∨ R)

Interestingly enough these are also substitution instances of (MP). Indeed, we have the following general theorem.

> **Theorem:**
> If argument form ***A**** is a substitution instance of ***A***, and argument form ***A***** is a substitution instance of ***A****, then ***A***** is a substitution instance of ***A***.

With the notion of substitution instance in hand, we are now in a position to solve the original problem. To say that argument form (MP*) is valid *in virtue of* modus ponens (MP) is not to say that (MP*) is identical to (MP); rather, it is to say that (MP*) is a substitution instance of (MP). The remaining question is whether the validity of (MP) ensures the validity of its substitution instances. This is answered by the following theorem.

> **Theorem:**
> If argument form ***A*** is valid,
> then every substitution instance of ***A*** is also valid.

The *rigorous* proof of this theorem is beyond the scope of introductory logic.

4. SIMPLE INFERENCE RULES

In the present section, we lay down the ground work for constructing our system of formal derivation, which we will call system SL (short for 'sentential logic'). At the heart of any derivation system is a set of *inference rules*. Each inference rule corresponds to a valid argument of sentential logic, although not every valid argument yields a corresponding inference rule. We select a subset of valid arguments to serve as inference rules.

But how do we make the selection? On the one hand, we want to be parsimonious. We want to employ as few inference rules as possible and still be able to generate all the valid argument forms. On the other hand, we want each inference rule to be simple, easy to remember, and intuitively obvious. These two desiderata actually push in opposite directions; the most parsimonious system is not the most intuitively clear; the most intuitively clear system is not the most parsimonious. Our particular choice will accordingly be a compromise solution.

We have to select from the infinitely-many valid argument forms of sentential logic a handful of very fertile ones, ones that will generate the rest. To a certain extent, the choice is arbitrary. It is very much like inventing a game – we get to make up the rules. On the other hand, the rules are not entirely arbitrary, because each rule must correspond to a valid argument form. Also, note that, even though we can choose the rules initially, once we have chosen, we must adhere to the ones we have chosen.

Every inference rule corresponds to a valid argument form of sentential logic. Note, however, that in granting the validity of an argument form (say, *modus ponens*), we mean to grant that specific argument form as well as *every* substitution instance.

In order to convey that each inference rule subsumes infinitely many argument forms, we will use an alternate font to formulate the inference rules; in particular, capital script letters ($\mathcal{A}$, $\mathcal{B}$, $\mathcal{C}$, etc.) will stand for arbitrary formulas of sentential logic.

Thus, for example, the rule of *modus ponens* will be written as follows, where $\mathcal{A}$ and $\mathcal{C}$ are *arbitrary* formulas of sentential logic.

(MP)

$$\begin{array}{l} \mathcal{A} \rightarrow \mathcal{C} \\ \mathcal{A} \\ \hline \mathcal{C} \end{array}$$

Given that the script letters '$\mathcal{A}$' and '$\mathcal{C}$' stand for arbitrary formulas, (MP) stands for infinitely many argument forms, all looking like the following.

(MP)

conditional	(antecedent) → [consequent]
antecedent	(antecedent)
———	———
consequent	[consequent]

Along the same lines, the rule *modus tollens* may be written as follows.

(MT)

$$\begin{array}{l} \mathcal{A} \rightarrow \mathcal{C} \\ \sim\mathcal{C} \\ \hline \sim\mathcal{A} \end{array}$$

(MT)

conditional	(antecedent) → [consequent]
literal negation of consequent	~[consequent]
———	———
literal negation of antecedent	~(antecedent)

Note: By 'literal negation of formula $\mathcal{A}$' is meant the formula that results from prefixing the formula $\mathcal{A}$ with a tilde. The literal negation of a formula always has exactly one more symbol than the formula itself.

In addition to (MP) and (MT), there are two other similar rules that we are going to adopt, given as follows.

(MTP1)

$$\begin{array}{l} \mathcal{A} \vee \mathcal{B} \\ \sim\mathcal{A} \\ \hline \mathcal{B} \end{array}$$

(MTP2)

$$\begin{array}{l} \mathcal{A} \vee \mathcal{B} \\ \sim\mathcal{B} \\ \hline \mathcal{A} \end{array}$$

This mode of reasoning is traditionally called *modus tollendo ponens*, which means the mode of affirming by denying. In each case, an affirmative conclusion is reached on the basis of a negative premise. The reader should verify, using truth-tables, that the simplest instances of these inference rules are in fact valid. The reader should also verify the intuitive validity of these forms of reasoning. MTP corresponds to the "process of elimination": one has a choice between two things, one eliminates one choice, leaving the other.

Before putting these four rules to work, it is important to point out two classes of errors that a student is liable to make.

• Errors of the First Kind

The four rules given above are to be carefully distinguished from argument forms that look similar but are clearly invalid. The following arguments are not instances of any of the above rules; worse, they are invalid.

Invalid!	Invalid!	Invalid!	Invalid!
P → Q Q ——— P	P → Q ~P ——— ~Q	P ∨ Q P ——— ~Q	P ∨ Q Q ——— ~P

These modes of inference are collectively known as *modus morons*, which means the mode of reasoning like a moron. It is easy to show that every one of them is invalid. You can use truth-tables, or you can construct counter-examples; either way, they are invalid.

• Errors of the Second Kind

Many valid arguments are not substitution instances of inference rules. This isn't too surprising. Some arguments, however, *look like* (but are not) substitution instances of inference rules. The following are examples.

Valid but not MT!	Valid but not MT!	Valid but not MTP!	Valid but not MTP!
~P → Q ~Q ——— P	P → ~Q Q ——— ~P	~P ∨ ~Q P ——— ~Q	~P ∨ ~Q Q ——— ~P

The following are corresponding correct applications of the rules.

MT	MT	MTP	MTP
~P → Q ~Q ——— ~~P	P → ~Q ~~Q ——— ~P	~P ∨ ~Q ~~P ——— ~Q	~P ∨ ~Q ~~Q ——— ~P

The natural question is, "aren't ~~P and P the same?" In asking this question, one might be thinking of arithmetic: for example, --2 and 2 are one and same *number*. But the corresponding *numerals* are not identical: the linguistic expression '--2' is not identical to the linguistic expression '2'. Similarly, the Roman *numeral* 'VII' is not identical to the Arabic *numeral* '7' even though both *numerals* denote the same *number*. Just like people, numbers have names; the names of numbers are numerals. We don't confuse people and their names. We shouldn't confuse numbers and their names (numerals).

Thus, the answer is that the formulas ~~P and P are not the same; they are as different as the Roman numeral 'VII' and the Arabic numeral '7'.

Another possible reason to think ~~P and P are the same is that they are *logically equivalent*, which may be shown using truth tables. This means they have the same truth-value no matter what. They have the same truth-value; does that mean they are the same? Of course not! That is like arguing from the premise that John and Mary are legally equivalent (meaning that they are equal under the law) to the conclusion that John and Mary are the same. Logical equivalence, like legal equivalence, is not identity.

Consider a very similar question whose answer revolves around the distinction between equality and identity: are four quarters and a dollar bill the same? The answer is, "yes and no". Four quarters are *monetarily equal* to a dollar bill, but they are definitely not identical. Quarters are made of metal, dollar bills are made of paper; they are physically quite different. For some purposes they are interchangeable; that does not mean they are the same.

The same can be said about ~~P and P. They have the same *value* (in the sense of truth-value), but they are definitely not identical. One has three symbols, the other only one, so they are not identical. More importantly, for our purposes, they have different forms – one is a negation; the other is atomic.

A derivation system in general, and inference rules in particular, pertain exclusively to the forms of the formulas involved.

In this respect, derivation systems are similar to coin-operated machines – vending machines, pay phones, parking meters, automatic toll booths, etc. A vending machine, for example, does not "care" what the *value* of a coin is. It only "cares" about the coin's form; it responds exclusively to the shape and weight of the coin. A penny worth one dollar to collectors won't buy a soft drink from a vending machine. Similarly, if the machine does not accept pennies, it is no use to put in 25 of them, even though 25 pennies have the same monetary value as a quarter. Similarly frustrating at times, a dollar bill is worthless when dealing with many coin-operated machines.

A derivation system is equally "stubborn"; it is blind to content, and responds exclusively to form. The fact that truth-tables tell us that P and ~~P are logically equivalent is irrelevant. If P is required by an inference-rule, then ~~P won't work, and if ~~P is required, then P won't work, just like 25 pennies won't buy a stick of gum from a vending machine. What one must do is first trade P for ~~P. We will have such conversion rules available.

5. SIMPLE DERIVATIONS

We now have four inference rules, MP, MT, MTP1, and MTP2. How do we utilize these in demonstrating other arguments of sentential logic are also valid? In order to prove (show, demonstrate) that an argument is valid, one derives its conclusion from its premises. We have already seen intuitive examples in an earlier section. We now redo these examples formally.

The first technique of derivation that we examine is called *simple derivation*. It is temporary, and will be replaced in the next section. However, it demonstrates the key intuitions about derivations.

Simple derivations are defined as follows.

> **Definition:**
> A *simple derivation* of conclusion $\mathcal{C}$ from premises $\mathcal{P}_1$, $\mathcal{P}_2$, ..., $\mathcal{P}_n$ is a list of formulas (also called lines) satisfying the following conditions.
>
> (1) the last line is $\mathcal{C}$;
> (2) every line (formula) is
>
> either:
>
> a premise (one of $\mathcal{P}_1$, $\mathcal{P}_2$, ..., $\mathcal{P}_n$),
>
> or:
>
> follows from previous lines according to an inference rule.

The basic idea is that in order to prove that an argument is valid, it is sufficient to construct a simple derivation of its conclusion from its premises. Rather than dwell on abstract matters of definition, it is better to deal with some examples by way of explaining the method of simple derivation.

Example 1

Argument: $P ; P \rightarrow Q ; Q \rightarrow R / R$

Simple Derivation:

(1)	P	Pr
(2)	$P \rightarrow Q$	Pr
(3)	$Q \rightarrow R$	Pr
(4)	Q	1,2,MP
(5)	R	3,4,MP

This is an example of a simple derivation. The last line is the conclusion; every line is either a premise or follows by a rule. The annotation to the right of each formula indicates the precise justification for the presence of the formula in the derivation.

There are two possible justifications at the moment; the formula is a premise (annotation: 'Pr'); the formula follows from previous formulas by a rule (annotation: line numbers, rule).

Example 2

Argument: $P \rightarrow Q$; $Q \rightarrow R$; $\sim R$ / $\sim P$

Simple Derivation:

(1)	$P \rightarrow Q$	Pr
(2)	$Q \rightarrow R$	Pr
(3)	$\sim R$	Pr
(4)	$\sim Q$	2,3,MT
(5)	$\sim P$	1,4,MT

Example 3

Argument: $\sim P$; $\sim P \rightarrow \sim R$; $Q \rightarrow R$ / $\sim Q$

Simple Derivation:

(1)	$\sim P$	Pr
(2)	$\sim P \rightarrow \sim R$	Pr
(3)	$Q \rightarrow R$	Pr
(4)	$\sim R$	1,2,MP
(5)	$\sim Q$	3,4,MT

These three examples take care of the examples from Section 2. The following one is more unusual.

Example 4

Argument: $(P \rightarrow Q) \rightarrow P$; $P \rightarrow Q$ / Q

Simple Derivation:

(1)	$(P \rightarrow Q) \rightarrow P$	Pr
(2)	$P \rightarrow Q$	Pr
(3)	P	1,2,MP
(4)	Q	2,3,MP

What is unusual about this one is that line (2) is used twice, in connection with MP, once as minor premise, once as major premise. One can appeal to the same line over and over again, if the need arises.

We conclude this section with examples of slightly longer simple derivations.

Example 5

Argument: $P \rightarrow (Q \vee R)$; $P \rightarrow {\sim}R$; P / Q

Simple Derivation:

(1)	$P \rightarrow (Q \vee R)$	Pr
(2)	$P \rightarrow {\sim}R$	Pr
(3)	P	Pr
(4)	${\sim}R$	2,3,MP
(5)	$Q \vee R$	1,3,MP
(6)	Q	4,5,MTP2

Example 6

Argument: ${\sim}P \rightarrow (Q \vee R)$; $P \rightarrow Q$; ${\sim}Q$ / R

Simple Derivation:

(1)	${\sim}P \rightarrow (Q \vee R)$	Pr
(2)	$P \rightarrow Q$	Pr
(3)	${\sim}Q$	Pr
(4)	${\sim}P$	2,3,MT
(5)	$Q \vee R$	1,4,MP
(6)	R	3,5,MTP1

Example 7

Argument: $(P \vee R) \vee (P \rightarrow Q)$; ${\sim}(P \rightarrow Q)$; $R \rightarrow (P \rightarrow Q)$ / P

Simple Derivation:

(1)	$(P \vee R) \vee (P \rightarrow Q)$	Pr
(2)	${\sim}(P \rightarrow Q)$	Pr
(3)	$R \rightarrow (P \rightarrow Q)$	Pr
(4)	$P \vee R$	1,2,MTP2
(5)	${\sim}R$	2,3,MT
(6)	P	4,5,MTP2

Example 8

Argument: $P \rightarrow {\sim}Q$; ${\sim}Q \rightarrow (R \,\&\, S)$; ${\sim}(R \,\&\, S)$; $P \vee T$ / T

Simple Derivation:

(1)	$P \rightarrow {\sim}Q$	Pr
(2)	${\sim}Q \rightarrow (R \,\&\, S)$	Pr
(3)	${\sim}(R \,\&\, S)$	Pr
(4)	$P \vee T$	Pr
(5)	${\sim}{\sim}Q$	2,3,MT
(6)	${\sim}P$	1,5,MT
(7)	T	4,6,MTP1

6. THE OFFICIAL INFERENCE RULES

So far, we have discussed only four inference rules: *modus ponens*, *modus tollens*, and the two forms of *modus tollendo ponens*. In the present section, we add quite a few more inference rules to our list.

Since the new rules will be given more pictorial, non-Latin, names, we are going to rename our original four rules in order to maintain consistency. Also, we are going to consolidate our original four rules into two rules.

In constructing the full set of inference rules, we would like to pursue the following overall plan. For each of the five connectives, we want two rules: on the one hand, we want a rule for "introducing" the connective; on the other hand, we want a rule for "eliminating" the connective. An introduction-rule is also called an *in-rule*; an elimination-rule is called an *out-rule*.

Also, it would be nice if the name of each rule is suggestive of what the rule does. In particular, the name should consist of two parts: (1) reference to the specific connective involved, and (2) indication whether the rule is an introduction (in) rule or an elimination (out) rule.

Thus, if we were to follow the overall plan, we would have a total of ten rules, listed as follows.

Ampersand-In	&I
Ampersand-Out	&O
Wedge-In	∨I
Wedge-Out	∨O
Double-Arrow-In	↔I
Double-Arrow-Out	↔O
*Arrow-In	→I
Arrow-Out	→O
*Tilde-In	~I
*Tilde-Out	~O

However, for reasons of simplicity of presentation, the general plan is not followed completely. In particular, there are three points of difference, which are marked by an asterisk. What we adopt instead, in the derivation system SL, are the following inference rules.

• INFERENCE RULES (INITIAL SET)

Ampersand-In (&I)	$\dfrac{\begin{array}{l}\mathcal{A}\\ \mathcal{B}\end{array}}{\mathcal{A} \,\&\, \mathcal{B}}$	$\dfrac{\begin{array}{l}\mathcal{A}\\ \mathcal{B}\end{array}}{\mathcal{B} \,\&\, \mathcal{A}}$
Ampersand-Out (&O)	$\dfrac{\mathcal{A} \,\&\, \mathcal{B}}{\mathcal{A}}$	$\dfrac{\mathcal{A} \,\&\, \mathcal{B}}{\mathcal{B}}$
Wedge-In (∨I)	$\dfrac{\mathcal{A}}{\mathcal{A} \vee \mathcal{B}}$	$\dfrac{\mathcal{A}}{\mathcal{B} \vee \mathcal{A}}$
Wedge-Out (∨O)	$\dfrac{\begin{array}{l}\mathcal{A} \vee \mathcal{B}\\ \sim\mathcal{A}\end{array}}{\mathcal{B}}$	$\dfrac{\begin{array}{l}\mathcal{A} \vee \mathcal{B}\\ \sim\mathcal{B}\end{array}}{\mathcal{A}}$
Double-Arrow-In (↔I)	$\dfrac{\begin{array}{l}\mathcal{A} \rightarrow \mathcal{B}\\ \mathcal{B} \rightarrow \mathcal{A}\end{array}}{\mathcal{A} \leftrightarrow \mathcal{B}}$	$\dfrac{\begin{array}{l}\mathcal{A} \rightarrow \mathcal{B}\\ \mathcal{B} \rightarrow \mathcal{A}\end{array}}{\mathcal{B} \leftrightarrow \mathcal{A}}$
Double-Arrow-Out (↔O)	$\dfrac{\mathcal{A} \leftrightarrow \mathcal{B}}{\mathcal{A} \rightarrow \mathcal{B}}$	$\dfrac{\mathcal{A} \leftrightarrow \mathcal{B}}{\mathcal{B} \rightarrow \mathcal{A}}$
Arrow-Out (→O)	$\dfrac{\begin{array}{l}\mathcal{A} \rightarrow \mathcal{B}\\ \mathcal{A}\end{array}}{\mathcal{B}}$	$\dfrac{\begin{array}{l}\mathcal{A} \rightarrow \mathcal{B}\\ \sim\mathcal{B}\end{array}}{\sim\mathcal{A}}$
Double Negation (DN)	$\dfrac{\mathcal{A}}{\sim\sim\mathcal{A}}$	$\dfrac{\sim\sim\mathcal{A}}{\mathcal{A}}$

A few notes may help clarify the above inference rules.

Notes

(1) Arrow-out (→O), the rule for decomposing conditional formulas, replaces both *modus ponens* and *modus tollens*.

(2) Wedge-out (∨O), the rule for decomposing disjunctions, replaces both forms of *modus tollendo ponens*.

(3) Double negation (DN) stands in place of both the tilde-in and the tilde-out rule.

(4) There is *no* arrow-in rule! [The rule for introducing arrow is not an inference rule but rather a show-rule, which is a different kind of rule, to be discussed later.]

(5) In each of the rules, $\mathcal{A}$ and $\mathcal{B}$ are arbitrary formulas of sentential logic. Each rule is short for infinitely many substitution instances.

(6) In each of the rules, the order of the premises is completely irrelevant.

(7) In the wedge-in (∨I) rule, the formula $\mathcal{B}$ is *any* formula whatsoever; it does not even have to be anywhere near the derivation in question!

There is one point that is extremely important, given as follows, which will be repeated as the need arises.

> Inference rules apply
> to whole lines,
> not to pieces of lines.

In other words, what are given above are not actually the inference rules themselves, but only pictures suggestive of the rules. The actual rules are more properly written as follows.

• INFERENCE RULES; OFFICIAL FORMULATION

> **Ampersand-In** (&I): If one has available *lines*, $\mathcal{A}$ and $\mathcal{B}$, then one is entitled to write down their conjunction, in one order $\mathcal{A}\&\mathcal{B}$, or the other order $\mathcal{B}\&\mathcal{A}$.

> **Ampersand-Out** (&O): If one has available a *line* of the form $\mathcal{A}\&\mathcal{B}$, then one is entitled to write down either conjunct $\mathcal{A}$ or conjunct $\mathcal{B}$.

Wedge-In (∨I): If one has available a *line* $\mathcal{A}$, then one is entitled to write down the disjunction of $\mathcal{A}$ with any formula $\mathcal{B}$, in one order $\mathcal{A}\vee\mathcal{B}$, or the other order $\mathcal{B}\vee\mathcal{A}$.

Wedge-Out (∨O): If one has available a *line* of the form $\mathcal{A}\vee\mathcal{B}$, and if one additionally has available a *line* which is the negation of the first disjunct, $\sim\mathcal{A}$, then one is entitled to write down the second disjunct, $\mathcal{B}$. Likewise, if one has available a *line* of the form $\mathcal{A}\vee\mathcal{B}$, and if one additionally has available a *line* which is the negation of the second disjunct, $\sim\mathcal{B}$, then one is entitled to write down the first disjunct, $\mathcal{A}$.

Double-Arrow-In (↔I): If one has available a *line* that is a conditional $\mathcal{A}\rightarrow\mathcal{B}$, and one additionally has available a line that is the converse $\mathcal{B}\rightarrow\mathcal{A}$, then one is entitled to write down either the biconditional $\mathcal{A}\leftrightarrow\mathcal{B}$ or the biconditional $\mathcal{B}\leftrightarrow\mathcal{A}$.

Double-Arrow-Out (↔O): If one has available a *line* of the form $\mathcal{A}\leftrightarrow\mathcal{B}$, then one is entitled to write down both the conditional $\mathcal{A}\rightarrow\mathcal{B}$ and its converse $\mathcal{B}\rightarrow\mathcal{A}$.

Arrow-Out (→O): If one has available a *line* of the form $\mathcal{A}\rightarrow\mathcal{B}$, and if one additionally has available a *line* which is the antecedent $\mathcal{A}$, then one is entitled to write down the consequent $\mathcal{B}$. Likewise, if one has available a *line* of the form $\mathcal{A}\rightarrow\mathcal{B}$, and if one additionally has available a *line* which is the negation of the consequent, $\sim\mathcal{B}$, then one is entitled to write down the negation of the antecedent, $\sim\mathcal{A}$.

Double Negation (DN): If one has available a *line* $\mathcal{A}$, then one is entitled to write down the double-negation $\sim\sim\mathcal{A}$. Similarly, if one has available a *line* of the form $\sim\sim\mathcal{A}$, then one is entitled to write down the formula $\mathcal{A}$.

The word 'available' is used in a technical sense that will be explained in a later section.

To this list, we will add a few further inference rules in a later section. They are not crucial to the derivation system; they merely make doing derivations more convenient.

7. SHOW-LINES AND SHOW-RULES; DIRECT DERIVATION

Having discussed simple derivations, we now begin the official presentation of the derivation system SL. In constructing system SL, we lay down a set of *system rules* – the rules of SL. It's a bit confusing: we have inference rules, already presented; now we have system rules as well. System rules are simply the official rules for constructing derivations, and include, among other things, all the inference rules.

For example, we have already seen two system rules, in effect. They are the two principles of simple derivation, which are now officially formulated as system rules.

System Rule 1 (The Premise Rule)

> At any point in a derivation, ***prior*** to the first show-line, any premise may be written down. The annotation is 'Pr'.

System Rule 2 (The Inference-Rule Rule)

> At any point in a derivation, a formula may be written down if it follows from previous available lines by an inference rule. The annotation cites the line numbers, and the inference rule, in that order.

System Rule 2 is actually short-hand for the list of all the inference rules, as formulated at the end of Section 6.

The next thing we do in elaborating system SL is to enhance the notion of simple derivation to obtain the notion of a direct derivation. This enhancement is quite simple; it even seems redundant, at the moment. But as we further elaborate system SL, this enhancement will become increasingly crucial. Specifically, we add the following additional system rule, which concerns a new kind of line, called a *show-line*, which may be introduced at any point in a derivation.

System Rule 3 (The Show-Line Rule)

> At *any* point in a derivation, one is entitled to write down the expression 'SHOW: $\mathcal{A}$',
> for *any* formula $\mathcal{A}$ *whatsoever.*

In writing down the line 'SHOW: $\mathcal{A}$', all one is saying is, "I will now *attempt* to show the formula $\mathcal{A}$". What the rule amounts to, then, is that at any point one is entitled to *attempt* to show anything one pleases. This is very much like saying that any citizen (over a certain age) is entitled to run for president. But rights are not guarantees; you can try, but you may not succeed.

Allowing show-lines changes the derivation system quite a bit, at least in the long run. However, at the current stage of development of system SL, there is generally only one reasonable kind of show-line. Specifically, one writes down 'SHOW: *C*', where *C* is the conclusion of the argument one is trying to prove valid. Later, we will see other uses of show-lines.

All derivations start pretty much the same way: one writes down all the premises, as permitted by System Rule 1; then one writes down 'SHOW: *C*' (where *C* is the conclusion), which is permitted by System Rule 3.

Consider the following example, which is the beginning of a derivation.

Example 1

(1)	(P ∨ Q) → ~R	Pr
(2)	P & T	Pr
(3)	R ∨ ~S	Pr
(4)	U → S	Pr
(5)	SHOW: ~U	???

These five lines may be regarded as simply stating the problem – we want to show one formula, given four others. I write '???' in the annotation column because this still needs explaining; more about this later.

Given the problem, we can construct what is very similar to a simple derivation, as follows.

(1)	(P ∨ Q) → ~R	Pr
(2)	P & T	Pr
(3)	R ∨ ~S	Pr
(4)	U → S	Pr
(5)	SHOW: ~U	???
(6)	P	2,&O
(7)	P ∨ Q	6,∨I
(8)	~R	1,7,→O
(9)	~S	3,8,∨O
(10)	~U	4,9,→O

Notice that, if we deleted the show-line, (5), the result is a simple derivation.

We are allowed to *try* to show anything. But how do we know when we have succeeded? In order to decide when a formula has in fact been shown, we need additional system rules, which we call "show-rules". The first show-rule is so simple it barely requires mentioning. Nevertheless, in order to make system SL completely clear and precise, we must make this rule explicit.

The first show-rule may be intuitively formulated as follows.

Direct Derivation (Intuitive Formulation)

If one is trying to show formula $\mathcal{A}$, and one actually obtains $\mathcal{A}$ as a later line, then one has succeeded.

The intuitive formulation is, unfortunately, not sufficiently precise for the purposes to which it will ultimately be put. So we formulate the following official system rule of derivation.

System Rule 4 (a show-rule)

Direct Derivation (DD)

If one has a show-line 'SHOW: $\mathcal{A}$', and one obtains $\mathcal{A}$ as a later **available** line, and there are no intervening uncancelled show-lines, then one is entitled to box and cancel 'SHOW: $\mathcal{A}$'. The annotation is 'DD'

As it is officially written, direct derivation is a very complicated rule. Don't worry about it now. The subtleties of the rule don't come into play until later.

For the moment, however, we do need to understand the idea of cancelling a show-line and boxing off the associated sub-derivation. Cancelling a show-line simply amounts to striking through the word 'SHOW', to obtain '~~SHOW~~'. This indicates that the formula has in fact been shown. Now the formula $\mathcal{A}$ can be used. The trade-off is that one must box off the associated derivation. No line inside a box can be further used. One, in effect, trades the derivation for the formula shown. More about this restriction later.

The intuitive content of direct derivation is pictorially presented as follows.

Direct Derivation (DD)

The box is of little importance right now, but later it becomes very important in helping organize very complex derivations, ones that involve several show-lines. For the moment, simply think of the box as a decoration, a flourish if you like, to celebrate having shown the formula.

Let us return to our original derivation problem. Completing it according to the strict rules yields the following.

(1)	(P ∨ Q) → ~R	Pr
(2)	P & T	Pr
(3)	R ∨ ~S	Pr
(4)	U → S	Pr
(5)	~~SHOW~~: ~U	DD
(6)	P	2,&O
(7)	P ∨ Q	6,∨I
(8)	~R	1,7,→O
(9)	~S	3,8,∨O
(10)	~U	4,9,→O

Note that 'SHOW' has been struck through, resulting in '~~SHOW~~'. Note the annotation for line (5); 'DD' indicates that the show-line has been cancelled in accordance with the show-rule Direct Derivation. Finally, note that every formula below the show-line has been boxed off.

Later, we will have other, more complicated, show-rules. For the moment, however, we just have direct derivation.

8. EXAMPLES OF DIRECT DERIVATIONS

In the present section, we look at several examples of direct derivations.

Example 1

(1)	~P → (Q ∨ R)	Pr
(2)	P → Q	Pr
(3)	~Q	Pr
(4)	~~SHOW~~: R	DD
(5)	~P	2,3,→O
(6)	Q ∨ R	1,5,→O
(7)	R	3,6,∨O

Example 2

(1)	P & Q	Pr
(2)	~~SHOW~~: ~~P & ~~Q	DD
(3)	P	1,&O
(4)	Q	1,&O
(5)	~~P	3,DN
(6)	~~Q	4,DN
(7)	~~P & ~~Q	5,6,&I

Example 3

(1)	P & Q	Pr
(2)	(Q ∨ R) → S	Pr
(3)	~~SHOW~~: P & S	DD
(4)	P	1,&O
(5)	Q	1,&O
(6)	Q ∨ R	5,∨I
(7)	S	2,6,→O
(8)	P & S	4,7,&I

Example 4

(1)	A & B	Pr
(2)	(A ∨ E) → C	Pr
(3)	D → ~C	Pr
(4)	~~SHOW~~: ~D	DD
(5)	A	1,&O
(6)	A ∨ E	5,∨I
(7)	C	2,6,→O
(8)	~~C	7,DN
(9)	~D	3,8,→O

Example 5

(1)	A & ~B	Pr
(2)	B ∨ (A → D)	Pr
(3)	(C & E) ↔ D	Pr
(4)	~~SHOW~~: A & C	DD
(5)	A	1,&O
(6)	~B	1,&O
(7)	A → D	2,6,∨O
(8)	D	5,7,→O
(9)	D → (C & E)	3,↔O
(10)	C & E	8,9,→O
(11)	C	10,&O
(12)	A & C	5,11,&I

Example 6

(1)	A → B	Pr
(2)	(A → B) → (B → A)	Pr
(3)	(A ↔ B) → A	Pr
(4)	~~SHOW~~: A & B	DD
(5)	B → A	1,2,→O
(6)	A ↔ B	1,5,↔I
(7)	A	3,6,→O
(8)	B	1,7,→O
(9)	A & B	7,8,&I

Example 7

(1)	~A & B	Pr
(2)	(C ∨ B) → (~D → A)	Pr
(3)	~D ↔ E	Pr
(4)	~~SHOW~~: ~E	DD
(5)	~A	1,&O
(6)	B	1,&O
(7)	C ∨ B	6,∨I
(8)	~D → A	2,7,→O
(9)	~~D	5,8,→O
(10)	E → ~D	3,↔O
(11)	~E	9,10,→O

NOTE: From now on, for the sake of typographical neatness, we will draw boxes in a purely skeletal fashion. In particular, we will only draw the left side of each box; the remaining sides of each box should be mentally filled in. For example, using skeletal boxes, the last two derivations are written as follows.

Example 6 (rewritten)

(1)	A → B	Pr
(2)	(A → B) → (B → A)	Pr
(3)	(A ↔ B) → A	Pr
(4)	~~SHOW~~: A & B	DD
(5)	│ B → A	1,2,→O
(6)	│ A ↔ B	1,5,↔I
(7)	│ A	3,6,→O
(8)	│ B	1,7,→O
(9)	│ A & B	7,8,&I

Example 7 (rewritten)

(1)	~A & B	Pr
(2)	(C ∨ B) → (~D → A)	Pr
(3)	~D ↔ E	Pr
(4)	SHOW: ~E	DD
(5)	│ ~A	1,&O
(6)	│ B	1,&O
(7)	│ C ∨ B	6,∨I
(8)	│ ~D → A	2,7,→O
(9)	│ ~~D	5,8,→O
(10)	│ E → ~D	3,↔O
(11)	│ ~E	9,10,→O

NOTE: In your own derivations, you can draw as much, or as little, of a box as you like, so long as you include at a minimum its left side. For example, you can use any of the following schemes.

SHOW: SHOW: SHOW: SHOW:

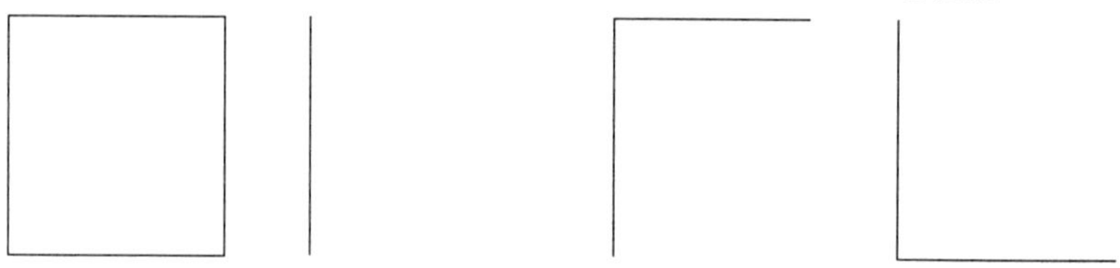

Finally, we end this section by rewriting the Direct Derivation Picture, in accordance with our minimal boxing scheme.

Direct Derivation (DD)

SHOW: $\mathcal{A}$	DD
│ o	
│ o	
│ o	
│ o	
│ o	
│ o	
│ o	
│ $\mathcal{A}$	

9. CONDITIONAL DERIVATION

So far, we only have one method by which to cancel a show-line – direct derivation. In the present section, we examine a new derivation method, which will enable us to prove valid a larger class of sentential arguments.

Consider the following argument.

(A) $P \rightarrow Q$
$Q \rightarrow R$
―――――
$P \rightarrow R$

This argument is valid, as can easily be demonstrated using truth-tables. Can we derive the conclusion from the premises? The following begins the derivation.

(1)	$P \rightarrow Q$	Pr
(2)	$Q \rightarrow R$	Pr
(3)	SHOW: $P \rightarrow R$	???
(4)	???	???

What formulas can we write down at line (4)? There are numerous formulas that follow from the premises according to the inference rules. But, not a single one of them makes any progress toward showing the conclusion $P \rightarrow R$. In fact, upon close examination, we see that we have no means at our disposal to prove this argument. We are stuck.

In other words, as it currently stands, derivation system SL is inadequate. The above argument is valid, by truth-tables, but it cannot be proven in system SL. Accordingly, system SL must be strengthened so as to allow us to prove the above argument. Of course, we don't want to make the system so strong that we can derive invalid conclusions, so we have to be careful, as usual.

How might we argue for such a conclusion? Consider a concrete instance of the argument form.

(I) if the gas tank gets a hole, then the car runs out of gas;
if the car runs out of gas, then the car stops;
therefore, if the gas tank gets a hole, then the car stops.

In order to argue for the conclusion of (I), it seems natural to argue as follows. First, suppose the premises are true, in order to show the conclusion. The conclusion says that

the car stops **if** the gas tank gets a hole

or in other words,

the car stops **supposing** the gas tank gets a hole.

So, suppose also that the antecedent,

the gas tank gets a hole,

is true. In conjunction with the first premise, we can infer the following by *modus ponens* (→O):

the car runs out of gas.

And from this in conjunction with the second premise, we can infer the following by *modus ponens* (→O).

the car stops

So supposing the antecedent (the gas tank gets a hole), we have deduced the consequent (the car stops). In other words, we have shown the conclusion – if the gas tank gets a hole, then the car stops.

The above line of reasoning is made formal in the following official derivation.

Example 1

(1)	H → R	Pr
(2)	R → S	Pr
(3)	~~SHOW~~: H → S	CD
(4)	| H	As
(5)	| ~~SHOW~~: S	DD
(6)	| | R	1,4,→O
(7)	| | S	2,6,→O

This new-fangled derivation requires explaining. First of all, there are two show-lines; in particular, one derivation is nested inside another derivation. This is because the original problem – showing H→S – is reduced to another problem, showing S assuming H. This procedure is in accordance with a new show-rule, called conditional derivation, which may be intuitively formulated as follows.

Conditional Derivation (Intuitive Formulation)

In order to show a conditional $\mathcal{A}\rightarrow\mathcal{C}$, it is sufficient to show the consequent $\mathcal{C}$, assuming the antecedent $\mathcal{A}$.

The official formulation of conditional derivation is considerably more complicated, being given by the following two system rules.

System Rule 5 (a show-rule)

> **Conditional Derivation** (CD)
>
> If one has a show-line of the form 'SHOW: $\mathcal{A}\rightarrow C$', and one has C as a later available line, and there are no subsequent uncancelled show-lines, then one is entitled to box and cancel 'SHOW: $\mathcal{A}\rightarrow C$'.
> The annotation is 'CD'

System Rule 6 (an assumption rule)

> If one has a show-line of the form 'SHOW: $\mathcal{A}\rightarrow C$', then one is entitled to write down the antecedent $\mathcal{A}$ on the very next line, as an assumption.
> The annotation is 'As'

It is probably easier to understand conditional derivation by way of the associated picture.

> **Conditional Derivation** (CD)
>
> ~~SHOW~~: $\mathcal{A}\rightarrow C$ CD
> | $\mathcal{A}$ As
> | ~~SHOW~~: C
> | |
> | |

This is supposed to depict the nature of conditional derivation; one shows a conditional $\mathcal{A}\rightarrow C$ by assuming its antecedent $\mathcal{A}$ and showing its consequent C.

In order to further our understanding of conditional derivation, we do a few examples.

Example 2

(1)	$P \to R$	Pr
(2)	$Q \to S$	Pr
(3)	~~SHOW~~: $(P \,\&\, Q) \to (R \,\&\, S)$	CD
(4)	│ $P \,\&\, Q$	As
(5)	│ ~~SHOW~~: $R \,\&\, S$	DD
(6)	│ │ P	4,&O
(7)	│ │ Q	4,&O
(8)	│ │ R	1,6,→O
(9)	│ │ S	2,7,→O
(10)	│ │ $R \,\&\, S$	8,9,&I

Example 3

(1)	$Q \to R$	Pr
(2)	$R \to (P \to S)$	Pr
(3)	~~SHOW~~: $(P \,\&\, Q) \to S$	CD
(4)	│ $P \,\&\, Q$	As
(5)	│ ~~SHOW~~: S	DD
(6)	│ │ P	4,&O
(7)	│ │ Q	4,&O
(8)	│ │ R	1,7,→O
(9)	│ │ $P \to S$	2,8,→O
(10)	│ │ S	6,9,→O

The above examples involve two show-lines; each one involves a direct derivation inside a conditional derivation. The following examples introduce a new twist – three show-lines in the same derivation, with a conditional derivation inside a conditional derivation.

Example 4

(1)	$(P \,\&\, Q) \to R$	Pr
(2)	~~SHOW~~: $P \to (Q \to R)$	CD
(3)	│ P	As
(4)	│ ~~SHOW~~: $Q \to R$	CD
(5)	│ │ Q	As
(6)	│ │ ~~SHOW~~: R	DD
(7)	│ │ │ $P \,\&\, Q$	3,5,&I
(8)	│ │ │ R	1,7,→O

Example 5

(1)	(P & Q) → R	Pr
(2)	~~SHOW~~: (P → Q) → (P → R)	CD
(3)	│ P → Q	As
(4)	│ ~~SHOW~~: P → R	CD
(5)	│ │ P	As
(6)	│ │ ~~SHOW~~: R	DD
(7)	│ │ │ Q	3,5,→O
(8)	│ │ │ P & Q	5,7,&I
(9)	│ │ │ R	1,8,→O

Needless to say, the depth of nesting is not restricted; consider the following example.

Example 6

(1)	(P & Q) → (R → S)	Pr
(2)	~~SHOW~~: R → [(P → Q) → (P → S)]	CD
(3)	│ R	As
(4)	│ ~~SHOW~~: (P → Q) → (P → S)	CD
(5)	│ │ P → Q	As
(6)	│ │ ~~SHOW~~: P → S	CD
(7)	│ │ │ P	As
(8)	│ │ │ ~~SHOW~~: S	DD
(9)	│ │ │ │ Q	5,7,→O
(10)	│ │ │ │ P & Q	7,9,&I
(11)	│ │ │ │ R → S	1,10,→O
(12)	│ │ │ │ S	3,11,→O

Irrespective of the complexity of the above problems, they are solved in the same systematic manner. At each point where we come across 'SHOW: $\mathcal{A}$→$\mathcal{C}$', we immediately write down two more lines – we assume the antecedent, $\mathcal{A}$, in order to (attempt to) show the consequent, $\mathcal{C}$.

That is all there is to it!

10. INDIRECT DERIVATION (FIRST FORM)

System SL is now a *complete* set of rules for sentential logic; every valid argument of sentential logic can be proved valid in system SL. System SL is also *consistent*, which is to say that no invalid argument can be proven in system SL. Demonstrating these two very important logical facts – that system SL is both complete and consistent – is well outside the scope of introductory logic. It rather falls under the scope of *metalogic*, which is studied in more advanced courses in logic.

Even though system SL is complete as it stands, we will nonetheless enhance it further, thereby sacrificing elegance in favor of convenience. Consider the following argument form.

(a1) P → Q
P → ~Q
———
~P

Using truth-tables, one can quickly demonstrate that (a1) is valid. What happens when we try to construct a derivation that proves it to be valid? Consider the following start.

(1)	P → Q	Pr
(2)	P → ~Q	Pr
(3)	SHOW: ~P	???
(4)	???	???

An *attempted* derivation, using DD and CD, might go as follows.

> Consider line (3), which is a negation. We cannot show it by conditional derivation; it's not a conditional! That leaves direct derivation. Well, the premises are both conditionals, so the appropriate rule is arrow-out. But arrow-out requires a minor premise. In the case of (1) we need P or ~Q; in the case of (2), we need P or ~~Q; none of these is available. We are stuck!

We are trying to show ~P, which says in effect that P is false. Let's try a sneaky approach to the problem. Just for the helluvit, let us *assume* the opposite of what we are trying to show, and see what happens. So right below 'SHOW: ~P', we write P as an assumption. That yields the following partial derivation.

(1)	P → Q	Pr
(2)	P → ~Q	Pr
(3)	SHOW: ~P	???
(4)	P	As??
(6)	Q	1,4,→O
(7)	~Q	1,5,→O
(8)	Q & ~Q	5,6,&I

We have gotten down to line (8) which is Q&~Q. From our study of truth-tables, we know that this formula is a self-contradiction; it is false no matter what. So we see that assuming P at line (4) leads to a very bizarre result, a self-contradiction at line (8).

So, we have shown, in effect, that if P is true, then so is Q&~Q, which means that we have shown P→(Q&~Q). To see this, let us rewrite the problem as follows. Notice especially the new show-line (4).

(1)	P → Q	Pr
(2)	P → ~Q	Pr
(3)	SHOW: ~P	???
(4)	~~SHOW~~: P → (Q & ~Q)	CD
(5)	│ P	As
(6)	│ ~~SHOW~~: Q & ~Q	DD
(7)	│ │ Q	1,5,→O
(8)	│ │ ~Q	2,5,→O
(9)	│ │ Q & ~Q	7,8,&I

This is OK as far as it goes, but it is still not complete; show-line (3) has not been cancelled yet, which is marked in the annotation column by '???'. Line (4) is permitted, by the show-line rule (we can try to show anything!). Lines (5) and (6) then are written down in accordance with conditional derivation. The remaining lines are completely ordinary.

So how do we complete the derivation? We are trying to show ~P; we have in fact shown P→(Q&~Q); in other words, we have shown that if P is true, then so is Q&~Q. But the latter can't be true, so neither can the former (by *modus tollens*). This reasoning can be made formal in the following part derivation.

(1)	P → Q	Pr
(2)	P → ~Q	Pr
(3)	SHOW: ~P	DD
(4)	~~SHOW~~: P → (Q & ~Q)	CD
(5)	│ P	As
(6)	│ ~~SHOW~~: Q & ~Q	DD
(7)	│ │ Q	1,5,→O
(8)	│ │ ~Q	2,5,→O
(9)	│ │ Q & ~Q	7,8,&I
(10)	~(Q & ~Q)	???
(11)	~P	4,10,→O

This is an OK derivation, except for line (10), which has no justification. At this stage in the elaboration of system SL, we could introduce a new system rule that allows one to write ~($\mathcal{A}$&~$\mathcal{A}$) at any point in a derivation. This rule would work perfectly well, but it is not nearly as tidy as what we do instead. We choose instead to abbreviate the above chain of reasoning considerably, by introducing a further show-rule, called indirect derivation, whose intuitive formulation is given as follows.

Indirect Derivation (First Form)

Intuitive Formulation

In order to show a negation $\sim\mathcal{A}$, it is sufficient to show *any contradiction*, assuming the un-negated formula, $\mathcal{A}$.

We must still provide the official formulation of indirect derivation, which as usual is considerably more complex; see below.

Recall that a contradiction is any formula whose truth table yields all F's in the output column. There are infinitely many contradictions in sentential logic. For this reason, at this point, it is convenient to introduce a new symbol into the vocabulary of sentential logic. In addition to the usual symbols – the letters, the connective symbols, and the parentheses – we introduce the symbol '**X**', in accordance with the following syntactic and semantic rules.

Syntactic Rule: **X** is a formula.

Semantic Rule: **X** is false no matter what.

[Alternatively, **X** is a "zero-place" logical connective, whose truth table always produces F.] In other words, **X** is a *generic contradiction*; it is equivalent to every contradiction.

With our new generic contradiction, we can reformulate Indirect Derivation as follows.

Indirect Derivation (First Form)

Second Formulation

In order to show a negation $\sim\mathcal{A}$, it is sufficient to show **X**, assuming the un-negated formula, $\mathcal{A}$.

In addition to the syntactic and semantic rules governing **X**, we also need inference rules; in particular, as with the other logical symbols, we need an elimination rule, and an introduction rule. These are given as follows.

Contradiction-In (**×**I)

$\mathcal{A}$
$\sim\mathcal{A}$
———
×

Contradiction-Out (**×**O)

×
———
$\mathcal{A}$

We will have little use for the elimination rule, **×**O; it is included simply for symmetry. By contrast, the introduction rule, **×**I, will be used extensively.

We are now in a position to write down the official formulation of indirect derivation of the first form (we discuss the second form in the next section).

System Rule 7 (a show rule)

Indirect Derivation (First Form)

If one has a show-line of the form 'SHOW: $\sim\mathcal{A}$', then if one has **×** as a later available line, and there are no subsequent uncancelled show-lines, then one is entitled to cancel 'SHOW: $\sim\mathcal{A}$' and box off all subsequent lines. The annotation is 'ID'.

System Rule 8 (an assumption rule)

If one has a show-line of the form 'SHOW: $\sim\mathcal{A}$', then one is entitled to write down the un-negated formula $\mathcal{A}$ on the *very next* line, as an assumption. The annotation is 'As'.

As with earlier rules, we offer a pictorial abbreviation of indirect derivation as follows.

Indirect Derivation (First Form)

~~SHOW~~: $\sim\mathcal{A}$	ID
| $\mathcal{A}$	As
| ~~SHOW~~: ✕	
| |	

With our new rules in hand, let us now go back and do our earlier derivation in accordance with the new rules.

Example 1

(1)	$P \to Q$	Pr
(2)	$P \to \sim Q$	Pr
(3)	~~SHOW~~: $\sim P$	ID
(4)	| P	As
(5)	| ~~SHOW~~: ✕	DD
(6)	| | Q	1,4,→O
(7)	| | $\sim Q$	2,4,→O
(8)	| | ✕	6,7,✕I

On line (3), we are trying to show $\sim$P, which is a negation, so we do it by ID This entails writing down P on the next line as an assumption, and writing down 'SHOW: ✕' on the following line. On line (8), we obtain ✕ from lines (6) and (7), applying our new rule ✕I.

Let's do another simple example.

Example 2

(1)	$P \to Q$	Pr
(2)	$Q \to \sim P$	Pr
(3)	~~SHOW~~: $\sim P$	ID
(4)	| P	As
(5)	| ~~SHOW~~: ✕	DD
(6)	| | Q	1,4,→O
(7)	| | $\sim P$	2,6,→O
(8)	| | ✕	4,7,✕I

In the previous two examples, ✕ is obtained from an atomic formula and its negation. Sometimes, ✕ comes from more complex formulas, as in the following examples.

Example 3

(1)	~(P ∨ Q)	Pr
(2)	~~SHOW~~: ~P	ID
(3)	│ P	As
(4)	│ ~~SHOW~~: **X**	DD
(5)	│ │ P ∨ Q	3,∨I
(6)	│ │ **X**	1,5,**X**I

Here, **X** comes by **X**I from P∨Q and ~(P∨Q).

Example 4

(1)	~(P & Q)	Pr
(2)	~~SHOW~~: P → ~Q	CD
(3)	│ P	As
(4)	│ ~~SHOW~~: ~Q	ID
(5)	│ │ Q	As
(6)	│ │ ~~SHOW~~: **X**	DD
(7)	│ │ │ P & Q	3,5,&I
(8)	│ │ │ **X**	1,7,**X**I

Here, **X** comes, by **X**I, from P&Q and ~(P&Q).

11. INDIRECT DERIVATION (SECOND FORM)

In addition to indirect derivation of the first form, we also add indirect derivation of the second form, which is very similar to the first form. Consider the following derivation problem.

(1)	P → Q	Pr
(2)	~P → Q	Pr
(3)	SHOW: Q	???

The same problem as before arises; we have no simple means of dealing with either premise. (3) is atomic, so we must show it by direct derivation, but that approach comes to a screeching halt!

Once again, let's do something sneaky (but completely legal!), and see where that leads.

(1)	P → Q	Pr
(2)	~P → Q	Pr
(3)	SHOW: Q	???
(4)	SHOW: ~~Q	???

We have written down an additional show-line (which is completely legal, remember). The new problem facing us – to show ~~Q – appears much more promising;

specifically, we are trying to show a negation, so we can attack it using indirect derivation, which yields the following part-derivation.

(1)	$P \to Q$	Pr
(2)	$\sim P \to Q$	Pr
(3)	SHOW: Q	???
(4)	~~SHOW~~: $\sim\sim Q$	ID
(5)	│ $\sim Q$	As
(6)	│ ~~SHOW~~: ✕	DD
(7)	│ │ $\sim P$	1,5,→O
(8)	│ │ $\sim\sim P$	2,5,→O
(9)	│ │ ✕	7,8,✕I

The derivation is not complete. Line (3) is not cancelled. We are trying to show Q; we have in fact shown $\sim\sim Q$. This is a near-hit because we can apply Double Negation to line (4) to get Q. This yields the following completed derivation.

(1)	$P \to Q$	Pr
(2)	$\sim P \to Q$	Pr
(3)	~~SHOW~~: Q	DD
(4)	│ ~~SHOW~~: $\sim\sim Q$	ID
(5)	│ │ $\sim Q$	As
(6)	│ │ ~~SHOW~~: ✕	DD
(7)	│ │ │ $\sim P$	1,5,→O
(8)	│ │ │ $\sim\sim P$	2,5,→O
(9)	│ │ │ ✕	7,8,✕I
(10)	│ Q	4,DN

This derivation presents something completely novel. Upon getting to line (9), we have shown $\sim\sim Q$, which is marked by cancelling the 'SHOW' and boxing off the associated derivation. We can now use the formula $\sim\sim Q$ in connection with the usual rules of inference. In this particular case, we apply double negation to obtain line (10). This is in accordance with the following principle.

> As soon as one cancels a show-line 'SHOW: $\mathcal{A}$', thus obtaining '~~SHOW~~: $\mathcal{A}$', the formula $\mathcal{A}$ is available, at least until the show-line itself gets boxed off.

In order to abbreviate the above derivation somewhat, we enhance the method of indirect derivation so as to include, in effect, the above double negation maneuver. The intuitive formulation of this rule is given as follows.

> **Indirect Derivation (Second Form)**
>
> Intuitive Formulation
>
> In order to show a formula $\mathcal{A}$, it is sufficient to show ✕, assuming its negation $\sim\mathcal{A}$.

As usual, the official formulation of the rule is more complex.

System Rule 9 (a show rule)

> **Indirect Derivation (Second Form)**
>
> If one has a show-line 'SHOW: $\mathcal{A}$', then if one has ✕ as a later available line, and there are no intervening uncancelled show lines, then one is entitled to cancel 'SHOW: $\mathcal{A}$' and box off all subsequent formulas. The annotation is 'ID'

System Rule 10 (an assumption rule)

> If one has a show-line 'SHOW: $\mathcal{A}$', then one is entitled to write down the negation $\sim\mathcal{A}$ on the *very next* line, as an assumption. The annotation is 'As'

As usual, we also offer a pictorial version of the rule.

> **Indirect Derivation (Second Form)**
>
> ~~SHOW~~: $\mathcal{A}$
> | $\sim\mathcal{A}$
> | ~~SHOW~~: ✕
> | |
> | |

With this new show-rule in hand, we can now rewrite our earlier derivation, as follows.

Example 1

(1)	P → Q	Pr
(2)	~P → Q	Pr
(3)	~~SHOW~~: Q	DD
(4)	| ~Q	As
(5)	| ~~SHOW~~: ✕	DD
(6)	| | ~P	1,4,→O
(7)	| | ~~P	2,4,→O
(8)	| | ✕	6,7,✕I

In this particular problem, ✕ is obtained by ✕I from ~P and ~~P.

Let's look at one more example of the second form of indirect derivation.

Example 2

(1)	~(P & ~Q)	Pr
(2)	~~SHOW~~: P → Q	CD
(3)	│ P	As
(4)	│ ~~SHOW~~: Q	ID
(5)	││ ~Q	As
(6)	││ ~~SHOW~~: ✕	DD
(7)	│││ P & ~Q	3,5,&I
(8)	│││ ✕	1,7,✕I

In this derivation we show P→Q by conditional derivation, which means we assume P and show Q. This is shown, in turn, by indirect derivation (second form), which means we assume ~Q to show **✕**. In this particular problem, **✕** is obtained by **✕**I from P&~Q and ~(P&~Q).

12. SHOWING DISJUNCTIONS USING INDIRECT DERIVATION

The second form of ID is very useful for showing atomic formulas, as demonstrated in the previous section. It is also useful for showing disjunctions. Consider the following derivation problem.

(1)	~P → Q	Pr
(2)	SHOW: P ∨ Q	???

We are asked to show a disjunction P∨Q. CD is not available because this formula is not a conditional. ID of the first form is not available because it is not a negation. DD is available but it does not work (except in conjunction with the double-negation maneuver). That leaves the second form of ID, which yields the following.

(1)	~P → Q	Pr
(2)	SHOW: P ∨ Q	ID
(3)	~(P ∨ Q)	As
(4)	SHOW: ✕	DD
(5)	???	

At this point, we are nearly stuck. We don't have the minor premise to deal with line (1), and we have no rule for dealing with line (3). So, what do we do? We can always write down a show-line of our own choosing, so we choose to write down 'SHOW: ~P'. This produces the following part-derivation.

Line	Formula	Justification
(1)	~P → Q	Pr
(2)	SHOW: P ∨ Q	ID
(3)	~(P ∨ Q)	As
(4)	SHOW: ✕	DD
(5)	~~SHOW~~: ~P	ID
(6)	│ P	As
(7)	│ ~~SHOW~~: ✕	DD
(8)	│ │ P ∨ Q	6,∨I
(9)	│ │ ✕	3,8,✕I
(10)	???	

We are still not finished, but now we have shown ~P, so we can use it (while it is still available). This enables us to complete the derivation as follows.

Line	Formula	Justification
(1)	~P → Q	Pr
(2)	~~SHOW~~: P ∨ Q	ID
(3)	│ ~(P ∨ Q)	As
(4)	│ ~~SHOW~~: ✕	DD
(5)	│ │ ~~SHOW~~: ~P	ID
(6)	│ │ │ P	As
(7)	│ │ │ ~~SHOW~~: ✕	DD
(8)	│ │ │ │ P ∨ Q	6,∨I
(9)	│ │ │ │ ✕	3,8,✕I
(10)	│ │ Q	1,5,→O
(11)	│ │ P ∨ Q	10,∨I
(12)	│ │ ✕	3,11,✕I

Lines 5-9 constitute a crucial, but completely routine, sub-derivation. Given how important, and yet how routine, this sub-derivation is, we now add a further inference-rule to our list. System SL is already complete as it stands, so we don't require this new rule. Adding it to system SL decreases its elegance. We add it purely for the sake of convenience.

The new rule is called tilde-wedge-out (~∨O). As its name suggests, it is a rule for breaking down formulas that are negations of disjunctions. It is pictorially presented as follows.

Tilde-Wedge-Out (~∨O)

$$\frac{\sim(\mathcal{A} \vee \mathcal{B})}{\sim\mathcal{A}} \qquad \frac{\sim(\mathcal{A} \vee \mathcal{B})}{\sim\mathcal{B}}$$

As with all inference rules, this rule applies exclusively to lines, not to parts of lines. In other words, the official formulation of the rule goes as follows.

> **Tilde-Wedge-Out** ($\sim\vee$O)
>
> If one has available a line of the form $\sim(\mathcal{A} \vee \mathcal{B})$, then one is entitled to write down both $\sim\mathcal{A}$ and $\sim\mathcal{B}$.

Once we have the new rule $\sim\vee$O, the above derivation is much, much simpler.

Example 1

(1)	$\sim P \rightarrow Q$	Pr
(2)	~~SHOW~~: $P \vee Q$	ID
(3)	| $\sim(P \vee Q)$	As
(4)	| ~~SHOW~~: ✕	DD
(5)	| | $\sim P$	3,$\sim\vee$O
(6)	| | $\sim Q$	3,$\sim\vee$O
(7)	| | Q	1,5,$\rightarrow$O
(8)	| | ✕	6,7,✕I

In the above problem, we show a disjunction using the second form of indirect derivation. This involves a general strategy for showing any disjunction, formulated as follows.

General Strategy for Showing Disjunctions

> If you have a show-line of the form 'SHOW: $\mathcal{A}\vee\mathcal{B}$', then use indirect derivation: first assume $\sim[\mathcal{A}\vee\mathcal{B}]$, then write down 'SHOW: ✕', then apply $\sim\vee$O to obtain $\sim\mathcal{A}$ and $\sim\mathcal{B}$, then proceed from there.

In cartoon form:

~~SHOW~~: $\mathcal{A} \vee \mathcal{B}$	ID
| $\sim[\mathcal{A} \vee \mathcal{B}]$	As
| ~~SHOW~~: ✕	
| | $\sim\mathcal{A}$	$\sim\vee$O
| | $\sim\mathcal{B}$	$\sim\vee$O

This particular strategy actually applies to any disjunction, simple or complex. In the previous example, the disjunction is simple (its disjuncts are atomic). In the next example, the disjunction is complex (its disjuncts are not atomic).

Example 2

(1)	(P ∨ Q) → (P & Q)	Pr
(2)	SHOW: (P & Q) ∨ (~P & ~Q)	ID
(3)	│ ~[(P & Q) ∨ (~P & ~Q)]	As
(4)	│ SHOW: ✕	DD
(5)	│ │ ~(P & Q)	3,~∨O
(6)	│ │ ~(~P & ~Q)	3,~∨O
(7)	│ │ ~(P ∨ Q)	1,5,→O
(8)	│ │ ~P	7,~∨O
(9)	│ │ ~Q	7,~∨O
(10)	│ │ ~P & ~Q	8,9,&I
(11)	│ │ ✕	6,10,✕I

The basic strategy is exactly like the previous problem. The only difference is that the formulas are more complex.

13. FURTHER RULES

In the previous section, we added the rule ~∨O to our list of inference rules. Although it is not strictly required, it does make a number of derivations much easier. In the present section, for the sake of symmetry, we add corresponding rules for the remaining two-place connectives; specifically, we add ~&O, ~→O, and ~↔O. That way, we have a rule for handling any negated molecular formula.

Also, we add one more rule that is sometimes useful, the Rule of Repetition.

The additional negation rules are given as follows.

Tilde-Ampersand-Out (~&O)

$$\frac{\sim(\mathcal{A} \,\&\, \mathcal{B})}{\mathcal{A} \rightarrow \sim\mathcal{B}}$$

Tilde-Arrow-Out (~→O)

$$\frac{\sim(\mathcal{A} \rightarrow \mathcal{C})}{\mathcal{A} \,\&\, \sim\mathcal{C}}$$

Tilde-Double-Arrow-Out (~↔O)

$$\frac{\sim(\mathcal{A} \leftrightarrow \mathcal{B})}{\sim\mathcal{A} \leftrightarrow \mathcal{B}}$$

The reader is urged to verify that these are all valid argument forms of sentential logic. There are other valid forms that could serve equally well as the rules in question. The choice is to a certain arbitrary. The advantage of the particular choice becomes more apparent in a later chapter on predicate logic.

Finally in this section, we officially present the Rule of Repetition.

Repetition (R)

$$\frac{\mathcal{A}}{\mathcal{A}}$$

In other words, if you have an available formula, $\mathcal{A}$, you can simply copy (repeat) it at any later time. See Problem #120 for an application of this rule.

14. SHOWING CONJUNCTIONS AND BICONDITIONALS

In the previous sections, strategies are suggested for showing various kinds of formulas, as follows.

Formula Type	Strategy
Conditional	Conditional Derivation
Negation	Indirect Derivation (1)
Atomic Formula	Indirect Derivation (2)
Disjunction	Indirect Derivation (2)

That leaves only two kinds of formulas – conjunctions and biconditionals. In the present section, we discuss the strategies for these kinds of formulas.

Strategy for Showing Conjunctions

If you have a show-line of the form 'SHOW: $\mathcal{A}$&$\mathcal{B}$', then write down two further show-lines. Specifically, first write down 'SHOW: $\mathcal{A}$' and complete the associated derivation, then write down 'SHOW: $\mathcal{B}$' and complete the associated derivation. Finally, apply &I, and cancel 'SHOW: $\mathcal{A}$&$\mathcal{B}$' by direct derivation.

This strategy is easier to see in its cartoon version.

~~SHOW~~: $\mathcal{A}$ & $\mathcal{B}$	DD
| ~~SHOW~~: $\mathcal{A}$	
| |	
| ~~SHOW~~: $\mathcal{B}$	
| |	
| $\mathcal{A}$ & $\mathcal{B}$	&I

There is a parallel strategy for biconditionals, given as follows.

Strategy for Showing Biconditionals

If you have a show-line of the form 'SHOW: $\mathcal{A}\leftrightarrow\mathcal{B}$', then write down two further show-lines. Specifically, first write down 'SHOW: $\mathcal{A}\rightarrow\mathcal{B}$' and complete the associated derivation, then write down 'SHOW: $\mathcal{B}\rightarrow\mathcal{A}$' and complete the associated derivation. Finally, apply ↔I and cancel 'SHOW: $\mathcal{A}\leftrightarrow\mathcal{B}$' by direct derivation.

The associated cartoon version is as follows.

SHOW: $\mathcal{A} \leftrightarrow \mathcal{B}$	DD
│ SHOW: $\mathcal{A} \rightarrow \mathcal{B}$	
│ │	
│ SHOW: $\mathcal{B} \rightarrow \mathcal{A}$	
│ │	
│ $\mathcal{A} \leftrightarrow \mathcal{B}$	$\leftrightarrow$I

We conclude this section by doing a few examples that use these two strategies.

Example 1

(1)	$(A \vee B) \rightarrow C$	Pr
(2)	SHOW: $(A \rightarrow C) \& (B \rightarrow C)$	DD
(3)	│ SHOW: $A \rightarrow C$	CD
(4)	│ │ A	As
(5)	│ │ SHOW: C	DD
(6)	│ │ │ $A \vee B$	4,$\vee$I
(7)	│ │ │ C	1,6,$\rightarrow$O
(8)	│ SHOW: $B \rightarrow C$	CD
(9)	│ │ B	As
(10)	│ │ SHOW: C	DD
(11)	│ │ │ $A \vee B$	9,$\vee$I
(12)	│ │ │ C	1,11,$\rightarrow$O
(13)	│ $(A \rightarrow C) \& (B \rightarrow C)$	3,8,&I

Example 2

(1)	$\sim P \rightarrow Q$	Pr
(2)	$Q \rightarrow \sim P$	Pr
(3)	SHOW: $P \leftrightarrow \sim Q$	DD
(4)	│ SHOW: $P \rightarrow \sim Q$	CD
(5)	│ │ P	As
(6)	│ │ SHOW: $\sim Q$	DD
(7)	│ │ │ $\sim\sim P$	5,DN
(8)	│ │ │ $\sim Q$	2,7,$\rightarrow$O
(9)	│ SHOW: $\sim Q \rightarrow P$	CD
(10)	│ │ $\sim Q$	As
(11)	│ │ SHOW: P	DD
(12)	│ │ │ $\sim\sim P$	1,10,$\rightarrow$O
(13)	│ │ │ P	12,DN
(14)	│ $P \leftrightarrow \sim Q$	4,9,$\leftrightarrow$I

Example 3

(1)	(P & Q) → ~R	Pr
(2)	Q → R	Pr
(3)	~~SHOW~~: P ↔ (P & ~Q)	DD
(4)	~~SHOW~~: P → (P & ~Q)	CD
(5)	P	As
(6)	~~SHOW~~: P & ~Q	DD
(7)	~~SHOW~~: ~Q	ID
(8)	Q	As
(9)	~~SHOW~~: ×	DD
(10)	P & Q	5,8,&I
(11)	~R	1,10,→O
(12)	R	2,8,→O
(13)	×	11,12,×I
(14)	P & ~Q	5,7,&I
(15)	~~SHOW~~: (P & ~Q) → P	CD
(16)	P & ~Q	As
(17)	~~SHOW~~: P	DD
(18)	P	16,&O
(19)	P ↔ (P & ~Q)	4,15,↔I

15. THE WEDGE-OUT STRATEGY

We now have a strategy for dealing with every kind of show-line, whether it be atomic, a negation, a conjunction, a disjunction, a conditional, or a biconditional.

One often runs into problems that do not immediately surrender to any of these strategies. Consider the following problem, partly completed.

(1)	(P → Q) ∨ (P → R)	Pr
(2)	SHOW: (P & ~Q) → R	CD
(3)	P & ~Q	As
(4)	SHOW: R	ID
(5)	~R	As
(6)	SHOW: ×	DD
(7)	P	3,&O
(8)	~Q	3,&O
(9)	???	???

Everything goes smoothly until we reach line (9), at which point we are stuck. The premise is a disjunction; so in order to decompose it by wedge-out, we need one of the minor premises; that is, we need either ~(P → Q) or ~(P → R). If we had, say, the first one, then we could proceed as follows.

(1)	$(P \rightarrow Q) \vee (P \rightarrow R)$	Pr
(2)	SHOW: $(P \,\&\, {\sim}Q) \rightarrow R$	CD
(3)	$P \,\&\, {\sim}Q$	As
(4)	SHOW: R	ID
(5)	${\sim}R$	As
(6)	SHOW: **×**	DD
(7)	P	3,&O
(8)	${\sim}Q$	3,&O
(9)	${\sim}(P \rightarrow Q)$	?????
(10)	$P \rightarrow R$	1,9,∨O
(11)	R	7,10,→O
(12)	**×**	5,11,**×**I

This is great, except for line (9), which is completely without justification! For this reason the derivation remains incomplete. However, if we could somehow get ~(P→Q), then the derivation could be legally completed. So what can we do? One thing is to try to *show* the needed formula. Remember, one can write down any show-line whatsoever. Doing this produces the following partly completed derivation.

(1)	$(P \rightarrow Q) \vee (P \rightarrow R)$	Pr
(2)	SHOW $(P \,\&\, {\sim}Q) \rightarrow R$	CD
(3)	$P \,\&\, {\sim}Q$	As
(4)	SHOW: R	ID
(5)	${\sim}R$	As
(6)	SHOW: **×**	DD
(7)	P	3,&O
(8)	${\sim}Q$	3,&O
(9)	~~SHOW~~: ${\sim}(P \rightarrow Q)$	ID
(10)	│ $P \rightarrow Q$	As
(11)	│ ~~SHOW~~: **×**	DD
(12)	│ │ Q	7,10,→O
(13)	│ │ **×**	8,12,**×**I

Notice that we have shown exactly what we needed, so we can use it to complete the derivation as follows.

Example 1

(1)	$(P \rightarrow Q) \vee (P \rightarrow R)$	Pr
(2)	~~SHOW~~: $(P \,\&\, {\sim}Q) \rightarrow R$	CD
(3)	$P \,\&\, {\sim}Q$	As
(4)	~~SHOW~~: R	ID
(5)	~R	As
(6)	~~SHOW~~: ✕	DD
(7)	P	3,&O
(8)	~Q	3,&O
(9)	~~SHOW~~: $\sim(P \rightarrow Q)$	ID
(10)	$P \rightarrow Q$	As
(11)	~~SHOW~~: ✕	DD
(12)	Q	7,10,→O
(13)	✕	8,12,✕I
(14)	$P \rightarrow R$	1,9,∨O
(15)	~P	5,14,→O
(16)	✕	7,15,✕I

The above derivation is an example of a general strategy, called the wedge-out strategy, which is formulated as follows.

> **Wedge-Out Strategy**
>
> If you have as an available line a disjunction $\mathcal{A} \vee \mathcal{B}$, then look for means to break it down using wedge-out. This requires having either $\sim\mathcal{A}$ or $\sim\mathcal{B}$. Look for ways to get one of these. If you get stuck, try to *show* one of them; i.e., write 'SHOW: $\sim\mathcal{A}$' or 'SHOW: $\sim\mathcal{B}$'.

In pictures, this strategy looks thus:

$\mathcal{A} \vee \mathcal{B}$	$\mathcal{A} \vee \mathcal{B}$
SHOW: $\mathcal{C}$	SHOW: $\mathcal{C}$
o	o
o	o
~~SHOW~~: $\sim\mathcal{A}$	~~SHOW~~: $\sim\mathcal{B}$
|	|
$\mathcal{B}$ ∨O	$\mathcal{A}$ ∨O
o	o
o	o
o	o

How does one decide which one to show; the rule of thumb (not absolutely reliable, however) is this:

> **Rule of Thumb**
>
> In the wedge-out strategy, the choice of which disjunct to attack is largely unimportant, so you might as well choose the first one.

Since the wedge-out strategy is so important, let's do one more example. Here the crucial line is line (7).

Example 2

(1)	(P & R) ∨ (Q & R)	Pr
(2)	~~SHOW~~: ~P → Q	CD
(3)	│ ~P	As
(4)	│ ~~SHOW~~: Q	ID
(5)	│ │ │ ~Q	As
(6)	│ │ │ ~~SHOW~~: ✕	DD
(7)	│ │ │ │ ~~SHOW~~: ~(P & R)	ID
(8)	│ │ │ │ │ P & R	As
(9)	│ │ │ │ │ ~~SHOW~~: ✕	DD
(10)	│ │ │ │ │ │ P	8,&O
(11)	│ │ │ │ │ │ ✕	3,10,✕I
(12)	│ │ │ │ Q & R	1,7,∨O
(13)	│ │ │ │ Q	12,&O
(14)	│ │ │ │ ✕	5,13,✕I

16. THE ARROW-OUT STRATEGY

There is one more strategy that we will examine, one that is very similar to the wedge-out strategy; the difference is that it pertains to conditionals.

> **Arrow-Out Strategy**
>
> If you have as an available line a conditional $\mathcal{A} \rightarrow \mathcal{C}$, then look for means to break it down using arrow-out. This requires having either $\mathcal{A}$ or $\sim\mathcal{C}$. Look for ways to get one of these. If you get stuck, try to *show* one of them; i.e., write 'SHOW: $\mathcal{A}$' or 'SHOW: $\sim\mathcal{C}$'.

In pictures:

$\mathcal{A} \rightarrow \mathcal{C}$	$\mathcal{A} \rightarrow \mathcal{C}$
SHOW: $\mathcal{B}$	SHOW: $\mathcal{B}$
o	o
o	o
~~SHOW~~: $\mathcal{A}$	~~SHOW~~: $\sim\mathcal{C}$
$\mathcal{C}$ →O	$\sim\mathcal{A}$ →O
o	o
o	o
o	o

The following is a derivation that employs the arrow-out strategy. The crucial line is line (5).

Example 1

(1)	(P → Q) → (P → R)	Pr
(2)	~~SHOW~~: (P & Q) → R	CD
(3)	P & Q	As
(4)	~~SHOW~~: R	DD
(5)	~~SHOW~~: P → Q	CD
(6)	P	As
(7)	~~SHOW~~: Q	DD
(8)	Q	3,&O
(9)	P → R	1,5,→O
(10)	P	3,&O
(11)	R	9,10,→O

17. SUMMARY OF THE SYSTEM RULES FOR SYSTEM SL

1. System Rule 1 (The Premise Rule)

> At any point in a derivation, ***prior*** to the first show-line, any premise may be written down. The annotation is 'Pr'

2. System Rule 2 (The Inference-Rule Rule)

> At any point in a derivation, a formula may be written down if it follows from previous available lines by an inference rule. The annotation cites the lines numbers, and the inference rule, in that order.

3. System Rule 3 (The Show-Line Rule)

> At *any* point in a derivation, one is entitled to write down the expression 'SHOW: $\mathcal{A}$',
> for *any* formula $\mathcal{A}$ *whatsoever*.

4. System Rule 4 (a show-rule)

> **Direct Derivation (DD)**
>
> If one has a show-line 'SHOW: $\mathcal{A}$', and one obtains $\mathcal{A}$ as a later available line, and there are no intervening uncancelled show-lines, then one is entitled to box and cancel 'SHOW: $\mathcal{A}$'. The annotation is 'DD'

5. System Rule 5 (a show-rule)

> **Conditional Derivation (CD)**
>
> If one has a show-line of the form 'SHOW: $\mathcal{A}\rightarrow\mathcal{C}$', and one has $\mathcal{C}$ as a later available line, and there are no subsequent uncancelled show-lines, then one is entitled to box and cancel 'SHOW: $\mathcal{A}\rightarrow\mathcal{C}$'. The annotation is 'CD'

6. System Rule 6 (an assumption rule)

> If one has a show-line of the form 'SHOW: $\mathcal{A}\rightarrow\mathcal{C}$', then one is entitled to write down the antecedent $\mathcal{A}$ on the very next line, as an assumption. The annotation is 'As'

7. System Rule 7 (a show rule)

> **Indirect Derivation (First Form)**
>
> If one has a show-line of the form 'SHOW: $\sim\mathcal{A}$', then if one has **×** as a later available line, and there are no intervening uncancelled show-lines, then one is entitled to box and cancel 'SHOW: $\sim\mathcal{A}$'. The annotation is 'ID'.

8. System Rule 8 (an assumption rule)

> If one has a show-line of the form 'SHOW: $\sim\mathcal{A}$', then one is entitled to write down the un-negated formula $\mathcal{A}$ on the *very next* line, as an assumption. The annotation is 'As'

9. System Rule 9 (a show rule)

> **Indirect Derivation (Second Form)**
>
> If one has a show-line 'SHOW: $\mathcal{A}$', then if one has **×** as a later available line, and there are no intervening uncancelled show lines, then one is entitled to box and cancel 'SHOW: $\mathcal{A}$'. The annotation is 'ID'

10. System Rule 10 (an assumption rule)

> If one has a show-line 'SHOW: $\mathcal{A}$', then one is entitled to write down the negation $\sim\mathcal{A}$ on the *very next* line, as an assumption. The annotation is 'As'

11. System Rule 11 (Definition of available formula)

Formula $\mathcal{A}$ in a derivation is ***available*** if and only if either $\mathcal{A}$ occurs (as a whole line!), but is not inside a box, or '~~SHOW~~: $\mathcal{A}$' occurs (as a whole line!), but is not inside a box.

12. System Rule 12 (definition of box-and-cancel)

To ***box and cancel*** a show-line 'SHOW: $\mathcal{A}$' is to strike through 'SHOW' resulting in '~~SHOW~~', and box off *all* lines below 'SHOW: $\mathcal{A}$' (which is to say all lines *at the time* the box-and-cancel occurs).

18. PICTORIAL SUMMARY OF THE RULES OF SYSTEM SL

INITIAL INFERENCE RULES

Ampersand-In (&I)

$$\frac{\mathcal{A} \quad \mathcal{B}}{\mathcal{A} \,\&\, \mathcal{B}} \qquad \frac{\mathcal{A} \quad \mathcal{B}}{\mathcal{B} \,\&\, \mathcal{A}}$$

Ampersand-Out (&O)

$$\frac{\mathcal{A} \,\&\, \mathcal{B}}{\mathcal{A}} \qquad \frac{\mathcal{A} \,\&\, \mathcal{B}}{\mathcal{B}}$$

Wedge-In (∨I)

$$\frac{\mathcal{A}}{\mathcal{A} \vee \mathcal{B}} \qquad \frac{\mathcal{A}}{\mathcal{B} \vee \mathcal{A}}$$

Wedge-Out (∨O)

$$\frac{\mathcal{A} \vee \mathcal{B} \quad {\sim}\mathcal{A}}{\mathcal{B}} \qquad \frac{\mathcal{A} \vee \mathcal{B} \quad {\sim}\mathcal{B}}{\mathcal{A}}$$

Double-Arrow-In (↔I)

$$\frac{\mathcal{A} \rightarrow \mathcal{B} \quad \mathcal{B} \rightarrow \mathcal{A}}{\mathcal{A} \leftrightarrow \mathcal{B}} \qquad \frac{\mathcal{A} \rightarrow \mathcal{B} \quad \mathcal{B} \rightarrow \mathcal{A}}{\mathcal{B} \leftrightarrow \mathcal{A}}$$

Double-Arrow-Out (↔O)

$$\frac{\mathcal{A} \leftrightarrow \mathcal{B}}{\mathcal{A} \rightarrow \mathcal{B}} \qquad \frac{\mathcal{A} \leftrightarrow \mathcal{B}}{\mathcal{B} \rightarrow \mathcal{A}}$$

Arrow-Out (→O)

$$\frac{\mathcal{A} \rightarrow \mathcal{C} \quad \mathcal{A}}{\mathcal{C}} \qquad \frac{\mathcal{A} \rightarrow \mathcal{C} \quad {\sim}\mathcal{C}}{{\sim}\mathcal{A}}$$

Double Negation (DN)

$$\frac{\mathcal{A}}{{\sim}{\sim}\mathcal{A}} \qquad \frac{{\sim}{\sim}\mathcal{A}}{\mathcal{A}}$$

ADDITIONAL INFERENCE RULES

Contradiction-In (✕I)

$$\frac{\mathcal{A} \quad {\sim}\mathcal{A}}{\times}$$

Contradiction-Out (✕O)

$$\frac{\times}{\mathcal{A}}$$

Tilde-Wedge-Out (~∨O)

$$\frac{\sim(\mathcal{A} \vee \mathcal{B})}{\sim\mathcal{A}} \qquad \frac{\sim(\mathcal{A} \vee \mathcal{B})}{\sim\mathcal{B}}$$

Tilde-Ampersand-Out (~&O)

$$\frac{\sim(\mathcal{A} \;\&\; \mathcal{B})}{\mathcal{A} \rightarrow \sim\mathcal{B}}$$

Tilde-Arrow-Out (~→O)

$$\frac{\sim(\mathcal{A} \rightarrow \mathcal{C})}{\mathcal{A} \;\&\; \sim\mathcal{C}}$$

Tilde-Double-Arrow-Out (~↔O)

$$\frac{\sim(\mathcal{A} \leftrightarrow \mathcal{B})}{\sim\mathcal{A} \leftrightarrow \mathcal{B}}$$

Repetition (R)

$$\frac{\mathcal{A}}{\mathcal{A}}$$

SHOW-RULES

Direct Derivation (DD)

~~SHOW~~: $\mathcal{A}$	DD
|	
| $\mathcal{A}$	

Conditional Derivation (CD)

~~SHOW~~: $\mathcal{A} \rightarrow C$	CD
| $\mathcal{A}$	As
| ~~SHOW~~: C	
| |	

Indirect Derivation (First Form)

~~SHOW~~: $\sim\mathcal{A}$	ID
| $\mathcal{A}$	As
| ~~SHOW~~: ×	
| |	

Indirect Derivation (Second Form)

~~SHOW~~: $\mathcal{A}$	ID
| $\sim\mathcal{A}$	As
| ~~SHOW~~: ×	
| |	

19. PICTORIAL SUMMARY OF STRATEGIES

~~SHOW~~: $\mathcal{A}$ & $\mathcal{B}$	DD
~~SHOW~~: $\mathcal{A}$	
~~SHOW~~: $\mathcal{B}$	
$\mathcal{A}$ & $\mathcal{B}$	&I

~~SHOW~~: $\mathcal{A} \rightarrow \mathcal{C}$	CD
A	As
~~SHOW~~: $\mathcal{C}$	

~~SHOW~~: $\mathcal{A} \vee \mathcal{B}$	ID
$\sim[\mathcal{A} \vee \mathcal{B}]$	As
~~SHOW~~: ✕	
$\sim\mathcal{A}$	$\sim\vee$O
$\sim\mathcal{B}$	$\sim\vee$O

~~SHOW~~: $\mathcal{A} \leftrightarrow \mathcal{B}$	DD
~~SHOW~~: $\mathcal{A} \rightarrow \mathcal{B}$	
~~SHOW~~: $\mathcal{B} \rightarrow \mathcal{A}$	
$\mathcal{A} \leftrightarrow \mathcal{B}$	$\leftrightarrow$I

~~SHOW~~: $\sim\mathcal{A}$	ID
| $\mathcal{A}$	As
| ~~SHOW~~: ×	
| |	

~~SHOW~~: A	ID
| $\sim$A	As
| ~~SHOW~~: ×	
| |	

Wedge-Out Strategy

> **Wedge-Out Strategy**
>
> If you have as an available line a disjunction $\mathcal{A}\vee\mathcal{B}$, then look for means to break it down using wedge-out. This requires having either $\sim\mathcal{A}$ or $\sim\mathcal{B}$. Look for ways to get one of these. If you get stuck, try to *show* one of them; i.e., write 'SHOW: $\sim\mathcal{A}$' or 'SHOW: $\sim\mathcal{B}$'.

$\mathcal{A}\vee\mathcal{B}$		$\mathcal{A}\vee\mathcal{B}$	
SHOW: $\mathcal{C}$		SHOW: $\mathcal{C}$	
o		o	
o		o	
~~SHOW~~: $\sim\mathcal{A}$		~~SHOW~~: $\sim\mathcal{B}$	
|		|	
$\mathcal{B}$	$\vee$O	$\mathcal{A}$	$\vee$O
o		o	
o		o	
o		o	

Arrow-Out Strategy

> If you have as an available line a conditional $\mathcal{A}\rightarrow\mathcal{B}$, then look for means to break it down using arrow-out. This requires having either $\mathcal{A}$ or $\sim\mathcal{B}$. Look for ways to get one of these. If you get stuck, try to *show* one of them; i.e., write 'SHOW: $\mathcal{A}$' or 'SHOW: $\sim\mathcal{B}$'.

$\mathcal{A}\rightarrow\mathcal{C}$	$\mathcal{A}\rightarrow\mathcal{C}$
SHOW: $\mathcal{B}$	SHOW: $\mathcal{B}$
o	o
o	o
~~SHOW~~: $\mathcal{A}$	~~SHOW~~: $\sim\mathcal{C}$
|	|
$\mathcal{C}$ $\rightarrow$O	$\sim\mathcal{A}$ $\rightarrow$O
o	o
o	o
o	o

20. EXERCISES FOR CHAPTER 5

EXERCISE SET A (Simple Derivation)

For each of the following arguments, construct a simple derivation of the conclusion (marked by '/') from the premises, using the simple rules MP, MT, MTP1, and MTP2.

(1) P ; P → Q ; Q → R ; R → S / S

(2) P → Q ; Q → R ; R → S ; ~S / ~P

(3) ~P ∨ Q ; ~Q ; P ∨ R / R

(4) P ∨ Q ; ~P ; Q → R / R

(5) P ; P → ~Q ; R → Q ; ~R → S / S

(6) P ∨ ~Q ; ~P ; R → Q ; ~R → S / S

(7) (P → Q) → P ; P → Q / Q

(8) (P → Q) → R ; R → P ; P → Q / Q

(9) (P → Q) → (Q → R) ; P → Q ; P / R

(10) ~P → Q ; ~Q ; R ∨ ~P / R

(11) ~P → (~Q ∨ R) ; P → R ; ~R / ~Q

(12) P → ~Q ; ~S → P ; ~~Q / ~~S

(13) P ∨ Q ; Q → R ; ~R / P

(14) ~P → (Q ∨ R) ; P → Q ; ~Q / R

(15) P → R ; ~P → (S ∨ R) ; ~R / S

(16) P ∨ ~Q ; ~R → ~~Q ; R → ~S ; ~~S / P

(17) (P → Q) ∨ (R → S) ; (P → Q) → R ; ~R / R → S

(18) (P → Q) → (R → S) ; (R → T) ∨ (P → Q) ; ~(R → T) / R → S

(19) ~R → (P ∨ Q) ; R → P ; (R → P) → ~P / Q

(20) (P → Q) ∨ R ; [(P → Q) ∨ R] → ~R ; (P → Q) → (Q → R) / ~Q

EXERCISE SET B (Direct Derivation)

Convert each of the simple derivations in Exercise Set A into a direct derivation; use the introduction-elimination rules.

EXERCISE SET C (Direct Derivation)

Directions for remaining exercises: For each of the following arguments, construct a derivation of the conclusion (marked by '/') from the premises, using the rules of System SL.

(21) P & Q ; P → (R & S) / Q & S

(22) P & Q ; (P ∨ R) → S / P & S

(23) P ; (P ∨ Q) → (R & S) ; (R ∨ T) → U / U

(24) P → Q ; P ∨ R ; ~Q / R & ~P

(25) P → Q ; ~R → (Q → S) ; R → T ; ~T & P / Q & S

(26) P → Q ; R ∨ ~Q ; ~R & S ; (~P & S) → T / T

(27) P ∨ ~Q ; ~R → Q ; R → ~S ; S / P

(28) P & Q ; (P ∨ T) → R ; S → ~R / ~S

(29) P & Q ; P → R ; (P & R) → S / Q & S

(30) P → Q ; Q ∨ R ; (R & ~P) → S ; ~Q / S

(31) P & Q / Q & P

(32) P & (Q & R) / (P & Q) & R

(33) P / P & P

(34) P / P & (P ∨ Q)

(35) P & ~P / Q

(36) P ↔ ~Q ; Q ; P ↔ ~S / S

(37) P & ~Q ; Q ∨ (P → S) ; (R & T) ↔ S / P & R

(38) P → Q ; (P → Q) → (Q → P) ; (P ↔ Q) → P / P & Q

(39) ~P & Q ; (R ∨ Q) → (~S → P) ; ~S ↔ T / ~T

(40) P & ~Q ; Q ∨ (R → S) ; ~V → ~P ; V → (S → R) ; (R ↔ S) → T ;
U ↔ (~Q & T) / U

EXERCISE SET D (Conditional Derivation)

(41) $(P \vee Q) \rightarrow R$ / $Q \rightarrow R$

(42) $Q \rightarrow R$ / $(P \& Q) \rightarrow (P \& R)$

(43) $P \rightarrow Q$ / $(Q \rightarrow R) \rightarrow (P \rightarrow R)$

(44) $P \rightarrow Q$ / $(R \rightarrow P) \rightarrow (R \rightarrow Q)$

(45) $(P \& Q) \rightarrow R$ / $P \rightarrow (Q \rightarrow R)$

(46) $P \rightarrow (Q \rightarrow R)$ / $(P \rightarrow Q) \rightarrow (P \rightarrow R)$

(47) $(P \& Q) \rightarrow R$ / $[(P \rightarrow Q) \rightarrow P] \rightarrow [(P \rightarrow Q) \rightarrow R]$

(48) $(P \& Q) \rightarrow (R \rightarrow S)$ / $(P \rightarrow Q) \rightarrow [(P \& R) \rightarrow S]$

(49) $[(P \& Q) \& R] \rightarrow S$ / $P \rightarrow [Q \rightarrow (R \rightarrow S)]$

(50) $(\sim P \& Q) \rightarrow R$ / $(\sim Q \rightarrow P) \rightarrow (\sim P \rightarrow R)$

EXERCISE SET E (Indirect Derivation – First Form)

(51) $P \rightarrow Q$; $P \rightarrow \sim Q$ / $\sim P$

(52) $P \rightarrow Q$; $Q \rightarrow \sim P$ / $\sim P$

(53) $P \rightarrow Q$; $\sim Q \vee \sim R$; $P \rightarrow R$ / $\sim P$

(54) $P \rightarrow R$; $Q \rightarrow \sim R$ / $\sim(P \& Q)$

(55) $P \& Q$ / $\sim(P \rightarrow \sim Q)$

(56) $P \& \sim Q$ / $\sim(P \rightarrow Q)$

(57) $\sim P$ / $\sim(P \& Q)$

(58) $\sim P \& \sim Q$ / $\sim(P \vee Q)$

(59) $P \leftrightarrow Q$; $\sim Q$ / $\sim(P \vee Q)$

(60) $P \& Q$ / $\sim(\sim P \vee \sim Q)$

(61) $\sim P \vee \sim Q$ / $\sim(P \& Q)$

(62) $P \vee Q$ / $\sim(\sim P \& \sim Q)$

(63) $P \rightarrow Q$ / $\sim(P \& \sim Q)$

(64) $P \rightarrow (Q \rightarrow \sim P)$ / $P \rightarrow \sim Q$

(65) $(P \& Q) \rightarrow R$ / $(P \& \sim R) \rightarrow \sim Q$

(66) $(P \& Q) \rightarrow \sim R$ / $P \rightarrow \sim(Q \& R)$

(67) $P \rightarrow (Q \rightarrow R)$ / $(Q \& \sim R) \rightarrow \sim P$

(68) $P \rightarrow \sim(Q \& R)$ / $(P \& Q) \rightarrow \sim R$

(69) $P \rightarrow \sim(Q \& R)$ / $(P \rightarrow Q) \rightarrow (P \rightarrow \sim R)$

(70) $P \rightarrow (Q \rightarrow R)$ / $(P \rightarrow \sim R) \rightarrow (P \rightarrow \sim Q)$

EXERCISE SET F (Indirect Derivation – Second Form)

(71) $P \to Q\ ;\ \sim P \to Q\ /\ Q$

(72) $P \vee Q\ ;\ P \to R\ ;\ Q \vee \sim R\ /\ Q$

(73) $\sim P \to R\ ;\ Q \to R\ ;\ P \to Q\ /\ R$

(74) $(P \vee \sim Q) \to (R\ \&\ \sim S)\ ;\ Q \vee S\ /\ Q$

(75) $(P \vee Q) \to (R \to S)\ ;\ (\sim S \vee T) \to (P\ \&\ R)\ /\ S$

(76) $\sim(P\ \&\ \sim Q)\ /\ P \to Q$

(77) $P \to (\sim Q \to R)\ /\ (P\ \&\ \sim R) \to Q$

(78) $P\ \&\ (Q \vee R)\ /\ \sim(P\ \&\ Q) \to R$

(79) $P \vee Q\ /\ Q \vee P$

(80) $\sim P \to Q\ /\ P \vee Q$

(81) $\sim(P\ \&\ Q)\ /\ \sim P \vee \sim Q$

(82) $P \to Q\ /\ \sim P \vee Q$

(83) $P \vee Q\ ;\ P \to R\ ;\ Q \to S\ /\ R \vee S$

(84) $\sim P \to Q\ ;\ P \to R\ /\ Q \vee R$

(85) $\sim P \to Q\ ;\ \sim R \to S\ ;\ \sim Q \vee \sim S\ /\ P \vee R$

(86) $(P\ \&\ \sim Q) \to R\ /\ P \to (Q \vee R)$

(87) $\sim P \to (\sim Q \vee R)\ /\ Q \to (P \vee R)$

(88) $P\ \&\ (Q \vee R)\ /\ (P\ \&\ Q) \vee R$

(89) $(P \vee Q)\ \&\ (P \vee R)\ /\ P \vee (Q\ \&\ R)$

(90) $(P \vee Q) \to (P\ \&\ Q)\ /\ (P\ \&\ Q) \vee (\sim P\ \&\ \sim Q)$

EXERCISE SET G (Strategies)

(91) $P \rightarrow (Q \& R)$ / $(P \rightarrow Q) \& (P \rightarrow R)$

(92) $(P \vee Q) \rightarrow R$ / $(P \rightarrow R) \& (Q \rightarrow R)$

(93) $(P \vee Q) \rightarrow (P \& Q)$ / $P \leftrightarrow Q$

(94) $P \leftrightarrow Q$ / $Q \leftrightarrow P$

(95) $P \leftrightarrow Q$ / $\sim P \leftrightarrow \sim Q$

(96) $P \leftrightarrow Q$; $Q \rightarrow \sim P$ / $\sim P \& \sim Q$

(97) $(P \rightarrow Q) \vee (\sim Q \rightarrow R)$ / $P \rightarrow (Q \vee R)$

(98) $P \vee Q$; $P \rightarrow \sim Q$ / $(P \rightarrow Q) \rightarrow (Q \& \sim P)$

(99) $P \vee Q$; $\sim(P \& Q)$ / $(P \rightarrow Q) \rightarrow \sim(Q \rightarrow P)$

(100) $P \vee Q$; $P \rightarrow \sim Q$ / $(P \& \sim Q) \vee (Q \& \sim P)$

(101) $(P \vee Q) \rightarrow (P \& Q)$ / $(\sim P \vee \sim Q) \rightarrow (\sim P \& \sim Q)$

(102) $P \& (Q \vee R)$ / $(P \& Q) \vee (P \& R)$

(103) $(P \& Q) \vee (P \& R)$ / $P \& (Q \vee R)$

(104) $P \vee (Q \& R)$ / $(P \vee Q) \& (P \vee R)$

(105) $(P \& Q) \vee [(P \& R) \vee (Q \& R)]$ / $P \vee (Q \& R)$

(106) $P \vee Q$; $P \vee R$; $Q \vee R$ / $[P \& Q] \vee [(P \& R) \vee (Q \& R)]$

(107) $(P \rightarrow Q) \vee (P \rightarrow R)$ / $P \rightarrow (Q \vee R)$

(108) $(P \rightarrow R) \vee (Q \rightarrow R)$ / $(P \& Q) \rightarrow R$

(109) $P \leftrightarrow (Q \& \sim P)$ / $\sim(P \vee Q)$

(110) $(P \& Q) \vee (\sim P \& \sim Q)$ / $P \leftrightarrow Q$

EXERCISE SET H (Miscellaneous)

(111) $P \rightarrow (Q \vee R)$ / $(P \rightarrow Q) \vee (P \rightarrow R)$

(112) $(P \leftrightarrow Q) \rightarrow R$ / $P \rightarrow (Q \rightarrow R)$

(113) $P \rightarrow (\sim Q \rightarrow R)$ / $\sim(P \rightarrow R) \rightarrow Q$

(114) $(P \& Q) \rightarrow R$ / $(P \rightarrow R) \vee (Q \rightarrow R)$

(115) $P \leftrightarrow \sim Q$ / $(P \& \sim Q) \vee (Q \& \sim P)$

(116) $(P \rightarrow \sim Q) \rightarrow R$ / $\sim(P \& Q) \rightarrow R$

(117) $P \leftrightarrow (Q \& \sim P)$ / $\sim P \& \sim Q$

(118) P / $(P \& Q) \vee (P \& \sim Q)$

(119) $P \leftrightarrow \sim P$ / Q

(120) $(P \leftrightarrow Q) \leftrightarrow R$ / $P \leftrightarrow (Q \leftrightarrow R)$

21. ANSWERS TO EXERCISES FOR CHAPTER 5

EXERCISE SET A

#1:

(1)	P	Pr
(2)	$P \rightarrow Q$	Pr
(3)	$Q \rightarrow R$	Pr
(4)	$R \rightarrow S$	Pr
(5)	Q	1,2,MP
(6)	R	3,5,MP
(7)	S	4,6,MP

#2:

(1)	$P \rightarrow Q$	Pr
(2)	$Q \rightarrow R$	Pr
(3)	$R \rightarrow S$	Pr
(4)	$\sim S$	Pr
(5)	$\sim R$	3,4,MT
(6)	$\sim Q$	2,5,MT
(7)	$\sim P$	1,6,MT

#3:

(1)	$\sim P \vee Q$	Pr
(2)	$\sim Q$	Pr
(3)	$P \vee R$	Pr
(4)	$\sim P$	1,2,MTP2
(5)	R	3,4,MTP1

#4:

(1)	$P \vee Q$	Pr
(2)	$\sim P$	Pr
(3)	$Q \rightarrow R$	Pr
(4)	Q	1,2,MTP1
(5)	R	3,4,MP

#5:

(1)	P	Pr
(2)	$P \rightarrow \sim Q$	Pr
(3)	$R \rightarrow Q$	Pr
(4)	$\sim R \rightarrow S$	Pr
(5)	$\sim Q$	1,2,MP
(6)	$\sim R$	3,5,MT
(7)	S	4,6,MP

#6:

(1)	$P \vee \sim Q$	Pr
(2)	$\sim P$	Pr
(3)	$R \rightarrow Q$	Pr
(4)	$\sim R \rightarrow S$	Pr
(5)	$\sim Q$	1,2,MTP1
(6)	$\sim R$	3,5,MT
(7)	S	4,6,MP

#7:

(1)	$(P \rightarrow Q) \rightarrow P$	Pr
(2)	$P \rightarrow Q$	Pr
(3)	P	1,2,MP
(4)	Q	2,3,MP

#8:

(1)	$(P \rightarrow Q) \rightarrow R$	Pr
(2)	$R \rightarrow P$	Pr
(3)	$P \rightarrow Q$	Pr
(4)	R	1,3,MP
(5)	P	2,4,MP
(6)	Q	3,5,MP

#9:

(1)	$(P \rightarrow Q) \rightarrow (Q \rightarrow R)$	Pr
(2)	$P \rightarrow Q$	Pr
(3)	P	Pr
(4)	$Q \rightarrow R$	1,2,MP
(5)	Q	2,3,MP
(6)	R	4,5,MP

#10:

(1)	$\sim P \rightarrow Q$	Pr
(2)	$\sim Q$	Pr
(3)	$R \vee \sim P$	Pr
(4)	$\sim\sim P$	1,2,MT
(5)	R	3,4,MTP2

#11:

(1)	$\sim P \rightarrow (\sim Q \vee R)$	Pr
(2)	$P \rightarrow R$	Pr
(3)	$\sim R$	Pr
(4)	$\sim P$	2,3,MT
(5)	$\sim Q \vee R$	1,4,MP
(6)	$\sim Q$	3,5,MTP2

#12:

(1)	$P \rightarrow \sim Q$	Pr
(2)	$\sim S \rightarrow P$	Pr
(3)	$\sim\sim Q$	Pr
(4)	$\sim P$	1,3,MT
(5)	$\sim\sim S$	2,4,MT

#13:

(1)	$P \vee Q$	Pr
(2)	$Q \rightarrow R$	Pr
(3)	$\sim R$	Pr
(4)	$\sim Q$	2,3,MT
(5)	P	1,4,MTP2

#14:

(1)	~P → (Q ∨ R)	Pr
(2)	P → Q	Pr
(3)	~Q	Pr
(4)	~P	2,3,MT
(5)	Q ∨ R	1,4,MP
(6)	R	3,5,MTP1

#15:

(1)	P → R	Pr
(2)	~P → (S ∨ R)	Pr
(3)	~R	Pr
(4)	~P	1,3,MT
(5)	S ∨ R	2,4,MP
(6)	S	3,6,MTP2

#16:

(1)	P ∨ ~Q	Pr
(2)	~R → ~~Q	Pr
(3)	R → ~S	Pr
(4)	~~S	Pr
(5)	~R	3,4,MT
(6)	~~Q	2,5,MP
(7)	P	1,6,MTP2

#17:

(1)	(P → Q) ∨ (R → S)	Pr
(2)	(P → Q) → R	Pr
(3)	~R	Pr
(4)	~(P → Q)	2,3,MT
(5)	R → S	1,4,MTP1

#18:

(1)	(P → Q) → (R → S)	Pr
(2)	(R → T) ∨ (P → Q)	Pr
(3)	~(R → T)	Pr
(4)	P → Q	2,3,MTP1
(5)	R → S	1,4,MP

#19:

(1)	~R → (P ∨ Q)	Pr
(2)	R → P	Pr
(3)	(R → P) → ~P	Pr
(4)	~P	2,3,MP
(5)	~R	2,4,MT
(6)	P ∨ Q	1,5,MP
(7)	Q	4,6,MTP1

#20:

(1)	(P → Q) ∨ R	Pr
(2)	[(P → Q) ∨ R] → ~R	Pr
(3)	(P → Q) → (Q → R)	Pr
(4)	~R	1,2,MP
(5)	P → Q	1,4,MTP2
(6)	Q → R	3,5,MP
(7)	~Q	4,6,MT

EXERCISE SETS B-H

#1:

(1)	P	Pr
(2)	P → Q	Pr
(3)	Q → R	Pr
(4)	R → S	Pr
(5)	~~SHOW~~: S	DD
(6)	│ Q	1,2,→O
(7)	│ R	3,6,→O
(8)	│ S	4,7,→O

#2:

(1)	P → Q	Pr
(2)	Q → R	Pr
(3)	R → S	Pr
(4)	~S	Pr
(5)	~~SHOW~~: ~P	DD
(6)	│ ~R	3,4,→O
(7)	│ ~Q	2,6,→O
(8)	│ ~P	1,7,→O

#3:

(1)	~P ∨ Q	Pr
(2)	~Q	Pr
(3)	P ∨ R	Pr
(4)	~~SHOW~~: R	DD
(5)	│ ~P	1,2,∨O
(6)	│ R	3,5,∨O

#4:

(1)	P ∨ Q	Pr
(2)	~P	Pr
(3)	Q → R	Pr
(4)	~~SHOW~~: R	DD
(5)	│ Q	1,2,∨O
(6)	│ R	3,5,→O

#5:

(1)	P	Pr
(2)	P → ~Q	Pr
(3)	R → Q	Pr
(4)	~R → S	Pr
(5)	~~SHOW~~: S	DD
(6)	│ ~Q	1,2,→O
(7)	│ ~R	3,6,→O
(8)	│ S	4,7,→O

#6:

(1)	P ∨ ~Q	Pr
(2)	~P	Pr
(3)	R → Q	Pr
(4)	~R → S	Pr
(5)	~~SHOW~~: S	DD
(6)	│ ~Q	1,2,∨O
(7)	│ ~R	3,6,→O
(8)	│ S	4,7,→O

#7:

(1)	(P → Q) → P	Pr
(2)	P → Q	Pr
(3)	~~SHOW~~: Q	DD
(4)	│ P	1,2,→O
(5)	│ Q	2,4,→O

#8:

(1)	(P → Q) → R	Pr
(2)	R → P	Pr
(3)	P → Q	Pr
(4)	~~SHOW~~: Q	DD
(5)	│ R	1,3,→O
(6)	│ P	2,5,→O
(7)	│ Q	3,6,→O

#9:

(1)	(P → Q) → (Q → R)	Pr
(2)	P → Q	Pr
(3)	P	Pr
(4)	~~SHOW~~: R	DD
(5)	│ Q → R	1,2,→O
(6)	│ Q	2,3,→O
(7)	│ R	5,6,→O

#10:

(1)	~P → Q	Pr
(2)	~Q	Pr
(3)	R ∨ ~P	Pr
(4)	~~SHOW~~: R	DD
(5)	│ ~~P	1,2,→O
(6)	│ R	3,5,∨O

#11:

(1)	~P → (~Q ∨ R)	Pr
(2)	P → R	Pr
(3)	~R	Pr
(4)	~~SHOW~~: ~Q	DD
(5)	│ ~P	2,3,→O
(6)	│ ~Q ∨ R	1,5,→O
(7)	│ ~Q	3,6,∨O

#12:

(1)	P → ~Q	Pr
(2)	~S → P	Pr
(3)	~~Q	Pr
(4)	~~SHOW~~: ~~S	DD
(5)	│ ~P	1,3,→O
(6)	│ ~~S	2,5,→O

#13:

(1)	P ∨ Q	Pr
(2)	Q → R	Pr
(3)	~R	Pr
(4)	~~SHOW~~: P	DD
(5)	│ ~Q	2,3,→O
(6)	│ P	1,5,∨O

#14:

(1)	~P → (Q ∨ R)	Pr
(2)	P → Q	Pr
(3)	~Q	Pr
(4)	~~SHOW~~: R	DD
(5)	│ ~P	2,3,→O
(6)	│ Q ∨ R	1,5,→O
(7)	│ R	3,6,∨O

#15:

(1)	P → R	Pr
(2)	~P → (S ∨ R)	Pr
(3)	~R	Pr
(4)	~~SHOW~~: S	DD
(5)	│ ~P	1,3,→O
(6)	│ S ∨ R	2,5,→O
(7)	│ S	3,6,∨O

#16:

(1)	P ∨ ~Q	Pr
(2)	~R → ~~Q	Pr
(3)	R → ~S	Pr
(4)	~~S	Pr
(5)	~~SHOW~~: P	DD
(6)	│ ~R	3,4,→O
(7)	│ ~~Q	2,6,→O
(8)	│ P	1,7,∨O

#17:

(1)	(P → Q) ∨ (R → S)	Pr
(2)	(P → Q) → R	Pr
(3)	~R	Pr
(4)	~~SHOW~~: R → S	DD
(5)	│ ~(P → Q)	2,3,→O
(6)	│ R → S	1,5,∨O

#18:

(1)	(P → Q) → (R → S)	Pr
(2)	(R → T) ∨ (P → Q)	Pr
(3)	~(R → T)	Pr
(4)	SHOW: R → S	DD
(5)	│ P → Q	2,3,∨O
(6)	│ R → S	1,5,→O

#19:

(1)	~R → (P ∨ Q)	Pr
(2)	R → P	Pr
(3)	(R → P) → ~P	Pr
(4)	SHOW: Q	DD
(5)	│ ~P	2,3,→O
(6)	│ ~R	2,5,→O
(7)	│ P ∨ Q	1,6,→O
(8)	│ Q	5,7,∨O

#20:

(1)	(P → Q) ∨ R	Pr
(2)	[(P → Q) ∨ R] → ~R	Pr
(3)	(P → Q) → (Q → R)	Pr
(4)	SHOW: ~Q	DD
(5)	│ ~R	1,2,→O
(6)	│ P → Q	1,5,∨O
(7)	│ Q → R	3,6,→O
(8)	│ ~Q	5,7,→O

#21:

(1)	P & Q	Pr
(2)	P → (R & S)	Pr
(3)	SHOW: Q & S	DD
(4)	│ P	1,&O
(5)	│ Q	1,&O
(6)	│ R & S	2,4,→O
(7)	│ S	6,&O
(8)	│ Q & S	5,7,&I

#22:

(1)	P & Q	Pr
(2)	(P ∨ R) → S	Pr
(3)	SHOW: P & S	DD
(4)	│ P	1,&O
(5)	│ P ∨ R	4,∨I
(6)	│ S	2,5,→O
(7)	│ P & S	4,6,&I

#23:

(1)	P	Pr
(2)	(P ∨ Q) → (R & S)	Pr
(3)	(R ∨ T) → U	Pr
(4)	SHOW: U	DD
(5)	│ P ∨ Q	1,∨I
(6)	│ R & S	2,5,→O
(7)	│ R	6,&O
(8)	│ R ∨ T	7,∨I
(9)	│ U	3,8,→O

#24:

(1)	P → Q	Pr
(2)	P ∨ R	Pr
(3)	~Q	Pr
(4)	SHOW: R & ~P	DD
(5)	│ ~P	1,3,→O
(6)	│ R	2,5,∨O
(7)	│ R & ~P	5,6,&I

#25:

(1)	P → Q	Pr
(2)	~R → (Q → S)	Pr
(3)	R → T	Pr
(4)	~T & P	Pr
(5)	SHOW: Q & S	DD
(6)	│ ~T	4,&O
(7)	│ ~R	3,6,→O
(8)	│ Q → S	2,7,→O
(9)	│ P	4,&O
(10)	│ Q	1,9,→O
(11)	│ S	8,10:→O
(12)	│ Q & S	10,11,&I

#26:

(1)	P → Q	Pr
(2)	R ∨ ~Q	Pr
(3)	~R & S	Pr
(4)	(~P & S) → T	Pr
(5)	SHOW: T	DD
(6)	│ ~R	3,&O
(7)	│ S	3,&O
(8)	│ ~Q	2,6,∨O
(9)	│ ~P	1,8,→O
(10)	│ ~P & S	7,9,&I
(11)	│ T	4,10,→O

#27:

(1)	P ∨ ~Q	Pr
(2)	~R → Q	Pr
(3)	R → ~S	Pr
(4)	S	Pr
(5)	SHOW: P	DD
(6)	| ~~S	4,DN
(7)	| ~R	3,6,→O
(8)	| Q	2,7,→O
(9)	| ~~Q	8,DN
(10)	| P	1,9,∨O

#28:

(1)	P & Q	Pr
(2)	(P ∨ T) → R	Pr
(3)	S → ~R	Pr
(4)	SHOW: ~S	DD
(5)	| P	1,&O
(6)	| P ∨ T	5,∨I
(7)	| R	2,6,→O
(8)	| ~~R	7,DN
(9)	| ~S	3,8,→O

#29:

(1)	P & Q	Pr
(2)	P → R	Pr
(3)	(P & R) → S	Pr
(4)	SHOW: Q & S	DD
(5)	| P	1,&O
(6)	| R	2,5,→O
(7)	| P & R	5,6,&I
(8)	| S	3,7,→O
(9)	| Q	1,&O
(10)	| Q & S	8,9,&I

#30:

(1)	P → Q	Pr
(2)	Q ∨ R	Pr
(3)	(R & ~P) → S	Pr
(4)	~Q	Pr
(5)	SHOW: S	DD
(6)	| ~P	1,4,→O
(7)	| R	2,4,∨O
(8)	| R & ~P	6,7,&I
(9)	| S	3,8,→O

#31:

(1)	P & Q	Pr
(2)	SHOW: Q & P	DD
(3)	| P	1,&O
(4)	| Q	1,&O
(5)	| Q & P	3,4,&I

#32:

(1)	P & (Q & R)	Pr
(2)	SHOW: (P & Q) & R	DD
(3)	| P	1,&O
(4)	| Q & R	1,&O
(5)	| Q	4,&O
(6)	| P & Q	3,5,&I
(7)	| R	4,&O
(8)	| (P & Q) & R	6,7,&I

#33:

(1)	P	Pr
(2)	SHOW: P & P	DD
(3)	| P & P	1,1,&I

#34:

(1)	P	Pr
(2)	SHOW: P & (P ∨ Q)	DD
(3)	| P ∨ Q	1,∨I
(4)	| P & (P ∨ Q)	1,3,&I

#35:

(1)	P & ~P	Pr
(2)	SHOW: Q	DD
(3)	| P	1,&O
(4)	| ~P	1,&O
(5)	| P ∨ Q	3,∨I
(6)	| Q	4,5,∨O

#36:

(1)	P ↔ ~Q	Pr
(2)	Q	Pr
(3)	P ↔ ~S	Pr
(4)	SHOW: S	DD
(5)	| P → ~Q	1,↔O
(6)	| ~~Q	2,DN
(7)	| ~P	5,6,→O
(8)	| ~S → P	3,↔O
(9)	| ~~S	7,8,→O
(10)	| S	9,DN

#37:

(1)	P & ~Q	Pr
(2)	Q ∨ (P → S)	Pr
(3)	(R & T) ↔ S	Pr
(4)	SHOW: P & R	DD
(5)	| P	1,&O
(6)	| ~Q	1,&O
(7)	| P → S	2,6,∨O
(8)	| S	5,7,→O
(9)	| S → (R & T)	3,↔O
(10)	| R & T	8,9,→O
(11)	| R	10:&O
(12)	| P & R	5,11,&I

#38:

(1)	P → Q	Pr
(2)	(P → Q) → (Q → P)	Pr
(3)	(P ↔ Q) → P	Pr
(4)	~~SHOW~~: P & Q	DD
(5)	│ Q → P	1,2,→O
(6)	│ P ↔ Q	1,5,↔I
(7)	│ P	3,6,→O
(8)	│ Q	1,7,→O
(9)	│ P & Q	7,8,&I

#39:

(1)	~P & Q	Pr
(2)	(R ∨ Q) → (~S → P)	Pr
(3)	~S ↔ T	Pr
(4)	~~SHOW~~: ~T	DD
(5)	│ Q	1,&O
(6)	│ R ∨ Q	5,∨I
(7)	│ ~S → P	2,6,→O
(8)	│ ~P	1,&O
(9)	│ ~~S	7,8,→O
(10)	│ T → ~S	3,↔O
(11)	│ ~T	9,10,→O

#40:

(1)	P & ~Q	Pr
(2)	Q ∨ (R → S)	Pr
(3)	~V → ~P	Pr
(4)	V → (S → R)	Pr
(5)	(R ↔ S) → T	Pr
(6)	U ↔ (~Q & T)	Pr
(7)	~~SHOW~~: U	DD
(8)	│ P	1,&O
(9)	│ ~~P	8,DN
(10)	│ ~~V	3,9,→O
(11)	│ V	10,DN
(12)	│ S → R	4,11,→O
(13)	│ ~Q	1,&O
(14)	│ R → S	2,13,∨O
(15)	│ R ↔ S	12,14,↔I
(16)	│ T	5,15,→O
(17)	│ ~Q & T	13,16,&I
(18)	│ (~Q & T) → U	6,↔O
(19)	│ U	17,18,→O

#41:

(1)	(P ∨ Q) → R	Pr
(2)	~~SHOW~~: Q → R	CD
(3)	│ Q	As
(4)	│ ~~SHOW~~: R	DD
(5)	│ │ P ∨ Q	3,∨I
(6)	│ │ R	1,5,→O

#42:

(1)	Q → R	Pr
(2)	~~SHOW~~: (P & Q) → (P & R)	CD
(3)	│ P & Q	As
(4)	│ ~~SHOW~~: P & R	DD
(5)	│ │ P	3,&O
(6)	│ │ Q	3,&O
(7)	│ │ R	1,6,→O
(8)	│ │ P & R	5,7,&I

#43:

(1)	P → Q	Pr
(2)	~~SHOW~~: (Q → R) → (P → R)	CD
(3)	│ Q → R	As
(4)	│ ~~SHOW~~: P → R	CD
(5)	│ │ P	As
(6)	│ │ ~~SHOW~~: R	DD
(7)	│ │ │ Q	1,5,→O
(8)	│ │ │ R	3,7,→O

#44:

(1)	P → Q	Pr
(2)	~~SHOW~~: (R → P) → (R → Q)	CD
(3)	│ R → P	As
(4)	│ ~~SHOW~~: R → Q	CD
(5)	│ │ R	As
(6)	│ │ ~~SHOW~~: Q	DD
(7)	│ │ │ P	3,5,→O
(8)	│ │ │ Q	1,7,→O

#45:

(1)	(P & Q) → R	Pr
(2)	~~SHOW~~: P → (Q → R)	CD
(3)	│ P	As
(4)	│ ~~SHOW~~: Q → R	CD
(5)	│ │ Q	As
(6)	│ │ ~~SHOW~~: R	DD
(7)	│ │ │ P & Q	3,5,&I
(8)	│ │ │ R	1,7,→O

#46:

(1)	P → (Q → R)	Pr
(2)	~~SHOW~~: (P → Q) → (P → R)	CD
(3)	│ P → Q	As
(4)	│ ~~SHOW~~: P → R	CD
(5)	│ │ P	As
(6)	│ │ ~~SHOW~~: R	DD
(7)	│ │ │ Q	3,5,→O
(8)	│ │ │ Q → R	1,5,→O
(9)	│ │ │ R	7,8,→O

#47:

(1)	(P & Q) → R	Pr
(2)	~~SHOW~~: [(P→Q)→P]→[(P→Q)→R]	CD
(3)	(P → Q) → P	As
(4)	~~SHOW~~: (P → Q) → R	CD
(5)	P → Q	As
(6)	~~SHOW~~: R	DD
(7)	P	3,5,→O
(8)	Q	5,7,→O
(9)	P & Q	7,8,&I
(10)	R	1,9,→O

#48:

(1)	(P & Q) → (R → S)	Pr
(2)	~~SHOW~~: (P → Q) → [(P & R) → S]	CD
(3)	P → Q	As
(4)	~~SHOW~~: (P & R) → S	CD
(5)	P & R	As
(6)	~~SHOW~~: S	DD
(7)	P	5,&O
(8)	Q	3,7,→O
(9)	P & Q	7,8,&I
(10)	R → S	1,9,→O
(11)	R	5,&O
(12)	S	10:11→O

#49:

(1)	[(P & Q) & R] → S	Pr
(2)	~~SHOW~~: P → [Q → (R → S)]	CD
(3)	P	As
(4)	~~SHOW~~: Q → (R → S)	CD
(5)	Q	As
(6)	~~SHOW~~: R → S	CD
(7)	R	As
(8)	~~SHOW~~: S	DD
(9)	P & Q	3,5,&I
(10)	(P & Q) & R	7,9,&I
(11)	S	1,10,→O

#50:

(1)	(~P & Q) → R	Pr
(2)	~~SHOW~~: (~Q → P) → (~P → R)	CD
(3)	~Q → P	As
(4)	~~SHOW~~: ~P → R	CD
(5)	~P	As
(6)	~~SHOW~~: R	DD
(7)	~~Q	3,5,→O
(8)	Q	7,DN
(9)	~P & Q	5,8,&I
(10)	R	1,9,→O

#51:

(1)	P → Q	Pr
(2)	P → ~Q	Pr
(3)	~~SHOW~~: ~P	ID
(4)	P	As
(5)	~~SHOW~~: ✕	DD
(6)	Q	1,4,→O
(7)	~Q	2,4,→O
(8)	✕	6,7,✕I

#52:

(1)	P → Q	Pr
(2)	Q → ~P	Pr
(3)	~~SHOW~~: ~P	ID
(4)	P	As
(5)	~~SHOW~~: ✕	DD
(6)	Q	1,4,→O
(7)	~~P	4,DN
(8)	~Q	2,7,→O
(9)	✕	6,8,✕I

#53:

(1)	P → Q	Pr
(2)	~Q ∨ ~R	Pr
(3)	P → R	Pr
(4)	~~SHOW~~: ~P	ID
(5)	P	As
(6)	~~SHOW~~: ✕	DD
(7)	Q	1,5,→O
(8)	~~Q	7,DN
(9)	~R	2,8,∨O
(10)	~P	3,9:→O
(11)	✕	5,10,✕I

#54:

(1)	P → R	Pr
(2)	Q → ~R	Pr
(3)	~~SHOW~~: ~(P & Q)	ID
(4)	P & Q	As
(5)	~~SHOW~~: ✕	DD
(6)	P	4,&O
(7)	Q	4,&O
(8)	R	1,6,→O
(9)	~R	2,7,→O
(10)	✕	8,9,✕I

#55:

(1)	P & Q	Pr
(2)	~~SHOW~~: ~(P → ~Q)	ID
(3)	P → ~Q	As
(4)	~~SHOW~~: ✕	DD
(5)	P	1,&O
(6)	Q	1,&O
(7)	~Q	3,5,→O
(8)	✕	6,7,✕I

#56:

(1)	P & ~Q	Pr
(2)	~~SHOW~~: ~(P → Q)	ID
(3)	│ P → Q	As
(4)	│ ~~SHOW~~: ✕	DD
(5)	│ │ P	1,&O
(6)	│ │ ~Q	1,&O
(7)	│ │ Q	3,5,→O
(8)	│ │ ✕	6,7,✕I

#57:

(1)	~P	Pr
(2)	~~SHOW~~: ~(P & Q)	ID
(3)	│ P & Q	As
(4)	│ ~~SHOW~~: ✕	DD
(5)	│ │ P	3,&O
(6)	│ │ ✕	1,5,✕I

#58:

(1)	~P & ~Q	Pr
(2)	~~SHOW~~: ~(P ∨ Q)	ID
(3)	│ P ∨ Q	As
(4)	│ ~~SHOW~~: ✕	DD
(5)	│ │ ~P	1,&O
(6)	│ │ ~Q	1,&O
(7)	│ │ Q	3,5,∨O
(8)	│ │ ✕	6,7,✕I

#59:

(1)	P ↔ Q	Pr
(2)	~Q	Pr
(3)	~~SHOW~~: ~(P ∨ Q)	ID
(4)	│ P ∨ Q	As
(5)	│ ~~SHOW~~: ✕	DD
(6)	│ │ P	2,4,∨O
(7)	│ │ P → Q	1,↔O
(8)	│ │ Q	6,7,→O
(9)	│ │ ✕	2,8,✕I

#60:

(1)	P & Q	Pr
(2)	~~SHOW~~: ~(~P ∨ ~Q)	ID
(3)	│ ~P ∨ ~Q	As
(4)	│ ~~SHOW~~: ✕	DD
(5)	│ │ P	1,&O
(6)	│ │ Q	1,&O
(7)	│ │ ~~P	5,DN
(8)	│ │ ~Q	3,7,∨O
(9)	│ │ ✕	6,8,✕I

#61:

(1)	~P ∨ ~Q	Pr
(2)	~~SHOW~~: ~(P & Q)	ID
(3)	│ P & Q	As
(4)	│ ~~SHOW~~: ✕	DD
(5)	│ │ P	3,&O
(6)	│ │ Q	3:&O
(7)	│ │ ~~P	5,DN
(8)	│ │ ~Q	1,7,∨O
(9)	│ │ ✕	6,8,✕I

#62:

(1)	P ∨ Q	Pr
(2)	~~SHOW~~: ~(~P & ~Q)	ID
(3)	│ ~P & ~Q	As
(4)	│ ~~SHOW~~: ✕	DD
(5)	│ │ ~P	3,&O
(6)	│ │ ~Q	3,&O
(7)	│ │ Q	1,5,∨O
(8)	│ │ ✕	6,7,✕I

#63:

(1)	P → Q	Pr
(2)	~~SHOW~~: ~(P & ~Q)	ID
(3)	│ P & ~Q	As
(4)	│ ~~SHOW~~: ✕	DD
(5)	│ │ P	3,&O
(6)	│ │ ~Q	3,&O
(7)	│ │ Q	1,5,→O
(8)	│ │ ✕	6,7,✕I

#64:

(1)	P → (Q → ~P)	Pr
(2)	~~SHOW~~: P → ~Q	CD
(3)	│ P	As
(4)	│ ~~SHOW~~: ~Q	ID
(5)	│ │ Q	As
(6)	│ │ ~~SHOW~~: ✕	DD
(7)	│ │ Q → ~P	1,3,→O
(8)	│ │ ~P	5,7,→O
(9)	│ │ ✕	3,8,✕I

#65:

(1)	(P & Q) → R	Pr
(2)	~~SHOW~~: (P & ~R) → ~Q	CD
(3)	│ P & ~R	As
(4)	│ ~~SHOW~~: ~Q	ID
(5)	│ │ Q	As
(6)	│ │ ~~SHOW~~: ✕	DD
(7)	│ │ │ P	3,&O
(8)	│ │ │ P & Q	5,7,&I
(9)	│ │ │ R	1,8,→O
(10)	│ │ │ ~R	3,&O
(11)	│ │ │ ✕	9,10,✕I

#66:

(1)	(P & Q) → ~R	Pr
(2)	~~SHOW~~: P → ~(Q & R)	CD
(3)	P	As
(4)	~~SHOW~~: ~(Q & R)	ID
(5)	Q & R	As
(6)	~~SHOW~~: ✕	DD
(7)	Q	5,&O
(8)	P & Q	3,7,&I
(9)	~R	1,8,→O
(10)	R	5,&O
(11)	✕	9,10,✕I

#67:

(1)	P → (Q → R)	Pr
(2)	~~SHOW~~: (Q & ~R) → ~P	CD
(3)	Q & ~R	As
(4)	~~SHOW~~: ~P	ID
(5)	P	As
(6)	~~SHOW~~: ✕	DD
(7)	Q → R	1,5,→O
(8)	Q	3,&O
(9)	R	7,8,→O
(10)	~R	3,&O
(11)	✕	9,10,✕I

#68:

(1)	P → ~(Q & R)	Pr
(2)	~~SHOW~~: (P & Q) → ~R	CD
(3)	P & Q	As
(4)	~~SHOW~~: ~R	ID
(5)	R	As
(6)	~~SHOW~~: ✕	DD
(7)	P	3,&O
(8)	Q	3,&O
(9)	Q & R	5,8,&I
(10)	~(Q & R)	1,7,→O
(11)	✕	9:10,✕I

#69:

(1)	P → ~(Q & R)	Pr
(2)	~~SHOW~~: (P → Q) → (P → ~R)	CD
(3)	P → Q	As
(4)	~~SHOW~~: P → ~R	CD
(5)	P	As
(6)	~~SHOW~~: ~R	ID
(7)	R	As
(8)	~~SHOW~~: ✕	DD
(9)	Q	3,5,→O
(10)	Q & R	7,9,&I
(11)	~(Q & R)	1,5,→O
(12)	✕	10,11,✕I

#70:

(1)	P → (Q → R)	Pr
(2)	~~SHOW~~: (P → ~R) → (P → ~Q)	CD
(3)	P → ~R	As
(4)	~~SHOW~~: P → ~Q	CD
(5)	P	As
(6)	~~SHOW~~: ~Q	ID
(7)	Q	As
(8)	~~SHOW~~: ✕	DD
(9)	Q → R	1,5,→O
(10)	~R	3,5,→O
(11)	~Q	9,10,→O
(12)	✕	7,11,✕I

#71:

(1)	P → Q	Pr
(2)	~P → Q	Pr
(3)	~~SHOW~~: Q	ID
(4)	~Q	As
(5)	~~SHOW~~: ✕	DD
(6)	~P	1,4,→O
(7)	~~P	2,4,→O
(8)	✕	6,7,✕I

#72:

(1)	P ∨ Q	Pr
(2)	P → R	Pr
(3)	Q ∨ ~R	Pr
(4)	~~SHOW~~: Q	ID
(5)	~Q	As
(6)	~~SHOW~~: ✕	DD
(7)	P	1,5,∨O
(8)	R	2,7,→O
(9)	~R	3,5,∨O
(10)	✕	8,9,✕I

#73:

(1)	~P → R	Pr
(2)	Q → R	Pr
(3)	P → Q	Pr
(4)	~~SHOW~~: R	ID
(5)	~R	As
(6)	~~SHOW~~: ✕	DD
(7)	~Q	2,5,→O
(8)	~~P	1,5,→O
(9)	P	8,DN
(10)	Q	3,9,→O
(11)	✕	7,10,✕I

#74:

(1)	(P ∨ ~Q) → (R & ~S)	Pr
(2)	Q ∨ S	Pr
(3)	~~SHOW~~: Q	ID
(4)	~Q	As
(5)	~~SHOW~~: ×	DD
(6)	P ∨ ~Q	4,∨I
(7)	R & ~S	1,6,→O
(8)	~S	7,&O
(9)	S	2,4,∨O
(10)	×	8,9,×I

#75:

(1)	(P ∨ Q) → (R → S)	Pr
(2)	(~S ∨ T) → (P & R)	Pr
(3)	~~SHOW~~: S	ID
(4)	~S	As
(5)	~~SHOW~~: ×	DD
(6)	~S ∨ T	4,∨I
(7)	P & R	2,6,→O
(8)	P	7,&O
(9)	P ∨ Q	8,∨I
(10)	R → S	1,9,→O
(11)	R	7,&O
(12)	S	10,11,→O
(13)	×	4,12,×I

#76:

(1)	~(P & ~Q)	Pr
(2)	~~SHOW~~: P → Q	CD
(3)	P	As
(4)	~~SHOW~~: Q	ID
(5)	~Q	As
(6)	~~SHOW~~: ×	DD
(7)	P & ~Q	3,5,&I
(8)	×	1,7,×I

#77:

(1)	P → (~Q → R)	Pr
(2)	~~SHOW~~: (P & ~R) → Q	CD
(3)	P & ~R	As
(4)	~~SHOW~~: Q	ID
(5)	~Q	As
(6)	~~SHOW~~: ×	DD
(7)	P	3,&O
(8)	~R	3,&O
(9)	~Q → R	1,7,→O
(10)	~~Q	8,9,→O
(11)	×	5,10,×I

#78:

(1)	P & (Q ∨ R)	Pr
(2)	~~SHOW~~: ~(P & Q) → R	CD
(3)	~(P & Q)	As
(4)	~~SHOW~~: R	ID
(5)	~R	As
(6)	~~SHOW~~: ×	DD
(7)	Q ∨ R	1,&O
(8)	Q	5,7,∨O
(9)	P	1,&O
(10)	P & Q	8,9,&I
(11)	×	3,10,×I

#79:

(1)	P ∨ Q	Pr
(2)	~~SHOW~~: Q ∨ P	ID
(3)	~(Q ∨ P)	As
(4)	~~SHOW~~: ×	DD
(5)	~Q	3,~∨O
(6)	~P	3,~∨O
(7)	Q	1,6,∨O
(8)	×	5,7,×I

#80:

(1)	~P → Q	Pr
(2)	~~SHOW~~: P ∨ Q	ID
(3)	~(P ∨ Q)	As
(4)	~~SHOW~~: ×	DD
(5)	~P	3,~∨O
(6)	~Q	3,~∨O
(7)	Q	1,5,→O
(8)	×	6,7,×I

#81:

(1)	~(P & Q)	Pr
(2)	~~SHOW~~: ~P ∨ ~Q	ID
(3)	~(~P ∨ ~Q)	As
(4)	~~SHOW~~: ×	DD
(5)	~~P	3,~∨O
(6)	~~Q	3,~∨O
(7)	P	5,DN
(8)	Q	6,DN
(9)	P & Q	7,8,&I
(10)	×	1,9,×I

#82:

(1)	P → Q	Pr
(2)	~~SHOW~~: ~P ∨ Q	ID
(3)	~(~P ∨ Q)	As
(4)	~~SHOW~~: ×	DD
(5)	~~P	3,~∨O
(6)	~Q	3,~∨O
(7)	P	5,DN
(8)	Q	1,7,→O
(9)	×	6,8,×I

#83:

(1)	$P \vee Q$	Pr
(2)	$P \to R$	Pr
(3)	$Q \to S$	Pr
(4)	~~SHOW~~: $R \vee S$	ID
(5)	│ $\sim(R \vee S)$	As
(6)	│ ~~SHOW~~: ×	DD
(7)	│ │ $\sim R$	5,$\sim\vee$O
(8)	│ │ $\sim S$	5,$\sim\vee$O
(9)	│ │ $\sim P$	2,7,$\to$O
(10)	│ │ $\sim Q$	3,8,$\to$O
(11)	│ │ Q	1,9,$\vee$O
(12)	│ │ ×	10,11,×I

#84:

(1)	$\sim P \to Q$	Pr
(2)	$P \to R$	Pr
(3)	~~SHOW~~: $Q \vee R$	ID
(4)	│ $\sim(Q \vee R)$	As
(5)	│ ~~SHOW~~: ×	DD
(6)	│ │ $\sim Q$	4,$\sim\vee$O
(7)	│ │ $\sim R$	4,$\sim\vee$O
(8)	│ │ $\sim\sim P$	1,6,$\to$O
(9)	│ │ P	8,DN
(10)	│ │ R	2,9,$\to$O
(11)	│ │ ×	7,10,×I

#85:

(1)	$\sim P \to Q$	Pr
(2)	$\sim R \to S$	Pr
(3)	$\sim Q \vee \sim S$	Pr
(4)	~~SHOW~~: $P \vee R$	ID
(5)	│ $\sim(P \vee R)$	As
(6)	│ ~~SHOW~~: ×	DD
(7)	│ │ $\sim P$	5,$\sim\vee$O
(8)	│ │ $\sim R$	5,$\sim\vee$O
(9)	│ │ Q	1,7,$\to$O
(10)	│ │ S	2,8,$\to$O
(11)	│ │ $\sim\sim Q$	9,DN
(12)	│ │ $\sim S$	3,11,$\vee$O
(13)	│ │ ×	10,12,×I

#86:

(1)	$(P \,\&\, \sim Q) \to R$	Pr
(2)	~~SHOW~~: $P \to (Q \vee R)$	CD
(3)	│ P	As
(4)	│ ~~SHOW~~: $Q \vee R$	ID
(5)	│ │ $\sim(Q \vee R)$	As
(6)	│ │ ~~SHOW~~: ×	DD
(7)	│ │ │ $\sim Q$	5,$\sim\vee$O
(8)	│ │ │ $P \,\&\, \sim Q$	3,7,&I
(9)	│ │ │ R	1,8,$\to$O
(10)	│ │ │ $\sim R$	5,$\sim\vee$O
(11)	│ │ │ ×	9,10,×I

#87

(1)	$\sim P \to (\sim Q \vee R)$	Pr
(2)	~~SHOW~~: $Q \to (P \vee R)$	CD
(3)	│ Q	As
(4)	│ ~~SHOW~~: $P \vee R$	ID
(5)	│ │ $\sim(P \vee R)$	As
(6)	│ │ ~~SHOW~~: ×	DD
(7)	│ │ │ $\sim P$	5,$\sim\vee$O
(8)	│ │ │ $\sim R$	5,$\sim\vee$O
(9)	│ │ │ $\sim Q \vee R$	1,7,$\to$O
(10)	│ │ │ $\sim Q$	8,9,$\vee$O
(11)	│ │ │ ×	3,10,×I

#88:

(1)	$P \,\&\, (Q \vee R)$	Pr
(2)	~~SHOW~~: $(P \,\&\, Q) \vee R$	ID
(3)	│ $\sim[(P \,\&\, Q) \vee R]$	As
(4)	│ ~~SHOW~~: ×	DD
(5)	│ │ $\sim(P \,\&\, Q)$	3,$\sim\vee$O
(6)	│ │ $\sim R$	3,$\sim\vee$O
(7)	│ │ P	1,&O
(8)	│ │ $Q \vee R$	1,&O
(9)	│ │ Q	6,8,$\vee$O
(10)	│ │ $P \,\&\, Q$	7,9,&I
(11)	│ │ ×	5,10,×I

#89:

(1)	$(P \vee Q) \,\&\, (P \vee R)$	Pr
(2)	~~SHOW~~: $P \vee (Q \,\&\, R)$	ID
(3)	│ $\sim[P \vee (Q \,\&\, R)]$	As
(4)	│ ~~SHOW~~: ×	DD
(5)	│ │ $\sim P$	3,$\sim\vee$O
(6)	│ │ $\sim(Q \,\&\, R)$	3,$\sim\vee$O
(7)	│ │ $P \vee Q$	1,&O
(8)	│ │ Q	5,7,$\vee$O
(9)	│ │ $P \vee R$	1,&O
(10)	│ │ R	5,9,$\vee$O
(11)	│ │ $Q \,\&\, R$	8,10,&I
(12)	│ │ ×	6,11×I

#90:

(1)	$(P \vee Q) \to (P \,\&\, Q)$	Pr
(2)	~~SHOW~~: $(P\&Q) \vee (\sim P \,\&\, \sim Q)$	ID
(3)	│ $\sim[(P \,\&\, Q) \vee (\sim P \,\&\, \sim Q)]$	As
(4)	│ ~~SHOW~~: ×	DD
(5)	│ │ $\sim(P \,\&\, Q)$	3,$\sim\vee$O
(6)	│ │ $\sim(\sim P \,\&\, \sim Q)$	3,$\sim\vee$O
(7)	│ │ $\sim(P \vee Q)$	1,5,$\to$O
(8)	│ │ $\sim P$	7,$\sim\vee$O
(9)	│ │ $\sim Q$	7,$\sim\vee$O
(10)	│ │ $\sim P \,\&\, \sim Q$	8,9,&I
(11)	│ │ ×	6,10,×I

#91:

(1)	P → (Q & R)	Pr
(2)	~~SHOW~~: (P → Q) & (P → R)	DD
(3)	│ ~~SHOW~~: P → Q	CD
(4)	│ │ P	As
(5)	│ │ ~~SHOW~~: Q	DD
(6)	│ │ │ Q & R	1,4,→O
(7)	│ │ │ Q	6,&O
(8)	│ ~~SHOW~~: P → R	CD
(9)	│ │ P	As
(10)	│ │ ~~SHOW~~: R	DD
(11)	│ │ │ Q & R	1,9,→O
(12)	│ │ │ R	11&O
(13)	│ (P → Q) & (P → R)	3,8,&I

#92:

(1)	(P ∨ Q) → R	Pr
(2)	~~SHOW~~: (P → R) & (Q → R)	DD
(3)	│ ~~SHOW~~: P → R	CD
(4)	│ │ P	As
(5)	│ │ ~~SHOW~~: R	DD
(6)	│ │ │ P ∨ Q	4,∨I
(7)	│ │ │ R	1,6,→O
(8)	│ ~~SHOW~~: Q → R	CD
(9)	│ │ Q	As
(10)	│ │ ~~SHOW~~: R	DD
(11)	│ │ │ P ∨ Q	9,∨I
(12)	│ │ │ R	1,11,→O
(13)	│ (P → R) & (Q → R)	3,8,&I

#93:

(1)	(P ∨ Q) → (P & Q)	Pr
(2)	~~SHOW~~: P ↔ Q	DD
(3)	│ ~~SHOW~~: P → Q	CD
(4)	│ │ P	As
(5)	│ │ ~~SHOW~~: Q	DD
(6)	│ │ │ P ∨ Q	4,∨I
(7)	│ │ │ P & Q	1,6,→O
(8)	│ │ │ Q	7,&O
(9)	│ ~~SHOW~~: Q → P	CD
(10)	│ │ Q	As
(11)	│ │ ~~SHOW~~: P	DD
(12)	│ │ │ P ∨ Q	10,∨I
(13)	│ │ │ P & Q	1,12,→O
(14)	│ │ │ P	13,&O
(15)	│ P ↔ Q	3,9,↔I

#94:

(1)	P ↔ Q	Pr
(2)	~~SHOW~~: Q ↔ P	DD
(3)	│ P → Q	1,↔O
(4)	│ Q → P	1,↔O
(5)	│ Q ↔ P	3,4,↔I

#95:

(1)	P ↔ Q	Pr
(2)	~~SHOW~~: ∼P ↔ ∼Q	DD
(3)	│ ~~SHOW~~: ∼P → ∼Q	CD
(4)	│ │ ∼P	As
(5)	│ │ ~~SHOW~~: ∼Q	DD
(6)	│ │ │ Q → P	1,↔O
(7)	│ │ │ ∼Q	4,6,→O
(8)	│ ~~SHOW~~: ∼Q → ∼P	CD
(9)	│ │ ∼Q	As
(10)	│ │ ~~SHOW~~: ∼P	DD
(11)	│ │ │ P → Q	1,↔O
(12)	│ │ │ ∼P	9,11,→O
(13)	│ ∼P ↔ ∼Q	3,8,↔I

#96:

(1)	P ↔ Q	Pr
(2)	Q → ∼P	Pr
(3)	~~SHOW~~: ∼P & ∼Q	DD
(4)	│ ~~SHOW~~: ∼P	ID
(5)	│ │ P	As
(6)	│ │ ~~SHOW~~: ✕	DD
(7)	│ │ │ P → Q	1,↔O
(8)	│ │ │ Q	5,7,→O
(9)	│ │ │ ∼P	2,8,→O
(10)	│ │ │ ✕	5,9,✕I
(11)	│ ~~SHOW~~: ∼Q	ID
(12)	│ │ Q	As
(13)	│ │ ~~SHOW~~: ✕	DD
(14)	│ │ │ Q → P	1,↔O
(15)	│ │ │ P	12,14,→O
(16)	│ │ │ ∼P	2,12,→O
(17)	│ │ │ ✕	15,16,✕I
(18)	│ ∼P & ∼Q	4,11,&I

#97:

(1)	(P → Q) ∨ (∼Q → R)	Pr
(2)	~~SHOW~~: P → (Q ∨ R)	CD
(3)	│ P	As
(4)	│ ~~SHOW~~: Q ∨ R	ID
(5)	│ │ ∼(Q ∨ R)	As
(6)	│ │ ~~SHOW~~: ✕	DD
(7)	│ │ │ ∼Q	5,∼∨O
(8)	│ │ │ ∼R	5,∼∨O
(9)	│ │ │ ~~SHOW~~: ∼(P → Q)	ID
(10)	│ │ │ │ P → Q	As
(11)	│ │ │ │ ~~SHOW~~: ✕	DD
(12)	│ │ │ │ │ Q	3,10,→O
(13)	│ │ │ │ │ ✕	7,12,✕I
(14)	│ │ │ ∼Q → R	1,9,∨O
(15)	│ │ │ R	7,14,→O
(16)	│ │ │ ✕	8,15,✕I

#98:

(1)	P ∨ Q	Pr
(2)	P → ~Q	Pr
(3)	SHOW: (P → Q) → (Q & ~P)	CD
(4)	P → Q	As
(5)	SHOW: Q & ~P	DD
(6)	SHOW: Q	ID
(7)	~Q	As
(8)	SHOW: ✕	DD
(9)	~P	4,7,→O
(10)	P	1,7,∨O
(11)	✕	9,10,✕I
(12)	SHOW: ~P	ID
(13)	P	As
(14)	SHOW: ✕	DD
(15)	Q	4,13,→O
(16)	~Q	2,13,→O
(17)	✕	15,16,✕I
(18)	Q & ~P	6,12,&I

#99:

(1)	P ∨ Q	Pr
(2)	~(P & Q)	Pr
(3)	SHOW: (P → Q) → ~(Q → P)	CD
(4)	P → Q	As
(5)	SHOW: ~(Q → P)	ID
(6)	Q → P	As
(7)	SHOW: ✕	DD
(8)	P → ~Q	2,~→O
(9)	SHOW: ~P	ID
(9)	P	As
(10)	SHOW: ✕	DD
(11)	Q	4,9,→O
(12)	~Q	8,9,→O
(13)	✕	11,12,✕I
(14)	Q	1,8,∨O
(15)	P	6,14,→O
(16)	✕	9,15,✕I

#100:

(1)	P ∨ Q	Pr
(2)	P → ~Q	Pr
(3)	SHOW: (P & ~Q) ∨ (Q & ~P)	ID
(4)	~[(P & ~Q) ∨ (Q & ~P)]	As
(5)	SHOW: ✕	DD
(6)	~(P & ~Q)	4,~∨O
(7)	~(Q & ~P)	4,~∨O
(8)	SHOW: ~P	ID
(9)	P	As
(10)	SHOW: ✕	DD
(11)	~Q	2,9,→O
(12)	P & ~Q	9,11,&I
(13)	✕	6,12,✕I
(14)	Q	1,8,∨O
(15)	Q & ~P	8,14,&I
(16)	✕	7,15,✕I

#101:

(1)	(P ∨ Q) → (P & Q)	Pr
(2)	SHOW: (~P ∨ ~Q) → (~P & ~Q)	CD
(3)	~P ∨ ~Q	As
(4)	SHOW: ~P & ~Q	DD
(5)	SHOW: ~P	ID
(6)	P	As
(7)	SHOW: ✕	DD
(8)	P ∨ Q	6,∨I
(9)	P & Q	1,8,→O
(10)	~~P	6,DN
(11)	~Q	3,10,∨O
(12)	Q	9,&O
(13)	✕	11,12,✕I
(14)	SHOW: ~Q	ID
(15)	Q	As
(16)	SHOW: ✕	DD
(17)	P ∨ Q	15,∨I
(18)	P & Q	1,17,→O
(19)	~~Q	15,DN
(20)	~P	3,19,∨O
(21)	P	18,&O
(22)	✕	20,21,✕I
(23)	~P & ~Q	5,14,&I

#102:

(1)	P & (Q ∨ R)	Pr
(2)	SHOW: (P & Q) ∨ (P & R)	ID
(3)	~[(P & Q) ∨ (P & R)]	As
(4)	SHOW: ✕	DD
(5)	~(P & Q)	3,~∨O
(6)	~(P & R)	3,~∨O
(7)	P	1,&O
(8)	Q ∨ R	1,&O
(9)	P → ~Q	5,~&O
(10)	P → ~R	6,~&O
(11)	~Q	7,9,→O
(12)	R	8,11,∨O
(13)	~R	7,10,→O
(14)	✕	12,13,✕I

#103:

(1)	(P & Q) ∨ (P & R)	Pr
(2)	~~SHOW~~: P & (Q ∨ R)	DD
(3)	~~SHOW~~: P	ID
(4)	~P	As
(5)	~~SHOW~~: ✕	DD
(6)	~~SHOW~~: ~(P & Q)	ID
(7)	P & Q	As
(8)	~~SHOW~~: ✕	DD
(9)	P	7,&O
(10)	✕	4,9,✕I
(13)	P & R	1,6,∨O
(14)	P	13,&O
(15)	✕	4,14,✕I
(16)	~~SHOW~~: Q ∨ R	ID
(17)	~(Q ∨ R)	As
(18)	~~SHOW~~: ✕	DD
(19)	~Q	17,~∨O
(20)	~R	17,~∨O
(21)	~~SHOW~~: ~(P & Q)	ID
(22)	P & Q	As
(23)	~~SHOW~~: ✕	DD
(24)	Q	22,&O
(25)	✕	19.24,✕I

#104:

(1)	P ∨ (Q & R)	Pr
(2)	~~SHOW~~: (P ∨ Q) & (P ∨ R)	DD
(3)	~~SHOW~~: P ∨ Q	ID
(4)	~(P ∨ Q)	As
(5)	~~SHOW~~: ✕	DD
(6)	~P	4,~∨O
(7)	~Q	4,~∨O
(8)	Q & R	1,6,∨O
(9)	Q	8,&O
(10)	✕	7,9,✕I
(11)	~~SHOW~~: P ∨ R	ID
(12)	~(P ∨ R)	As
(13)	~~SHOW~~: ✕	DD
(14)	~P	12,~∨O
(15)	~R	12,~∨O
(16)	Q & R	1,14,∨O
(17)	Q	16,&O
(18)	✕	15,17,✕I
(19)	(P ∨ Q) & (P ∨ R)	3,11,&I

#105:

(1)	(P&Q) ∨ [(P&R) ∨ (Q&R)]	Pr
(2)	~~SHOW~~: P ∨ (Q & R)	ID
(3)	~[P ∨ (Q & R)]	As
(4)	~~SHOW~~: ✕	DD
(5)	~P	3,~∨O
(6)	~(Q & R)	3,~∨O
(7)	~~SHOW~~: ~(P & Q)	ID
(8)	P & Q	As
(9)	~~SHOW~~: ✕	DD
(10)	P	8,&O
(11)	✕	5,10,✕I
(12)	(P & R) ∨ (Q & R)	1,7,∨O
(13)	P & R	6,12,∨O
(14)	P	13,&O
(15)	✕	5,14,✕I

#106:

(1)	P ∨ Q	Pr
(2)	P ∨ R	Pr
(3)	Q ∨ R	Pr
(4)	~~SHOW~~: (P&Q)∨[(P&R)∨(Q&R)]	ID
(5)	~{(P&Q)∨[(P&R)∨(Q&R)]}	As
(6)	~~SHOW~~: ✕	DD
(7)	~(P & Q)	5,~∨O
(8)	~[(P & R) ∨ (Q & R)]	5,~∨O
(8)	~(P & R)	8,~∨O
(9)	~(Q & R)	8,~∨O
(10)	P → ~Q	7,~&O
(11)	P → ~R	8,~&O
(12)	Q → ~R	9,~&O
(13)	~~SHOW~~: ~P	ID
(14)	P	As
(15)	~~SHOW~~: ✕	DD
(16)	~Q	10,14,→O
(17)	~R	11,14,→O
(18)	R	3,16,∨O
(19)	✕	17,18,✕I
(20)	Q	1,13,∨O
(21)	R	2,13,∨O
(22)	~R	12,20,→O
(23)	✕	21,22,✕I

#107:

(1)	(P → Q) ∨ (P → R)	Pr
(2)	~~SHOW~~: P → (Q ∨ R)	CD
(3)	P	As
(4)	~~SHOW~~: Q ∨ R	ID
(5)	~(Q ∨ R)	As
(6)	~~SHOW~~: ✕	DD
(7)	~Q	5,~∨O
(8)	~R	5,~∨O
(9)	~~SHOW~~: ~(P → Q)	ID
(10)	P → Q	As
(11)	~~SHOW~~: ✕	DD
(12)	Q	3,10,→O
(13)	✕	7,12,✕I
(14)	P → R	1,9 ∨O
(15)	R	3,14,→O
(16)	✕	8,15,✕I

#108:

(1)	(P → R) ∨ (Q → R)	Pr
(2)	~~SHOW~~: (P & Q) → R	CD
(3)	P & Q	As
(4)	~~SHOW~~: R	ID
(5)	~R	As
(6)	~~SHOW~~: ✕	DD
(7)	~~SHOW~~: ~(P → R)	ID
(8)	P → R	As
(9)	~~SHOW~~: ✕	DD
(10)	P	3,&O
(11)	R	8,10,→O
(12)	✕	5,11,✕I
(13)	Q → R	1,7,∨O
(14)	Q	3,&O
(15)	R	13,14,→O
(16)	✕	5,15,✕I

#109:

(1)	P ↔ (Q & ~P)	Pr
(2)	~~SHOW~~: ~(P ∨ Q)	ID
(3)	P ∨ Q	As
(4)	~~SHOW~~: ✕	DD
(5)	P → (Q & ~P)	1,↔O
(6)	~~SHOW~~: P	ID
(7)	~P	As
(8)	~~SHOW~~: ✕	DD
(9)	Q	3,7,∨O
(10)	Q & ~P	7,9,&I
(11)	(Q & ~P) → P	1,→O
(12)	P	10,12,→O
(13)	✕	7,12,✕I
(14)	Q & ~P	5,6,→O
(15)	~P	14,&O
(16)	✕	6,15,✕I

#110:

(1)	(P & Q) ∨ (~P & ~Q)	Pr
(2)	~~SHOW~~: P ↔ Q	DD
(3)	~~SHOW~~: P → Q	CD
(4)	P	As
(5)	~~SHOW~~: Q	ID
(6)	~Q	As
(7)	~~SHOW~~: ✕	DD
(8)	~~SHOW~~: ~(P & Q)	ID
(9)	P & Q	As
(10)	~~SHOW~~: ✕	DD
(11)	Q	9,&O
(12)	✕	6,11,✕I
(13)	~P & ~Q	1,8,∨O
(14)	~P	13,&O
(15)	✕	4,14,✕I
(16)	~~SHOW~~: Q → P	CD
(17)	Q	As
(18)	~~SHOW~~: P	ID
(19)	~P	As
(20)	~~SHOW~~: ✕	DD
(21)	~~SHOW~~: ~(P & Q)	ID
(22)	P & Q	As
(23)	~~SHOW~~: ✕	DD
(24)	P	22,&O
(25)	✕	19,24,✕I
(26)	~P & ~Q	1,21,&O
(27)	~Q	26,&O
(28)	✕	17,27,✕I
(29)	P ↔ Q	3,16,↔I

#111:

(1)	P → (Q ∨ R)	Pr
(2)	~~SHOW~~: (P → Q) ∨ (P → R)	ID
(3)	~[(P → Q) ∨ (P → R)]	As
(4)	~~SHOW~~: ✕	DD
(5)	~(P → Q)	3,~∨O
(6)	~(P → R)	3,~∨O
(7)	P & ~Q	5,~→O
(8)	P & ~R	6,~→O
(9)	P	7,&O
(10)	~Q	7,&O
(11)	~R	8,&O
(12)	Q ∨ R	1,9,→O
(13)	R	10,12,∨O
(14)	✕	11,13,✕I

#112:

(1)	(P ↔ Q) → R	Pr
(2)	~~SHOW~~: P → (Q → R)	CD
(3)	P	As
(4)	~~SHOW~~: Q → R	CD
(5)	Q	As
(6)	~~SHOW~~: R	DD
(7)	~R	As
(8)	~~SHOW~~: ✕	DD
(9)	~(P ↔ Q)	1,7,→O
(10)	~P ↔ Q	9,~↔O
(11)	Q → ~P	9,↔O
(12)	~P	5,11,→O
(13)	✕	3,12,✕I

#113:

(1)	P → (~Q → R)	Pr
(2)	~~SHOW~~: ~(P → R) → Q	CD
(3)	~(P → R)	As
(4)	~~SHOW~~: Q	ID
(5)	~Q	As
(6)	~~SHOW~~: ✕	DD
(7)	P & ~R	3,~→O
(8)	P	7,&O
(9)	~R	7,&O
(10)	~Q → R	1,8,→O
(11)	R	5,10,→O
(12)	✕	9,11,✕I

#114:

(1)	(P & Q) → R	Pr
(2)	~~SHOW~~: (P → R) ∨ (Q → R)	ID
(3)	~[(P → R) ∨ (Q → R)]	As
(4)	~~SHOW~~: ✕	DD
(5)	~(P → R)	3,~∨O
(6)	~(Q → R)	3,~∨O
(7)	P & ~R	5,~→O
(8)	Q & ~R	6,~→O
(9)	P	7,&O
(10)	~R	7,&O
(11)	Q	7,&O
(12)	P & Q	9,11,&I
(13)	R	1,12,→O
(14)	✕	10,13,✕I

#115:

(1)	P ↔ ~Q	Pr
(2)	~~SHOW~~: (P & ~Q) ∨ (Q & ~P)	ID
(3)	~[(P & ~Q) ∨ (Q & ~P)]	As
(4)	~~SHOW~~: ✕	DD
(5)	~(P & ~Q)	3,~∨O
(6)	~(Q & ~P)	3,~∨O
(7)	P → ~~Q	5,~&O
(8)	Q → ~~P	6,~&O
(9)	P → ~Q	1,↔O
(10)	~Q → P	1,↔O
(11)	~~SHOW~~: P	ID
(12)	~P	As
(13)	~~SHOW~~: ✕	DD
(14)	~~Q	7,10,→O
(15)	~~~P	12,DN
(16)	~Q	7,15,→O
(17)	✕	14,16,✕I
(18)	~~Q	7,11,→O
(19)	~Q	9,11,→O
(20)	✕	18,19,✕I

#116:

(1)	(P → ~Q) → R	Pr
(2)	~~SHOW~~: ~(P & Q) → R	CD
(3)	~(P & Q)	As
(4)	~~SHOW~~: R	DD
(5)	P → ~Q	3,~&O
(6)	R	1,5,→O

#117:

(1)	P ↔ (Q & ~P)	Pr
(2)	~~SHOW~~: ~P & ~Q	DD
(3)	~~SHOW~~: ~P	ID
(4)	P	As
(5)	~~SHOW~~: ✕	DD
(6)	P → (Q & ~P)	1,↔O
(7)	Q & ~P	4,6,→O
(8)	~P	7,&O
(9)	✕	4,8,✕I
(10)	~~SHOW~~: ~Q	ID
(11)	Q	As
(12)	~~SHOW~~: ✕	DD
(13)	Q & ~P	3,11,&I
(14)	(Q & ~P) → P	1,↔O
(15)	P	13,14,→O
(16)	✕	3,15,✕I
(17)	~P & ~Q	3,10,&I

#118:

(1)	P	Pr
(2)	~~SHOW~~: (P & Q) ∨ (P & ~Q)	ID
(3)	~[(P & Q) ∨ (P & ~Q)]	As
(4)	~~SHOW~~: ✕	DD
(5)	~(P & Q)	3,~∨O
(6)	~(P & ~Q)	3,~∨O
(7)	P → ~Q	5,~&O
(8)	P → ~~Q	6,~&O
(9)	~Q	1,7,→O
(10)	~~Q	1,8,→O
(11)	✕	9,10,✕I

#119:

(1)	P ↔ ~P	Pr
(2)	~~SHOW~~: Q	ID
(3)	~Q	As
(4)	~~SHOW~~: ✕	DD
(5)	P → ~P	1,↔O
(6)	~P → P	1,↔O
(7)	~~SHOW~~: P	ID
(8)	~P	As
(9)	~~SHOW~~: ✕	DD
(10)	P	6,8,→O
(11)	✕	8,10,✕I
(12)	~P	5,7,→O
(13)	✕	7,12,✕I

#120:

(1)	(P ↔ Q) ↔ R	Pr
(2)	~~SHOW~~: P ↔ (Q ↔ R)	DD
(3)	~~SHOW~~: P → (Q ↔ R)	CD
(4)	P	As
(5)	~~SHOW~~: Q ↔ R	DD
(6)	~~SHOW~~: Q → R	CD
(7)	Q	As
(8)	~~SHOW~~: R	DD
(9)	~~SHOW~~: P → Q	CD
(10)	P	As
(11)	~~SHOW~~: Q	DD
(12)	Q	7,R
(13)	~~SHOW~~: Q → P	CD
(14)	Q	As
(15)	~~SHOW~~: P	DD
(16)	P	4,R
(17)	P ↔ Q	9,13,↔I
(18)	(P ↔ Q) → R	1,↔O
(19)	R	17,18,→O
(20)	~~SHOW~~: R → Q	CD
(21)	R	As
(22)	~~SHOW~~: Q	DD
(23)	R → (P ↔ Q)	1,↔O
(24)	P ↔ Q	21,23→O
(25)	P → Q	24,↔O
(26)	Q	4,25,→O
(27)	Q ↔ R	6,20,↔I
(28)	~~SHOW~~: (Q ↔ R) → P	CD
(29)	Q ↔ R	As
(30)	~~SHOW~~: P	ID
(31)	~P	As
(32)	~~SHOW~~: ✕	DD
(33)	~~SHOW~~: P → Q	CD
(34)	P	As
(35)	~~SHOW~~: Q	ID
(36)	~Q	As
(37)	~~SHOW~~: ✕	DD
(38)	✕	31,34,✕I
(39)	~~SHOW~~: Q → P	CD
(40)	Q	As
(41)	~~SHOW~~: P	DD
(42)	Q → R	29,↔O
(43)	R	40,42,→O
(44)	R → (P ↔ Q)	1,↔O
(45)	P ↔ Q	43,44,→O
(46)	Q → P	45,↔O
(47)	P	40,46,→O
(48)	P ↔ Q	33,39,↔I
(49)	(P ↔ Q) → R	1,↔O
(50)	R	48,49,→O
(51)	R → Q	29,↔O
(52)	Q	50,51,→O
(53)	P	39,52,→O
(54)	✕	31,53,✕I
(55)	P ↔ (Q ↔ R)	3,28,↔I

6 TRANSLATIONS IN MONADIC PREDICATE LOGIC

1. INTRODUCTION

As we have noted in earlier chapters, the validity of an argument is a function of its form, as opposed to its specific content. On the other hand, as we have also noted, the form of a statement or an argument is not absolute, but rather depends upon the level of logical analysis we are pursuing.

We have already considered two levels of logical analysis – syllogistic logic, and sentential logic. Whereas syllogistic logic considers quantifier expressions (e.g., 'all', 'some') as the sole logical terms, sentential logic considers statement connectives (e.g., 'and', 'or') as the sole logical connectives. Thus, these branches of logic analyze logical form quite differently from one another.

Predicate logic subsumes both syllogistic logic and sentential logic; in particular, it considers both quantifier expressions and statement connectives as logical terms. It accordingly represents a deeper level of logical analysis. As a consequence of the deeper logical analysis, numerous arguments that are not valid, either relative to syllogistic logic, or relative to sentential logic, turn out to be valid relative to predicate logic. Consider the following argument.

(A) if at least one person will show up, then we will meet
Adams will show up
/ we will meet

First of all, argument (A) is not a syllogism, so it is not a valid syllogism. Next, if we symbolize (A) in sentential logic, we obtain something like the following.

(F) $P \rightarrow M$
A
$/ M$

Here 'P' stands for 'at least one person will show up', 'A' stands for 'Adams will show up', and 'M' stands for 'we will meet'. It is easy to show (using truth tables) that (F) is not a valid sentential logic form.

Nevertheless, argument (A) is valid (intuitively, at least). What this means is that the formal techniques of sentential logic are not fully adequate to characterize the validity of arguments. In particular, (A) has further logical structure that is not captured by sentential logic. So, what we need is a further technique for uncovering the additional structure of (A) that reveals that it is indeed valid. This technique is provided by predicate logic.

2. THE SUBJECT-PREDICATE FORM OF ATOMIC STATEMENTS

Recall the distinction in sentential logic between the following sentences.

(1) Jay and Kay are Sophomores
(2) Jay and Kay are roommates

Whereas the former is equivalent to a conjunction, namely,

(1*) Jay is a Sophomore and Kay is a Sophomore,

the latter is an atomic statement, having no structure from the viewpoint of sentential logic. In particular, whereas in (1) 'and' is used *conjunctively* to assert something about Jay and Kay individually, in (2) 'and' is used *relationally* – to assert that a certain relation holds between Jay and Kay.

In predicate logic, we are able to uncover the additional logical structure of (2); indeed, we are able to uncover the additional logical structure of (1) as well. In particular, we are able to display atomic formulas as consisting of a predicate and one or more subjects.

Consider the atomic statements that compose (1).

(3) Jay is a Sophomore
(4) Kay is a Sophomore

Each of these consists of two grammatical components: a *subject* and a *predicate*. In (3), the subject is 'Jay', and the predicate is '...is a sophomore'; in (4), the subject is 'Kay', and the predicate is the same, '...is a sophomore'.

Next, consider the sole atomic statement in (2), which is (2) itself.

(5) Jay and Kay are roommates

This may be paraphrased as follows.

(5*) Jay is a roommate of Kay

Unlike (3) and (4), this sentence has two grammatical subjects – 'Jay' and 'Kay'. In addition to the subjects, there is also a predicate – '...is a roommate of...'

The basic idea in the three examples so far is that an atomic sentence can be grammatically analyzed into a predicate and one or more subjects. In order to further emphasize this point, let us consider a slightly more complicated example, involving several subjects in addition to a single predicate.

(6) Chris is sitting between Jay and Kay

Once more 'and' is used relationally rather than conjunctively; in particular (6) is not a conjunction, but is rather atomic. In this case, the predicate is fairly complex:

...is sitting between...and...,

and there are three grammatical subjects:

Chris, Jay, Kay

We now state the first principle of predicate logic.

> In predicate logic, every atomic sentence consists of one *predicate* and one or more *subjects*.

3. PREDICATES

Every predicate has a degree, which is a number. If a predicate has degree one, we call it a *one-place predicate*; if it has degree two, we call it a *two-place predicate*; and so forth.

In principle, for every number n, there are predicates of degree n, (i.e., n-place predicates). However, we are going to concentrate primarily on 1-place, 2-place, and 3-place predicates, in that order of emphasis.

To say that a predicate is a one-place predicate is to say that it takes a single grammatical subject. In other words, a one-place predicate forms a statement when combined with a *single* subject. The following are examples.

___ is clever
___ is a Sophomore
___ sleeps soundly
___ is very unhappy

Each of these is a 1-place predicate, because it takes a single term to form a statement; thus, for example, we obtain the following statements.

Jay is clever
Kay is a Sophomore
Chris sleeps soundly
Max is very unhappy

On the other hand, a *two-place predicate* takes two grammatical subjects, which is to say that it forms a statement when combined with two names. The following are examples.

___ is taller than ___
___ is south of ___
___ admires ___
___ respects ___
___ is a cousin of ___

Thus, for example, using various pairs of individual names, we obtain the following statements.

> Jones is taller than Smith
> New York is south of Boston
> Jay admires Kay
> Kay respects Jay
> Jay is a cousin of Kay

Finally, a *three-place predicate* takes three grammatical subjects, which is to say that it forms a statement when combined with three names. The following are examples.

> ___ is between ___ and ___
> ___ is a child of ___ and ___
> ___ is the sum of ___ and ___
> ___ borrowed ___ from ___
> ___ loaned ___ to ___
> ___ recommended ___ to ___

Thus, for example, we may obtain the following statements from these predicates.

> New York is between Boston and Philadelphia
> Chris is a child of Jay and Kay
> 11 is the sum of 4 and 7
> Jay borrowed this pen from Kay
> Kay loaned this pen to Jay
> Kay recommended the movie "Casablanca" to Jay

One-place predicates (also called monadic predicates) may be thought of as denoting *properties* (e.g., the property of being tall), whereas multi-place (1,2,3-place) predicates (also called polyadic predicates) may be thought of as denoting *relations* (e.g., the relation between two things when one is taller than the other).

Sometimes, the study of predicate logic is formally divided into *monadic predicate logic* (also called property logic) and *polyadic predicate logic* (also called relational logic). In this text, we do not *formally* divide the subject in this way. On the other hand, we deal primarily with monadic predicate logic in the present chapter, leaving polyadic predicate logic for the next chapter.

4. SINGULAR TERMS

Predicate logic analyzes every atomic sentence into a predicate and one or more subjects. In the present section, we examine the latter in a little more detail. In the previous section, the alert reader probably noticed that diverse sorts of expressions were substituted into the blanks of the predicates. Not only did we use names of people, but we also used numerals (which are names of numbers), the name of a movie, and even a demonstrative noun phrase '*this* pen'.

These are all examples of *singular terms* (also called individual terms), which include four sorts of expressions, among others.

(1) proper nouns
(2) definite descriptions
(3) demonstrative noun phrases
(4) pronouns

Examples of proper nouns include the following.

Jay, Kay, Chris, etc.
George Washington, John F. Kennedy, etc.
Paris, London, New York, etc.
Jupiter, Mars, Venus, etc.
1, 2, 3, 23, 45, etc.

Examples of definite descriptions that are singular terms include the following.

the largest river in the world
James Joyce's last book
the president of the U.S.
the square root of 2
the first person to finish the exam

Examples of demonstrative noun phrases that are singular terms include the following.

the person over there
this person, this pen, etc.
that person, that pen, etc.

The use of demonstrative noun phrases generally involves *pointing*, either explicitly or implicitly.

Examples of pronouns that are singular terms are basically all the third person singular pronouns,

he, she, it, him, her,

as well as "wh" expressions such as

who, whom, which (that), what, when, where, why.

Having seen various examples of singular terms, it is equally important to see examples of noun-like expressions that do not qualify as singular terms. These might be called, by analogy, *plural terms*.

Examples of Plural Terms

the people who play for the New York Yankees
the five smartest persons in the class
James Joyce's books
the European cities
the natural numbers
the people standing over there
they, them, these, those

Note carefully that many people use 'they' and 'them' as singular pronouns. Consider the following example.

(?1) I have a date tonight with a music major; I am meeting *them* at the concert hall.

One's response to hearing the word 'them' should be "exactly how many people do you have a date with?", or "is your date a schizophrenic?" More than likely, your date is a man, in which case your date is a "him", or is a woman, in which case your date is a "her". Unless your date consists of several people, it is not a "them".

Another very common example in which 'they/them/their' is used (incorrectly) as a singular pronoun is the following.

(?2) Everyone in the class likes *their* roommate.

In times long past, literate people thought that 'he', 'him', and 'his' had a use as singular third person *neutral* pronouns. In those care-free times, when men were men (and so were women!), the grammatically correct formulation of (?2) would have been the following.

(*2) Everyone in the class likes *his* roommate.

Nowadays, in the U.S. at least, many literate people reject the neutrality of 'he', 'him', and 'his' and accordingly insist on rewriting the above sentence in the following (slightly stilted) manner.

(!2) Everyone in the class likes *his or her* roommate.

Notwithstanding the fact that illiterate people use 'they', 'them', and 'their' as singular pronouns, these words are in fact *plural* pronouns, as can quickly be seen by examining the following two sentences.

(1) they are tall (plural verb form)
(2) they is tall (singular verb form)

A singular term refers to a single individual – a person, place, thing, event, etc., although perhaps a complex one, like IBM, or a very complex one, like the

Renaissance. In order to decide whether a noun phrase qualifies as a singular term, the simplest thing to do is to check whether the noun phrase can be used properly with the singular verb form 'is'. If the noun phrase requires the plural form 'are', then it is not a singular term, but is rather a plural term.

Let us conclude by stating a further, very important, principle of the grammar of predicate logic.

> In predicate logic,
> every subject is a *singular term*.

5. ATOMIC FORMULAS

Having discussed the manner in which every atomic sentence of predicate logic is decomposed into a predicate and (singular) subject(s), we now introduce the symbolic apparatus by which the form of such a sentence is formally displayed.

In sentential logic, you will recall, atomic sentences are abbreviated by upper case letters of the Roman alphabet. The fact that they are symbolized by letters reflects the fact that they are regarded as having no further logical structure. By contrast, in predicate logic, every atomic sentence is analyzed into its constituents, being its predicate and its subject or subjects.

In order to distinguish these constituents, we adopt a particular notational convention, which is simple if not entirely intuitive. This convention is presented as follows.

> (1) Predicates are symbolized by upper case letters.
>
> (2) Singular terms are symbolized by lower case letters.
>
> (3) Every atomic sentence is symbolized by juxtaposing the associated subject and predicate letters.
>
> (4) In particular, in each atomic sentence, the predicate letter goes first and is followed by the subject letter(s).

The following are examples.

Expression	Abbreviation
Predicates:	
___ is tall	T
___ is a Freshman	F
___ respects____	R
___ is a cousin of___	C
___ is between ___ and ___	B
Singular Terms:	
Jay	j
Kay	k
New York City	n
Jupiter	j
the tallest person in class	t
the movie "Casablanca"	c
Sentences:	
Jay is tall	Tj
Kay is a Freshman	Fk
Jay respects Kay	Rjk
Kay is a cousin of Jay	Ckj
Chris is between Jay and Kay	Bcjk

From occasion to occasion, different predicates can be abbreviated by the same letter; likewise for singular terms. However, in any given context (a statement or argument), one must be careful to use different letters to abbreviate different names. The letter 'j' can stand for 'Jay' or for 'Jupiter', but if 'Jay' and 'Jupiter' appear in the same statement or argument, then we cannot use 'j' to abbreviate both of them; for example, we might use 'j' for 'Jay' and 'u' for 'Jupiter'.

Notice that we use lower case letters to abbreviate all singular terms, including definite descriptions. Unlike proper nouns, definite descriptions have further logical structure, and this further structure is revealed and examined in more advanced branches of logic. However, for the purposes of intro logic, definite descriptions have no further logical structure; they are simply singular terms, and are accordingly abbreviated simply by lower case letters.

6. VARIABLES AND PRONOUNS

So far we have concentrated on singular terms that might be called *constants*. In addition to constants there are also *variables*. Variables play the same role in predicate logic that (singular third-person) pronouns play in ordinary language; specifically, they are used for cross-referencing inside a sentence or larger linguistic unit. Furthermore, variables play the same role in predicate logic that variables play in symbolic arithmetic (called algebra in high school); specifically, they enable us to refer to individuals (e.g., individual numbers), without referring to any particular individual (number). This is very useful, as we shall see shortly, in making general claims.

Concerning symbolization, whereas we use the lower case letters 'a', 'b', 'c', ..., 'w' as constants, we use the remaining lower case letters 'x, `y', 'z' as variables. If it turns out that we need more than 26 constants or variables, then we will subscript these with numerals to obtain, for example, 'a_2', 'y_3', 'z_{50}', etc. Thus, in principle, there are infinitely many constants and variables.

In order to avoid using subscripted variables, we also reserve the right to "requisition" constants to use as variables, if the need should arise. So, for example, if we need six variables, but only a few constants, then we will "draft" 'u', 'v', and 'w' into service as variables. If this should happen, it will be explicitly announced. For the most part, however, in intro logic we need only three variables, and there is no need to recruit constants.

When we combine a predicate with one or more singular terms, we obtain a *formula* of predicate logic. When one or more of these singular terms is a variable, we obtain an *open formula*. Open formulas of predicate logic correspond to open sentences of natural language.

Consider the following sentences of arithmetic.

(1)	2 is even	Et
(2)	3 is larger than 4	Ltf
(3)	it is even	Ex
(4)	this is larger than that	Lxy

Whereas (1) and (2) are closed sentences, and their symbolizations, to the right, are closed formulas, (3) and (4) are open sentences, and their symbolizations are open formulas.

So, what is the difference between open and closed sentences, anyway? The difference can be described by saying that, whereas (1) and (2) express propositions and are accordingly true or false, (3) and (4) do not (by themselves) express propositions and are accordingly neither true nor false.

On the other hand (this is the tricky part!), even though it does not autonomously express a proposition, an open sentence can be *used* to assert a proposition – specifically, by uttering it while "pointing" at a particular object or objects. If we "point" at the number two (insofar as that is possible), and say "it/this/that is even", then we have asserted the proposition that the number two is even; indeed, we have asserted a true proposition. Similarly, when we successively point at the number two and the number five, and say "this is larger than that", then we have asserted the proposition that two is larger than five; we have asserted a proposition, but a false proposition.

A closed sentence, by contrast, can be used to assert a proposition, even without having to point. If I say "two is even", I need not point at the number two in order to assert a proposition; the sentence does it for me.

One way to describe the difference between open and closed sentences is to say that, unlike closed sentences, open sentences are essentially *indexical* in character, which is to say that their use essentially involves *pointing*. (Here, think of the *index* finger, as used for pointing.) This pointing can be fairly straightforward, but it can also be oblique and subtle. This pointing can also be either *external* or *internal* to the sentence in which the indexical (i.e., pointing) expression occurs.

For example, in the sentence about the date with the music major, the pronoun refers to (points at) something external; the 'he or she' refers to the particular person about whom the speaker is talking. By contrast, in the sentence about roommates, the 'his or her' refers, not externally to a particular person, but rather internally to the expression 'everyone'.

Another use of internal pointing involves the following indexical expressions.

(1) the former
(2) the latter
(3) the party of the first part
(4) the party of the second part

The latter two expressions (an example of pointing!) are used almost exclusively in legal documents, and we will not examine them any further. The former two expressions, on the other hand, are important expressions in logic. If I refer to a music major and a business major, in that order, then if I say "the *former* respects the *latter*", I am saying that the music major respects the business major. If I say instead "he respects her", then it is not clear who respects whom. Thus, the words 'former' and 'latter' are useful substitutes for ordinary pronouns.

We conclude this section by announcing yet another principle of the grammar of predicate logic.

> In an atomic formula,
> every subject is either
> a *constant* or a *variable*.

7. COMPOUND FORMULAS

We have now described the atomic formulas of predicate logic; every such formula consists of an n-place predicate letter followed by n singular terms, each one being either a constant or a variable. The atomic formulas of predicate logic play exactly the same role that atomic formulas play in sentential logic; in particular, they can be combined with connectives to form molecular formulas.

We already know how to construct molecular formulas from atomic formulas in sentential logic. This skill carries over directly to predicate logic, the rules being precisely the same. If we have a formula, we can form its negation; if we have two formulas, we can form their conjunction, disjunction, conditional, and biconditional. The only difference is that the simple statements we begin with are not simply letters, as in sentential logic, but are rather combinations of predicate letters and singular terms.

The following are examples of compound statements in predicate logic, followed by their symbolizations.

(1)	if Jay is a Freshman, then Kay is a Freshman	Fj → Fk
(2)	Kay is not a Freshman	~Fk
(3)	neither Jay nor Kay is a Freshman	~Fj & ~Fk
(4)	Jay respects Kay, but Kay does not respect Jay	Rjk & ~Rkj

Next, we note that either (or both) of the proper nouns 'Jay' and 'Kay' can be replaced by pronouns. Correspondingly, either (or both) of the constants 'j' and 'k' can be replaced by variables (for example, 'x' and 'y'). We accordingly obtain various open sentences (formulas). For example, taking (1), we can construct the following open statements and associated open formulas.

(1)	if Jay is a Freshman, then Kay is a Freshman	Fj → Fk
(1a)	if Jay is a Freshman, then she is a Freshman	Fj → Fy
(1b)	if he is a Freshman, then Kay is a Freshman	Fx → Fk
(1c)	if he is a Freshman, then she is a Freshman	Fx → Fy

8. QUANTIFIERS

We have already seen that compound formulas can be constructed using the connectives of sentential logic. In addition to these truth-functional connectives, predicate logic has additional compound forming expressions – namely, the quantifiers.

Quantifiers are linguistic expressions denoting *quantity* in some form. Examples of quantifiers in English include the following.

every, all, each, both, any, either
some, most, many, several, a few
none, neither
at least one, at least two, etc.
at most one, at most two, etc.
exactly one, exactly two, etc.

These expressions are typically combined with noun phrases to produce sentences, such as the following.

every Freshman is clever
at least one Sophomore is clever
no Senior is clever
many Sophomores are clever
several Juniors are clever

In addition to these quantifier expressions, there are also derivative expressions, contractions, involving 'thing' and 'one'.

everyone, everything, someone, something, no one, nothing

These yield sentences such as the following

everyone is clever
everything is clever
someone is clever
something is clever
no one is clever
nothing is clever

Recall that there are numerous statement connectives in English, but in sentential logic we concentrate on just a few, logically fruitful, ones. Similarly, even though there are numerous quantifier expressions in English, in predicate logic we concentrate only on a couple of them, given as follows.

every
at least one

Not only do we concentrate on these two quantifier concepts, we render them very general, as follows.

everything is such that...
there is at least one thing such that...
at least one thing is such that...

Although these expressions are somewhat stilted (much like the official expression for negation 'it is not true that...'), they are sufficiently general to be used in a much wider variety of contexts than more colloquial quantifier expressions.

If this is not stilted enough, we must add one further feature to the above quantifiers, in order to obtain the official quantifiers of predicate logic. Recall that a pronoun can point internally, and in particular, it can point at a quantifier expression in the sentence. In the sentence

everyone likes his/her roommate

the pronoun 'his/her' points at the quantifier 'everyone'. But what if the sentence in question has more than one quantifier? Consider the following.

everyone knows someone who respects his/her mother

This sentence is ambiguous, because it isn't clear what the pronoun 'his/her' points at. This sentence might be paraphrased in either of the following ways.

everyone knows someone who respects *the former*'s mother
everyone knows someone who respects *the latter*'s mother

The additional feature needed by the quantifiers above is an *index*, in order to allow clear and consistent cross-referencing inside of sentences in which they appear. Since we are using variables as pronouns, it is convenient to use the very same symbolic devices as quantifier indices as well.

Thus, every quantifier comes with an index (a variable) attached to it. We thus obtain the following quantifier expressions.

everything x is such that...
everything y is such that...
everything z is such that...

there is at least one thing x such that...
there is at least one thing y such that...
there is at least one thing z such that...

These are symbolized respectively as follows.

∀x ∀y ∀z
∃x ∃y ∃z

Historically, the upside-down 'A' derives from the word 'all', and the backwards 'E' derives from the word 'exist'. Whereas the expressions '∀x', '∀y', '∀z' are called *universal quantifiers*, the expressions '∃x', '∃y', '∃z' are called *existential quantifiers*.

For every variable, there are two quantifiers, a universal quantifier, and an existential quantifier. Grammatically, a quantifier is a one-place connective, just like negation ~. In other words, we have the following grammatical principle.

> If **F** is a formula, then so are all the following.
>
> ∀x**F**, ∀y**F**, ∀z**F**
> ∃x**F**, ∃y**F**, ∃z**F**

Of course, in forming the compound formula, the outer parentheses (if any) of the formula **F** must be restored before prefixing the quantifier. This is just like negation. We will see examples of this later.

We now have the official quantifier expressions of predicate logic. How do they combine with other formulas to make quantified formulas? The basic idea (but not the whole story) is that one begins with an open formula involving (say) the variable 'x', and one prefixes '∀x' to obtain a universally quantified formula, or one prefixes '∃x' to obtain an existentially quantified formula.

For example, we can begin with the following open formula,

Fx: x is fascinating (it is fascinating),

and prefix either '∀x' or '∃x' to obtain the following formulas.

∀xFx: everything [x] is such that it [x] is fascinating

∃xFx: there is at least one thing [x] such that it [x] is fascinating

In each case, I have divided the sentence into a quantifier and an open formula. The variables are placed in parentheses, since they are not really part of the English sentence; rather, they are used to cross-reference the pronoun 'it'. In particular, the fact that 'x' is used for both the quantifier and the pronoun indicates that 'it' points back at (cross-references) the quantifier expression.

This is the simplest case, one in which the open formula $\mathcal{A}$ is atomic. It can also be molecular; it can even be a quantified formula (a great deal more about this in the next chapter). The following are all examples of open formulas involving 'x' together with the resulting quantified formulas. Notice the appearance of the parentheses in (2) and (3).

	Open Formula:	**Universal Formula:**	**Existential Formula:**
(1)	~Fx	∀x~Fx	∃x~Fx
(2)	Fx & Gx	∀x(Fx & Gx)	∃x(Fx & Gx)
(3)	Fx → Gx	∀x(Fx → Gx)	∃x(Fx → Gx)
(4)	Rxj	∀xRxj	∃xRxj
(5)	∃yRxy	∀x∃yRxy	∃x∃yRxy

The pairs to the right are all examples of quantified formulas, universal formulas and existential formulas respectively. These can in turn be combined using any of the sentential logic connectives, to obtain (e.g.) the following compound formulas.

(6)	∀x~Fx ∨ ∀x(Fx & Gx)	disjunction
(7)	~∀xRxj; ~∃xRxj; ~∀x∃yRxy; ~∃x∃yRxy	negations
(8)	∀xRxj → ∀x∃yRxy; ∃xRxj → ∃x∃yRxy	conditionals

At this stage, the important thing is not necessarily to be able to read the above formulas, but to be able to recognize them as formulas. Toward this end, keep in clear sight the rules of formula formation in predicate logic, which are sketched as follows.

Definition of Formula in Predicate Logic:

Atomic Formulas:

(1) If P is a predicate letter of degree n, then P followed by n singular terms is an atomic formula.

(2) Nothing else is an atomic formula.

Formulas:

(1) Every atomic formula is a formula.

(2) If $\mathcal{A}$ is a formula, then so is $\sim\mathcal{A}$.

(3) If $\mathcal{A}$ and $\mathcal{B}$ are formulas then so are the following.

$(\mathcal{A} \,\&\, \mathcal{B})$
$(\mathcal{A} \vee \mathcal{B})$
$(\mathcal{A} \rightarrow \mathcal{B})$
$(\mathcal{A} \leftrightarrow \mathcal{B})$

(4) If $\mathcal{A}$ is a formula, then so are the following.

$\forall x\mathcal{A}$, $\forall y\mathcal{A}$, $\forall z\mathcal{A}$, etc.
$\exists x\mathcal{A}$, $\exists y\mathcal{A}$, $\exists z\mathcal{A}$, etc.

(5) Nothing else is a formula.

9. COMBINING QUANTIFIERS WITH NEGATION

As noted at the end of the previous section, any formula can be prefixed by either a universal quantifier or an existential quantifier, just as any formula can be prefixed by negation, and the result is another formula.

In the present section, we concentrate on the way in which negation interacts with quantifiers.

Let us start with the following open formula.

(1) Px it is perfect

Then let us quantify it both universally and existentially, as follows.

(2) ∀xPx everything is such that
it is perfect

(3) ∃xPx at least one thing is such that
it is perfect

These can in turn be negated, yielding the following formulas.

(4) ~∀xPx it is not true that
everything is such that
it is perfect

(5) ~∃xPx it is not true that
at least one thing is such that
it is perfect

Before considering more colloquial paraphrases of the above sentences, let us consider an alternative tack. Let us first negate 'Px' to obtain the following.

(6) ~Px it is not true that
it is perfect

The latter sentence may be paraphrased as either of the following.

it is not perfect
It is imperfect

Many adjectives have ready-made negations (happy/unhappy, friendly/unfriendly, possible/impossible); most adjectives, however, do not have natural negations. On the other hand, we can always produce the negation of any adjective simply by prefixing 'non-' in front of the adjective.

Now, let us take the negated formula '~Px' and quantify in the two ways, which yields the following.

(7) ∀x~Px everything is such that
it is not true that
it is perfect

everything is such that
it is not perfect

everything is such that
it is imperfect

(8) ∃x~Px at least one thing is such that
it is not true that
it is perfect

at least one thing is such that
it is not perfect

at least one thing is such that
it is imperfect

Having written down all the simple formulas involving negation and quantifiers, let us now consider the idiomatic rendering of these sentences. First, to say

everything is such that it is perfect

is equivalent to saying

everything has a certain property – it is perfect.

These two sentences are simply verbose ways of saying

everything is perfect.

Similarly, to say

at least one thing is such that it is perfect,

which is an alternative to

there is at least one thing such that it is perfect,

is equivalent to saying

at least one thing has a certain property – it is perfect.

These two sentences are simply verbose ways of saying

at least one thing is perfect.

The latter sentence, in turn, can be thought of as one way of rendering precise the following.

something is perfect

Along similar lines, recall the way that the negation operator works; the official form of negation involves prefixing 'it is not true that' in front of the sentence in question. Thus, for example, one obtains the following.

it is not true that it is perfect

Recall that this is equivalent to the following more colloquial expression.

it is not perfect

The advantage of the verbose forms of negation and quantification is grammatical generality; we can always produce the official negation or quantification of a sentence, but we cannot always easily produce the colloquial negation or quantification.

For example, consider the following.

everything is such that
    it is not true that
        it is perfect,

which is equivalent to

everything is such that
    it is not perfect.

Following the above line of reasoning concerning colloquial quantification, the natural paraphrase of this is the following.

everything is not perfect

Unfortunately, the placement of 'not' in this sentence makes it unclear whether it modifies 'is' or 'perfect'; accordingly, this sentence is ambiguous in meaning between the following pair of sentences.

everything isn't perfect
    (i.e., not everything is perfect)

everything is non-perfect

These are not equivalent; if, some things are perfect and some things are not, the first is true, but the second is false.

The original sentence,

everything is such that it is not perfect,

says that everything has the property of being non-perfect (imperfect), or

everything is non-perfect (imperfect).

To say that everything is non-perfect (imperfect) is equivalent to saying

nothing is perfect,

which is much stronger than

not everything is perfect.

The latter sentence is a colloquial paraphrase of

it is not true that everything is perfect,

which is a colloquial paraphrase of

it is not true that
    everything is such that
        it is perfect.

This is precisely formula (4) above.

Now, if not everything is perfect, then there is at least one thing that isn't perfect, and conversely. To say the latter, we write

at least one thing is such that it is not perfect,

which is formula (8) above.

Finally, consider formula (5)

~∃xPx   it is not true that
  at least one thing is such that
    it is perfect

which is equivalent to

it is not true that
  at least one thing is perfect.

The number of things that are perfect is either zero, one, two, three, etc. To say that at least one thing is perfect is to say that the number of perfect things is at least one, that is, the number is not zero. To say that this is not true is to say that the number of perfect things is zero, which is to say

nothing is perfect.

Thus, we basically have six colloquial sentences.

(c1) everything is perfect
(c2) something is perfect (i.e., at least one thing is perfect)
(c3) everything is imperfect

(c4) something is imperfect
(c5) not everything is perfect
(c6) nothing is perfect

These correspond to the following formulas of predicate logic.

(f1) ∀xPx
(f2) ∃xPx
(f3) ∀x~Px
(f4) ∃x~Px
(f5) ~∀xPx
(f6) ~∃xPx

As noted earlier, two pairs of formulas are equivalent. In particular:

∀x~Px   [everything is imperfect]

is equivalent to

~∃xPx   [nothing is perfect],

and

$\exists x{\sim}Px$ [something is imperfect]

is equivalent to

${\sim}\forall xPx$ [not everything is perfect].

These are instances of two very general equivalences, which may be stated as follows.

$$\begin{aligned} {\sim}\forall x &= \exists x{\sim} \\ {\sim}\exists x &= \forall x{\sim} \end{aligned}$$

What this means is that for any formula $\mathcal{A}$, however complex, we have the following.

> ${\sim}\forall x\mathcal{A}$ is equivalent to $\exists x{\sim}\mathcal{A}$.
>
> ${\sim}\exists x\mathcal{A}$ is equivalent to $\forall x{\sim}\mathcal{A}$.

In order to understand them better, it might be worthwhile to compare these two equivalences with their counterparts in sentential logic – deMorgan's laws. In their simplest form, these laws of logic are stated as follows.

(dM1) ${\sim}(\mathcal{A}\&\mathcal{B})$ is equivalent to ${\sim}\mathcal{A}\vee{\sim}\mathcal{B}$.
(dM2) ${\sim}(\mathcal{A}\vee\mathcal{B})$ is equivalent to ${\sim}\mathcal{A}\&{\sim}\mathcal{B}$.

But there are more general forms as well, given as follows.

(M1) ${\sim}(\mathcal{A}_1\&\mathcal{A}_2\&..\&\mathcal{A}_n)$ is equivalent to $\mathcal{A}_1\vee{\sim}\mathcal{A}_2\vee...\vee{\sim}\mathcal{A}_n$
(M2) ${\sim}(\mathcal{A}_1\vee\mathcal{A}_2\vee...\vee\mathcal{A}_n)$ is equivalent to ${\sim}\mathcal{A}_1\&{\sim}\mathcal{A}_2\&...\&{\sim}\mathcal{A}_n$

In other words, the negation of any conjunction, however long, is equivalent to a corresponding disjunction of negations, and similarly, the negation of any disjunction, however long, is equivalent to a corresponding conjunction of negations.

But what does this have to do with universal and existential quantifiers. Well, imagine for a moment there are exactly two things in the universe – call them a and b, respectively. In such a universe, which is very small, every universally quantified statement is equivalent to a conjunction, and every existentially quantified statement is equivalent to a disjunction. In particular, we have the following.

everything is F :: a is F, and b is F

something is F :: a is F, and/or b is F

Or, in formulas:

$\forall xFx$ :: $Fa \& Fb$

$\exists xFx$ :: $Fa \vee Fb$

Similarly, if there are exactly three things in the universe (a, b, c), then we have the following equivalences.

everything is F :: a is F, and b is F, and c is F

something is F :: a is F, and/or b is F, and/or c is F

Or, in formulas:

∀xFx :: Fa & Fb & Fc

∃xFx :: Fa ∨ Fb ∨ Fc

This can be generalized to any (finite) number of things in the universe; for every universally/ existentially quantified statement, there is a corresponding conjunction/ disjunction of suitable length.

Having seen what the equivalence looks like in general, let us concentrate on the simplest non-trivial version – a universe with just two things (a and b) in it.

Next, let us consider what happens when we combine quantifiers with negation? First, the simplest.

everything is not-F :: a is not F and b is not F

something is not-F :: a is not F and/or b is not F

Or, in formulas:

∀x~Fx :: ~Fa & ~Fb

∃x~Fx :: ~Fa ∨ ~Fb

Negating the quantified statements yields:

not everything is F :: not(a is F and b is F)

nothing is F :: not something is F :: not(a is F and/or b is F)

Or, in formulas:

~∀xFx :: ~(Fa & Fb)

~∃xFx :: ~(Fa ∨ Fb)

Finally, we obtain the following chain of equivalences.

~∀xFx :: ~(Fa & Fb) :: ~Fa ∨ ~Fb :: ∃x~Fx

~∃xFx :: ~(Fa ∨ Fb) :: ~Fa & ~Fb :: ∀x~Fx

The same procedure can be carried out with three, or four, or any number of, individuals.

Note: In the previous example, the formula $\mathcal{A}$ is simple, being Fx. In general, $\mathcal{A}$ may be complex – for example, it might be the formula (Fx→Gx).

Then $\sim\mathcal{A}$ is the negation of the entire formula, which is $\sim$(Fx→Gx). (Notice that the parentheses are optional in the conditional, but not in its negation.)

10. SYMBOLIZING THE STATEMENT FORMS OF SYLLOGISTIC LOGIC

Recall that the statement forms of syllogistic logic are given as follows.

(f1) all A are B
(f2) some A are B
(f3) no A are B
(f4) some A are not B

These are all stated in the plural form. In order to translate these into predicate logic, the first thing we must do is to convert each plural form into the corresponding closest singular form.

(s1)	every A is B	[every A is a B]
(s2)	some A is B	[some A is a B]
(s3)	no A is B	[no A is a B]
(s4)	some A is not B	[some A is not a B]

Examples of sentences in these forms are given as follows.

(e1) every astronaut is brave
(e2) some astronaut is brave
(e3) no astronaut is brave
(e4) some astronaut is not brave

Note that the simple predicate 'is brave' can be replaced by the longer expression 'is a brave person'.

The next thing we must do is to convert the specific quantifier expressions 'every/some/no A' into the corresponding expressions involving general quantifiers 'every/some/thing is such that...'

Consider (s1); to say

every A is a B

is to say

everything that is A is B,

or if we have persons exclusively in mind,

everyone who is A is B.

For example, we could read the latter as follows.

everyone who is an astronaut is brave

We know how to formalize ‘everything (everyone) is B’.

everything is such that it is B $\forall xBx$

But we don't want to say that everything is B, just every A is B. How do we add the clause ‘that (who) is A’? Let us try the following paraphrases.

everything is B *provided* it is A

everything is such that it is B *provided* it is A

Now we are getting somewhere, since this sentence divides as follows.

everything is such that
it is B provided it is A

Adding the crucial pronoun indices (variables), we obtain the following.

everything x is such that
x is B provided x is A

Recall ‘$\mathcal{B}$ provided $\mathcal{A}$’ is equivalent to ‘$\mathcal{B}$ if $\mathcal{A}$’, which is equivalent to ‘if $\mathcal{A}$, then $\mathcal{B}$’, which is symbolized $\mathcal{A}\rightarrow\mathcal{B}$. Thus, the above sentence is symbolized as follows:

$$\forall x(Ax \rightarrow Bx).$$

Note carefully the parentheses around the conditional; it's OK to omit them when the formula stands by itself, but when it goes into making a larger formula, the outer parentheses must be restored. The same thing happens when we negate a conditional.

Of course, the corresponding formula without parentheses,

$$\forall xAx \rightarrow Bx,$$

is also a formula of predicate logic, just as $\sim A\rightarrow B$ is a formula of sentential logic. Both are conditionals. The latter says ‘if not A, then B’, in contrast to ‘it is not true that if A then B’, which is the reading of $\sim(A\rightarrow B)$. The most accurate translation of the predicate logic formula, which is logically equivalent to

$$\forall xAx \rightarrow By,$$

reads as follows.

if everything is A, then *this* is B,

where ‘this’ points at something external to the sentence. This is a perfectly good piece of English, but it is definitely not the same as saying that every A is B.

Next, let us consider (s2) above. To say

some A is B,

for example, to say

some astronaut is brave,

is to say

there is at least one A that (who) is also B,

which is equivalent to

there is at least one A *and* it (he/she) is also B.

Notice that the pronoun 'it' points internally at 'at least one A'.

We know how to say

there is at least one A.
there is at least one thing such that it is A
∃xAx

How do we add the clause 'that is also B' or 'and it is also B'? Well, we are saying that the thing in question is A, and we are saying *in addition* that it is B, so we are saying that it is A *and* it is B, which gives us the following.

there is at least one thing such that
it is A
and it is B

This is symbolized as follows.

∃x(Ax & Bx)

Notice once again that the outer parentheses are restored before the quantifier is prefixed. If we were to drop the parentheses, we obtain

∃xAx & Bx,

which is logically equivalent to

∃xAx & By,

which may be read

something is A, and *this* is B,

where 'this' points externally at whatever the person using this sentence is pointing toward. Although this is a perfectly good formula of predicate logic, it says something entirely different from 'some A is B'

Next, let us consider (s3) above. To say

no A is B,

for example,

no astronaut is brave,

is to *deny* that there is at least one A who is B. In other words, it is the negation of 'some A is B', and is accordingly symbolized as follows,

$$\sim\exists x(Ax \ \& \ Bx),$$

which is literally read as

it is not true that
  there is at least one thing such that
    it is A and it is B

Recall that $\sim\exists x\mathbb{F}$ is equivalent to $\forall x\sim\mathbb{F}$, for any formula $\mathbb{F}$. In the above case, $\mathbb{F}$ is the formula (Ax&Bx), we have the following equivalence.

$$\sim\exists x(Ax \ \& \ Bx) \ :: \ \forall x\sim(Ax \ \& \ Bx)$$

But, in sentential logic, we have the following equivalence (check the truth table!)

$$\sim(\mathcal{A} \ \& \ \mathcal{B}) \ :: \ \mathcal{A} \rightarrow \sim\mathcal{B}$$

So, putting these together, we obtain the following equivalence.

$$\sim\exists x(Ax \ \& \ Bx) \ :: \ \forall x(Ax \rightarrow \sim Bx)$$

Thus, we have an alternative way of formulating 'no A is B':

$$\forall x(Ax \rightarrow \sim Bx),$$

which is read literally as

everything is such that
  if it is A
    then it is not B

Finally, let us consider (s4) above. To say

some A is not B

is to say

there is at least one A and it is not B,

which is symbolized very much the same way as 'some A is B',

$$\exists x(Ax \ \& \ \sim Bx),$$

which is read literally as follows.

there is at least one thing such that
  it is A and it is not B

Let us compare this with the following negation,

not every A is B,

which is symbolized just like

it is not true that every A is B,

thus:

$\sim\forall x(Ax \rightarrow Bx)$,

whose literal reading is

it is not true that
every thing is such that
if it is A then it is B.

Recall that $\sim\forall x\mathbb{F}$ is equivalent to $\exists x\sim\mathbb{F}$, for any formula $\mathbb{F}$; in the above case $\mathbb{F}$ is the formula $(Ax \rightarrow Bx)$ – notice the parentheses – so we obtain the following equivalence.

$\sim\forall x(Ax \rightarrow Bx) :: \exists x\sim(Ax \rightarrow Bx)$

But recall the following equivalence of sentential logic.

$\sim(\mathcal{A} \rightarrow \mathcal{B}) :: \mathcal{A} \,\&\, \sim\mathcal{B}$

Thus, we have the following equivalence of predicate logic.

$\sim\forall x(Ax \rightarrow Bx) :: \exists x(Ax \,\&\, \sim Bx)$

In other words, to say

not every A is B

is the same as to say

some A is not B.

For example, the following in effect say the same thing.

not every astronaut is brave
some astronaut is not brave

11. SUMMARY OF THE BASIC QUANTIFIER TRANSLATION PATTERNS SO FAR EXAMINED

Before continuing, it is a good idea to review the basic patterns of translation that we have examined so far. These are given as follows.

Simple Quantification Plus Negation

(1)	everything is B	∀xBx
(2)	something is B	∃xBx
(3)	nothing is B	~∃xBx
(4)	something is non-B	∃x~Bx
(5)	everything is non-B	∀x~Bx
(6)	not everything is B	~∀xBx

Syllogistic Forms Plus Negation

(7)	every A is B	∀x(Ax → Bx)
(8)	some A is B	∃x(Ax & Bx)
(9)	no A is B	~∃x(Ax & Bx)
(10)	some A is not B	∃x(Ax & ~Bx)
(11)	every A is a non-B	∀x(Ax → ~Bx)
(12)	not every A is B	~∀x(Ax → Bx)

In addition to these, it is important to keep the following logical equivalences in mind when doing translations into predicate logic.

Basic Logical Equivalences

(1)	~∃xAx :: ∀x~Ax
(2)	~∀xAx :: ∃x~Ax
(3)	~∃x(Ax & Bx) :: ∀x(Ax → ~Bx)
(4)	~∀x(Ax → Bx) :: ∃x(Ax & ~Bx)

In looking over the above patterns, one might wonder why the following is not a correct translation:

(1) every A is B $\forall x(Ax \& Bx)$ WRONG!!!

The correct translation is given as follows.

(2) every A is B $\forall x(Ax \rightarrow Bx)$ RIGHT!!!

Remember there simply is no general symbol-by-symbol translation between colloquial English and the language of predicate logic; in the correct translation (2), no symbol in the formula corresponds to the 'is' in the colloquial sentence, and no symbol in the colloquial English sentence corresponds to '$\rightarrow$' in the formula.

The erroneous nature of (1) becomes apparent as soon as we translate the formula into English, which goes as follows.

everything is such that
it is A *and* it is B

For example,

everything is such that
it is an astronaut *and* it is brave

In other words,

everything is an astronaut who is brave,

or equivalently,

everything is a brave astronaut.

This is also equivalent to:

everything is an astronaut *and* everything is brave.

Needless to say, this does *not* say the same thing as:

every astronaut is brave.

O.K., arrow works when we have 'every A is B', but ampersand does not work. So, why doesn't arrow work just as well in the corresponding statement 'some A is B'? Why isn't the following a correct translation?

(3) some A is B $\exists x(Ax \rightarrow Bx)$ WRONG!!!

As noted above, the correct translation is:

(4) some A is B $\exists x(Ax \& Bx)$ RIGHT!!!

Once again, please note that there is no symbol-by-symbol translation between the colloquial English form and the predicate logic formula.

Let's see what happens when we translate the formula of (3) into English; the straight translation yields the following:

(3t) there is at least one thing such that
if it is A **then** it is B.

Does this say that some A is B? No! In fact, it is not clear what it says. If the conditional *were* subjunctive, rather than truth-functional, then (3t) might correspond to the following colloquial subjunctive sentence.

there is someone who
would be brave if he *were* an astronaut

From this, it surely does not follow that there is even a single brave astronaut, or even a single astronaut. To make this clear, consider the following analogous sentence.

some Antarctican is brave

Here, let us understand 'Antarctican' to mean a permanent citizen of Antarctica. This sentence must be carefully distinguished from the following.

there is someone who
would be brave if he/she were Antarctican

To say that some Antarctican is brave to say that there is at least one Antarctican who is brave, from which it obviously follows that there is at least one Antarctican. The sentence 'some Antarctican is brave' logically implies 'at least one Antarctican exists'.

By contrast, the sentence 'there is someone who would be brave if *he/she* were Antarctican' does not imply that any Antarctican exists. Whether there is such a person who would be brave were he/she to become an Antarctican, I really couldn't say, but I suspect it is probably true. It takes a brave person to live in Antarctica.

When we take if-then as a subjunctive conditional, we see very quickly that $\exists x(Ax \rightarrow Bx)$ simply does not say that some A is B. What happens if we insist that if-then is truth-functional? In that case, the sentence $\exists x(Ax \rightarrow Bx)$ is automatically true, so long as we can find someone who is *not* Antarctican!

Suppose that Smith is *not* Antarctican. Then the sentence

Smith is Antarctican

is false, and hence the conditional sentence

if Smith is Antarctican, then Smith is brave

is true! Why? Because of the truth table for if-then! But if Smith is such that if he is Antarctican then he is brave, then at least one person is such that if he is Antarctican then he is brave. Thus, the following existential sentence is true.

there is someone such that
if he is Antarctican,
then he is brave

We conclude this section by presenting the following rule of thumb about how symbolizations *usually* go. Of course, in saying that it is a rule of thumb, all one means is that it works quite often, not that it works always.

> **Rule of Thumb** (not absolute)
>
> If one has a universal formula,
> then the connective immediately "beneath"
> the universal quantifier
> is a conditional.
>
> If one has an existential formula,
> then the connective immediately "beneath"
> the existential quantifier
> is a conjunction.

The slogan that goes with this reads as follows:

> UNIVERSAL-CONDITIONAL
>
> EXISTENTIAL-CONJUNCTION

Remember! This is just a rule of thumb! There are numerous exceptions, which will be presented in subsequent sections.

12. FURTHER TRANSLATIONS INVOLVING SINGLE QUANTIFIERS

In the previous section, we saw how one can formulate the statement forms of syllogistic logic in terms of predicate logic. However, the expressive power of predicate logic is significantly greater than syllogistic logic. Syllogistic patterns are a very tiny fraction of the statement forms that can be formulated in predicate logic.

In the next three sections (Sections 12-14), we are going to explore numerous patterns of predicate logic that all have one thing in common with what we have so far examined. Specifically, they all involve exactly one quantifier. More specifically still, each one has one of the following forms.

(f1) $\forall x\mathcal{A}$
(f2) $\exists x\mathcal{A}$
(f3) $\sim\forall x\mathcal{A}$
(f4) $\sim\exists x\mathcal{A}$

In particular, either the main connective is a quantifier, or the main connective is negation, and the next connective is a quantifier.

We have already seen the simplest examples of these forms, in Sections 10 and 11.

∃xBx	something is B
∀xBx	everything is B
~∃xBx	nothing is B
~∀xBx	not everything is B
∃x~Bx	something is non-B
∀x~Bx	everything is non-B

We can also formulate sentences that have an overall form like one of the above, but which have more complicated formulas in place of 'Bx'. The following are examples.

(1) everything is both A and B
(2) everything is either A or B
(3) everything is A but not B
(4) something is both A and B
(5) something is either A or B
(6) something is A but not B
(7) nothing is both A and B
(8) nothing is either A or B
(9) nothing is A but not B

How do we translate these sorts of sentences into predicate logic? One way is first to notice that the *overall* forms of these sentences may be written and symbolized, respectively, as follows.

(o1)	everything is J	∀xJx
(o2)	everything is K	∀xKx
(o2)	everything is L	∀xLx
(o3)	something is J	∃xJx
(o4)	something is K	∃xKx
(o2)	something is L	∃xLx
(o5)	nothing is J	~∃xJx
(o6)	nothing is K	~∃xKx
(o2)	nothing is L	~∃xLx

Here, the pseudo-atomic formulas Jx, Kx, and Lx are respectively short for the more complex formulas, given as follows.

Jx	::	(Ax & Bx)
Kx	::	(Ax ∨ Bx)
Lx	::	(Ax & ~Bx)

Note the appearance of the outer parentheses. Substituting in accordance with these equivalences, we obtain the following translations of the above sentences.

(t1) $\forall x(Ax \& Bx)$
(t2) $\forall x(Ax \vee Bx)$
(t3) $\forall x(Ax \& \sim Bx)$
(t4) $\exists x(Ax \& Bx)$
(t5) $\exists x(Ax \vee Bx)$
(t6) $\exists x(Ax \& \sim Bx)$
(t7) $\sim\exists x(Ax \& Bx)$
(t8) $\sim\exists x(Ax \vee Bx)$
(t9) $\sim\exists x(Ax \& \sim Bx)$

The following paraphrase chains may help to see how one might go about producing the symbolization.

(c1) everything is both A and B

everything is such that
it is both A and B

everything is such that
it is A and it is B

$\forall x(Ax \& Bx)$

(c2) everything is either A or B

everything is such that
it is either A or B

everything is such that
it is A or it is B

$\forall x(Ax \vee Bx)$

(c3) everything is A but not B

everything is such that
it is A but not B

everything is such that
it is A and it is not B

$\forall x(Ax \& \sim Bx)$

(c4) something is both A and B

there is at least one thing such that
it is both A and B

there is at least one thing such that
it is A and it is B

$\exists x(Ax \& Bx)$

You will recall, of course, that 'something is both A and B' is logically equivalent to 'some A is B', as noted in the previous sections.

(c5) something is either A or B

there is at least one thing such that
it is either A or B

there is at least one thing such that
it is A or it is B

$\exists x(Ax \vee Bx)$

(c6) something is A but not B

there is at least one thing such that
it is A but not B

there is at least one thing such that
it is A and it is not B

$\exists x(Ax \& {\sim}Bx)$

(c7) nothing is both A and B

it is not true that something is both A and B

it is not true that
there is at least one thing such that
it is both A and B

it is not true that
there is at least one thing such that
it is A and it is B

${\sim}\exists x(Ax \& Bx)$

(c8) nothing is either A or B

it is not true that something is either A or B

it is not true that there is at least one thing such that
it is either A or B

it is not true that there is at least one thing such that
it is A or it is B

$\sim\exists x(Ax \vee Bx)$

(c9) nothing is A but not B

it is not true that something A but not B

it is not true that there is at least one thing such that
it is A but not B

it is not true that there is at least one thing such that
it is A and it is not B

$\sim\exists x(Ax \ \& \sim Bx)$

In the next section, we will further examine these kinds of sentences, but will introduce a further complication.

13. CONJUNCTIVE COMBINATIONS OF PREDICATES

So far, we have concentrated on formulas that have at most two predicates. In the present section, we drop that restriction and discuss formulas with three or more predicates. However, for the most part, we will concentrate on conjunctive combinations of predicates.

Consider the following sentences (which pertain to a fictional group of people, called Bozonians, who inhabit the fictional country of Bozonia).

E:
(e1) every Adult Bozonian is a Criminal
(e2) every Adult Criminal is a Bozonian
(e3) every Criminal Bozonian is an Adult
(e4) every Adult is a Criminal Bozonian
(e5) every Bozonian is an Adult Criminal
(e6) every Criminal is an Adult Bozonian

S:

(s1) some Adult Bozonian is a Criminal
(s2) some Adult Criminal is a Bozonian
(s3) some Criminal Bozonian is an Adult
(s4) some Adult is a Criminal Bozonian
(s5) some Bozonian is an Adult Criminal
(s6) some Criminal is an Adult Bozonian

N:

(n1) no Adult Bozonian is a Criminal
(n2) no Adult Criminal is a Bozonian
(n3) no Criminal Bozonian is an Adult
(n4) no Adult is a Criminal Bozonian
(n5) no Bozonian is an Adult Criminal
(n6) no Criminal is an Adult Bozonian

The predicate terms have been capitalized for easy spotting. The official predicates are as follows.

A: ...is an adult
B: ...is a Bozonian
C: ...is a criminal

You will notice that every sentence above involves at least one of the following predicate combinations.

AB: adult Bozonian (Bozonian adult)
AC: adult criminal (criminal adult)
BC: Bozonian criminal (criminal Bozonian)

In these particular cases, the predicates combine in the simplest manner possible – i.e., *conjunctively*. In other words, the following are equivalences for the complex predicates.

x is an Adult Bozonian	::	x is an Adult and x is a Bozonian
x is an Adult Criminal	::	x is an Adult and x is a Criminal
x is a Bozonian Criminal	::	x is a Bozonian and x is a Criminal

The above predicates combine *conjunctively*; this is *not* a universal feature of English, as evidenced by the following examples.

x is an alleged criminal
x is a putative solution
x is imitation leather
x is an expectant mother
x is an experienced sailor; x is an experienced hunter
x is a large whale; x is a small whale
x is a large shrimp; x is a small shrimp
x is a deer hunter; x is a shrimp fisherman

For example, an alleged criminal is not a criminal who is alleged; indeed an alleged criminal need not be a criminal at all. Similarly, an expectant mother need not be a mother at all.

By contrast, an experienced sailor is a sailor, but not a sailor who is generally experienced. Similarly, an experienced hunter is a hunter, but not a hunter who is generally experienced. In each case, the person is not experienced in general, but rather is experienced *at* a particular thing (sailing, hunting).

Along the same lines, a large whale is a whale, and a large shrimp is a shrimp, but neither is generally large; neither is nearly as large as a small ocean, let alone a small planet, or a small galaxy.

Finally a deer hunter is not a deer who hunts, but someone or something that hunts deer, and a shrimp fisherman is not a shrimp who fishes but someone who fishes for shrimp.

I am sure that the reader can come up with numerous other examples of predicates that don't combine conjunctively.

Sometimes, a predicate combination is ambiguous between a conjunctive and a non-conjunctive reading. The following is an example.

x is a Bostonian Cabdriver

This has a conjunctive reading.

x is a Bostonian who drives a cab
(perhaps in Boston, perhaps elsewhere)

But it also has a non-conjunctive reading.

x is a person who drives a cab in Boston
(who lives perhaps in Boston, perhaps elsewhere)

Another example, which seems to engender confusion is the following.

x is a male chauvinist

This has a conjunctive reading,

x is a male and x is a chauvinist,

which means

x is a male who is excessively (and blindly) patriotic (loyal).

However, this is not what is usually meant by the phrase 'male chauvinist'. As originally intended by the author of this phrase, a male chauvinist need not be male, and a male chauvinist need not be a chauvinist. Rather, a male chauvinist is a person (male or female) who is excessively (and blindly) loyal *in respect to* the alleged superiority of men to women.

It is important to realize that many predicates don't combine conjunctively. Nonetheless, we are going to concentrate exclusively on ones that do, for the sake of simplicity. When there are two readings of a predicate combination, we will opt for the conjunctive reading, and ignore the non-conjunctive reading.

Now, let's go back to the original problem of paraphrasing the various sentences concerning adults, Bozonians, and criminals. We do two examples from each group, in each case by presenting a paraphrase chain.

(e1) every Adult Bozonian is a Criminal

every AB is C

everything is such that

if it is AB, then it is C

everything is such that
if it is A and it is B, then it is C

$\forall x([Ax \,\&\, Bx] \rightarrow Cx)$

(e4) every Adult is a Bozonian Criminal

every A is BC

everything is such that
if it is A, then it is BC

everything is such that
if it is A, then it is B and it is C

$\forall x(Ax \rightarrow [Bx \,\&\, Cx])$

(s3) some Criminal Bozonian is an Adult

some CB is A

there is at least one thing such that
it is CB, and it is A

there is at least one thing such that
it is C and it is B, and it is A

$\exists x([Cx \,\&\, Bx] \,\&\, Ax)$

(s5) some Bozonian is an Adult Criminal

some B is AC

there is at least one thing such that
    it is B, and it is AC

there is at least one thing such that
    it is B, and it is A and it is C

∃x(Bx & [Ax & Cx]).

(n3) no Criminal Bozonian is an Adult

no CB is A

it is not true that some CB is A

it is not true that
    there is at least one thing such that
        it is CB, and it is A

it is not true that
    there is at least one thing such that
        it is C and it is B, and it is A

~∃x([Cx & Bx] & Ax).

(n6) no Criminal is an Adult Bozonian

no C is AB

it is not true that some C is AB

it is not true that
    there is at least one thing such that
        it is C, and it is AB

it is not true that
    there is at least one thing such that
        it is C, and it is A and it is B

~∃x(Cx & [Ax & Bx]).

The reader is invited to symbolize the remaining sentences from the above groups.

We can further complicate matters by adding an additional predicate letter (say) 'D', which symbolizes (say) '___ is deranged'. Consider the following two examples.

(e1) every Deranged Adult is a Criminal Bozonian
(e2) no Adult Bozonian is a Deranged Criminal

The symbolizations go as follows.

(s1) ∀x([Dx & Ax] → [Cx & Bx])
(s2) ~∃x([Ax & Bx] & [Dx & Cx])

Another possible complication concerns internal negations in the sentences. The following are examples, together with their step-wise paraphrases.

(1) every Adult who is not Bozonian is a Criminal

every A who is not B is C

everything is such that
  if it is an A who is not B,
    then it is C

everything is such that
  if it is A and it is not B,
    then it is C

∀x([Ax & ~Bx] → Cx)

(2) some Adult Bozonian is not a Criminal

some AB is not C

there is at least one thing such that
  it is AB, and it is not C

there is at least one thing such that
  it is A and it is B, and it is not C

∃x([Ax & Bx] & ~Cx)

(3) some Bozonian is an Adult who is not a Criminal

some B is an A who is not a C

there is at least one thing such that
    it is B, and it is an A who is not C

there is at least one thing such that
    it is B, and it is A and it is not C

∃x(Bx & [Ax & ~Cx])

(4) no Adult who is not a Bozonian is a Criminal

no A who is not B is C

it is not true that
    some A who is not B is C

it is not true that
    there is at least one thing such that
        it is an A who is not B, and it is C

it is not true that
    there is at least one thing such that
        it is A and it is not B, and it is C

~∃x([Ax & ~Bx] & Cx)

14. SUMMARY OF BASIC TRANSLATION PATTERNS FROM SECTIONS 12 AND 13

Forms With Only Two Predicates

(1)	everything is both A and B	$\forall x(Ax \& Bx)$
(2)	everything is A but not B	$\forall x(Ax \& \sim Bx)$
(3)	everything is either A or B	$\forall x(Ax \vee Bx)$

(1)	something is both A and B	$\exists x(Ax \& Bx)$
(2)	something is A but not B	$\exists x(Ax \& \sim Bx)$
(3)	something is either A or B	$\exists x(Ax \vee Bx)$

(1)	nothing is both A and B	$\sim\exists x(Ax \& Bx)$
(2)	nothing is A but not B	$\sim\exists x(Ax \& \sim Bx)$
(3)	nothing is either A or B	$\sim\exists x(Ax \vee Bx)$

Simple Conjunctive Combinations

(1)	every AB is C	$\forall x([Ax \& Bx] \rightarrow Cx)$
(2)	some AB is C	$\exists x([Ax \& Bx] \& Cx)$
(3)	some AB is not C	$\exists x([Ax \& Bx] \& \sim Cx)$
(4)	no AB is C	$\sim\exists x([Ax \& Bx] \& Cx)$

(5)	every A is BC	$\forall x(Ax \rightarrow [Bx \& Cx])$
(6)	some A is BC	$\exists x(Ax \& [Bx \& Cx])$
(7)	some A is not BC	$\exists x(Ax \& \sim[Bx \& Cx])$
(8)	no A is BC	$\sim\exists x(Ax \& [Bx \& Cx])$

Conjunctive Combinations Involving Negations

(1)	every A that is not B is C	$\forall x([Ax \& \sim Bx] \rightarrow Cx)$
(2)	some A that is not B is C	$\exists x([Ax \& \sim Bx] \& Cx)$
(3)	some A that is not B is not C	$\exists x([Ax \& \sim Bx] \& \sim Cx)$
(4)	no A that is not B is C	$\sim\exists x([Ax \& \sim Bx] \& Cx)$

(5)	every A is B but not C	$\forall x(Ax \rightarrow [Bx \& \sim Cx])$
(6)	some A is B but not C	$\exists x(Ax \& [Bx \& \sim Cx])$
(7)	no A is B but not C	$\sim\exists x(Ax \& [Bx \& \sim Cx])$

15. 'ONLY'

The standard quantifiers of predicate logic are 'every' and 'at least one'. We have already seen how to paraphrase various non-standard quantifiers into standard form. In particular, we paraphrase 'all' as 'every', 'some' as 'at least one', and 'no' as 'not at least one'.

In the present section, we examine another non-standard quantifier, 'only'; in particular, we show how it can be paraphrased using the standard quantifiers. In a later section, we examine a subtle variant – 'the only'. But for the moment let us concentrate on 'only' by itself.

The basic quantificational form for 'only' is:

only $\mathbb{A}$ are $\mathbb{B}$.

Examples include:

(1) only Men are NFL football players
(2) only Citizens are Voters

Occasionally, signs use 'only' as in:

employees only
members only
passenger cars only

These can often be paraphrased as follows.

(3) only Employees are Allowed
(4) only Members are Allowed
(5) only Passenger cars are Allowed

What is, in fact, allowed (or disallowed) depends on the context. Generally, signs employing 'only' are intended to *exclude* certain things, specifically things that *fail* to have a certain property (being an employee, being a member, being a passenger car, etc.).

Before dealing with the quantifier 'only', let us recall a similar expression in sentential logic – namely, 'only if'. In particular, recall that

A only if B

may be paraphrased as

not A if *not* B,

which in standard form is written

if not B, then not A $[\sim B \rightarrow \sim A]$

In other words, 'only' modifies 'if' by introducing two negations. The word 'if' always introduces the antecedent, and the word 'only' modifies 'if' by adding two negations in the appropriate places.

When combined with the connective 'if', the word 'only' behaves as a special sort of double-negative modifier. When 'only' acts as a quantifier, it behaves in a similar, double-negative, manner. Recall the signs involving 'only'; they are intended to *exclude* persons who *fail* to have a certain property.

Indeed, we can paraphrase 'only $\mathcal{A}$ are $\mathcal{B}$' in at least two very different ways involving double-negatives.

First, we can paraphrase 'only $\mathcal{A}$ are $\mathcal{B}$' using the negative quantifier 'no', as follows

(o) only $\mathbb{A}$ are $\mathbb{B}$

(p) *no non* $\mathbb{A}$ are $\mathbb{B}$

Strictly speaking, 'non' is not an English word, but simply a prefix; properly speaking, we should write the following.

(p*) *no non*-$\mathbb{A}$ are $\mathbb{B}$

However, the hyphen will generally be dropped, simply to avoid clutter in our intermediate symbolizations.

Thus, the following is the "skeletal" paraphrase:

only = no non [only = no non-]

However, in various colloquial examples, the following more "meaty" paraphrase is more suitable.

(p) *no one* who is *not* $\mathbb{A}$ is $\mathbb{B}$

So, for example, (1)-(4) may be paraphrased as follows.

(p1) no one who isn't a Man is an NFL football player
(p2) no one who isn't a Citizen is a Voter
(p3) no one who isn't an Employee is Allowed
(p4) no one who isn't a Member is Allowed

Next, we turn to symbolization. First the general form is:

(o) only $\mathbb{A}$ are $\mathbb{B}$,

which is paraphrased:

(p) no non $\mathbb{A}$ are $\mathbb{B}$ [no non $\mathbb{A}$ is $\mathbb{B}$]

This is symbolized as follows.

(s) $\sim\exists x(\sim\mathbb{A}x \;\&\; \mathbb{B}x)$

Similarly, (o1)-(o5) are symbolized as follows.

(s1) $\sim\exists x(\sim Mx \;\&\; Nx)$
(s2) $\sim\exists x(\sim Cx \;\&\; Vx)$
(s3) $\sim\exists x(\sim Ex \;\&\; Ax)$
(s4) $\sim\exists x(\sim Mx \;\&\; Ax)$
(s5) $\sim\exists x(\sim Px \;\&\; Ax)$

The quickest way to paraphrase 'only' is using the equivalence

ONLY = NO NON

An alternative paraphrase technique uses 'all'/'every' plus two occurrences of 'non'/'not', as follows

(o) only $\mathbb{A}$ are $\mathbb{B}$
(p1) all non $\mathbb{A}$ are non $\mathbb{B}$
(p2) every non $\mathbb{A}$ is non $\mathbb{B}$
(p3) everyone who is *not* $\mathbb{A}$ is *not* $\mathbb{B}$

These are symbolized as follows.

(s) $\forall x(\sim\mathbb{A}x \to \sim\mathbb{B}x)$

So, for example, (1)-(5) may be paraphrased as follows.

(p1) everyone who isn't a Man isn't an NFL football player
(p2) everyone who isn't a Citizen isn't a Voter
(p3) everyone who isn't an Employee isn't Allowed
(p4) everyone who isn't a Member isn't Allowed
(p5) everyone who isn't (driving) a Passenger car isn't Allowed

These in turn are symbolized as follows.

(s1) $\forall x(\sim Mx \to \sim Nx)$
(s2) $\forall x(\sim Cx \to \sim Vx)$
(s3) $\forall x(\sim Ex \to \sim Ax)$
(s4) $\forall x(\sim Mx \to \sim Ax)$
(s5) $\forall x(\sim Px \to \sim Ax)$

The two approaches above are equivalent, since the following is an equivalence of predicate logic.

$$\sim\exists x(\sim\mathbb{A}x \;\&\; \mathbb{B}x) \;::\; \forall x(\sim\mathbb{A}x \to \sim\mathbb{B}x)$$

To see this equivalence, first recall the following quantificational equivalence:

$$\sim\exists x\mathcal{F} \;::\; \forall x\sim\mathcal{F}$$

And recall the following sentential equivalence:

$$\sim(\sim\mathcal{A} \,\&\, \mathcal{B}) \;::\; \sim\mathcal{A} \rightarrow \sim\mathcal{B}$$

Accordingly,

$$\sim\exists x(\sim\mathbb{A}x \,\&\, \mathbb{B}x) \;::\; \forall x\sim(\sim\mathbb{A}x \,\&\, \mathbb{B}x)$$

And

$$\sim(\sim\mathbb{A}x \,\&\, \mathbb{B}x) \;::\; (\sim\mathbb{A}x \rightarrow \sim\mathbb{B}x)$$

So

$$\sim\exists x(\sim\mathbb{A}x \,\&\, \mathbb{B}x) \;::\; \forall x(\sim\mathbb{A}x \rightarrow \sim\mathbb{B}x)$$

There is still another sentential equivalence:

$$\sim\mathcal{A} \rightarrow \sim\mathcal{B} \;::\; \mathcal{B} \rightarrow \mathcal{A}$$

So

$$\sim\mathbb{B}x \rightarrow \sim\mathbb{A}x \;::\; (\mathbb{B}x \rightarrow \mathbb{A}x)$$

So

$$\sim\exists x(\sim\mathbb{A}x \,\&\, \mathbb{B}x) \;::\; \forall x(\mathbb{B}x \rightarrow \mathbb{A}x)$$

This equivalence enables us to provide yet another paraphrase and symbolization of 'only $\mathbb{A}$ are $\mathbb{B}$', as follows.

(o) only $\mathbb{A}$ are $\mathcal{B}$
(p) all $\mathbb{B}$ are $\mathbb{A}$
(s) $\forall x(\mathbb{B}x \rightarrow \mathbb{A}x)$

The latter symbolization is admitted in intro logic, just as $P\rightarrow Q$ is admitted as a symbolization of 'P only if Q', in addition to the official $\sim Q\rightarrow\sim P$. The problem is that the non-negative construals of 'only' statements sound funny (even wrong, to some people) in English.

In short, our official paraphrase/symbolization goes as follows.

(o) only $\mathbb{A}$ are $\mathbb{B}$

(p1) *no non* $\mathbb{A}$ is $\mathbb{B}$
(s1) $\sim\exists x(\sim\mathbb{A}x \,\&\, \mathbb{B}x)$

(p2) every *non* $\mathbb{A}$ is *non* $\mathbb{B}$
(s2) $\forall x(\sim\mathbb{A}x \rightarrow \sim\mathbb{B}x)$

Note carefully, however, for the sake of having a single form, the *former* paraphrase/symbolization will be used *exclusively* in the answers to the exercises.

16. AMBIGUITIES INVOLVING 'ONLY'

Having discussed the basic 'only' statement forms, we now move to examples involving more than two predicates. As it turns out, adding a third predicate can complicate matters.

Consider the following example.

(e1) only Poisonous Snakes are Dangerous

Let us assume that 'poisonous' combines conjunctively, so that a poisonous snake is simply a snake that is poisonous, even though a poisonous snake is quite different from a poisonous mushroom (a mushroom's bite is not very deadly!) Granting this simplifying assumption, we have the following paraphrase.

x is a Poisonous Snake :: x is Poisonous and x is a Snake

Now, if we follow the pattern of paraphrase suggested in the previous section, we obtain the following paraphrase.

(p1) no non Poisonous Snakes are Dangerous
no non Poisonous Snake is Dangerous

Unfortunately, the scope of 'non' is ambiguous. For the sentence

(1) x is a non poisonous snake

has two different readings, and hence two different symbolizations.

(r1) x is a non-poisonous snake
(r2) x is a non(poisonous snake)
x is not a poisonous snake

(s1) $\sim Px \& Sx$
(s2) $\sim(Px \& Sx)$

On one reading, to be a non poisonous snake is to be a snake that is not poisonous. On the other reading, to be a non poisonous snake is simply to be *anything but* a poisonous snake.

Our original sentence, and its paraphrase,

only poisonous snakes are dangerous
no non poisonous snakes are dangerous

are correspondingly ambiguous between the following readings.

$\sim\exists x(\sim[Px \& Sx] \& Dx)$

there is no thing x such that
x is not a Poisonous Snake,
but x is Dangerous

$\sim\exists x([\sim Px \& Sx] \& Dx)$

there is no thing x such that
x is a nonPoisonous Snake,
but x is Dangerous

To see that the original sentence really is ambiguous, consider the following four (very short) paragraphs.

(1) Few snakes are dangerous. In fact,
only poisonous snakes are dangerous.

(2) Few reptiles are dangerous. In fact,
only poisonous snakes are dangerous.

(3) Few animals are dangerous. In fact,
only poisonous snakes are dangerous.

(4) Few things are dangerous. In fact,
only poisonous snakes are dangerous.

In each paragraph, when we get to the second sentence, it is clear what the topic is – snakes, reptiles, animals, or things in general. What the topic is helps to determine the meaning of the second sentence.

For example, in the first paragraph, by the time we get to the second sentence, it is clear that we are talking *exclusively* about snakes, and not things in general. In particular, the sentence does not say whether there are any dangerous tigers, or dangerous mushrooms.

By contrast, in the fourth paragraph, the first sentence makes it clear that we are talking about things in general, so the second sentence is intended to exclude from the class of dangerous things anything that is not a poisonous snake.

An alternative method of clarifying the topic of the sentence is to rewrite the four sentences as follows.

(1) only Poisonous Snakes are Dangerous snakes
(2) only Poisonous Snakes are Dangerous reptiles
(3) only Poisonous Snakes are Dangerous animals
(4) only Poisonous Snakes are Dangerous things

These may be straightforwardly paraphrased and symbolized as follows.

(0) only A are B
no non A is B
$\sim\exists x(\sim Ax \& Bx)$

(1) only PS are DS
no non(PS) is DS
$\sim\exists x(\sim[Px \& Sx] \& [Dx \& Sx])$

(2) only PS are DR
no non(PS) is DR
$\sim\exists x(\sim[Px \& Sx] \& [Dx \& Rx])$

(3) only PS are DA
no non(PS) is DA
$\sim\exists x(\sim[Px \& Sx] \& [Dx \& Ax])$

(4) only PS are D
no non(PS) is D
$\sim\exists x(\sim[Px \& Sx] \& Dx)$

If we prefer to use the 'every' paraphrase of 'only', then the paraphrase and symbolization goes as follows.

(0) only A are B
every non A is non B
$\forall x(\sim Ax \rightarrow \sim Bx)$

(1) only PS are DS
every non(PS) is non(DS)
$\forall x(\sim[Px \& Sx] \rightarrow \sim[Dx \& Sx])$

(2) only PS are DR
every non(PS) is non(DR)
$\forall x(\sim[Px \& Sx] \rightarrow \sim[Dx \& Rx])$

(3) only PS arc DA
every non(PS) is non(DA)
$\forall x(\sim[Px \& Sx] \rightarrow \sim[Dx \& Ax])$

(4) only PS are D
every non(PS) is non-D
$\forall x(\sim[Px \& Sx] \rightarrow \sim Dx)$

17. 'THE ONLY'

The subtleties of 'only' are further complicated by combining it with the word 'the' to produce 'the only'.

[Still more complications arise when 'the' is combined with 'only' ('all') to produce 'only the' ('all the'); however, we are only going to deal with 'the only'.]

The nice thing about 'the only' is that it enables us to make 'only' statements without the kind of ambiguity seen in the previous section. Recall that only poisonous snakes are dangerous

is ambiguous between any of the following (among others):

only poisonous snakes are dangerous snakes
only poisonous snakes are dangerous reptiles
only poisonous snakes are dangerous animals
only poisonous snakes are dangerous things

These four propositions can also be expressed using 'the only', as follows.

(1) *the only* dangerous **snakes** are poisonous **snakes**
or: poisonous **snakes** are *the only* dangerous **snakes**

(2) *the only* dangerous **reptiles** are poisonous **snakes**
or: poisonous **snakes** are *the only* dangerous **reptiles**

(3) *the only* dangerous **animals** are poisonous **snakes**
or: poisonous **snakes** are *the only* dangerous **animals**

(4) *the only* dangerous **things** are poisonous **snakes**
or: poisonous **snakes** are *the only* dangerous **things**

The general form of these is:

the only AB are CD
or: CD are the only AB

Here, 'AB' and 'BC' are conjunctively-combined predicates.

Certain simplifications occasionally occur. For example, B and D may be the same predicate, or B may be the vacuous predicate 'is a thing' (which is never explicitly symbolized, since everything is a thing!).

The paraphrase and symbolization of 'the only' statements follows a pattern similar to the paraphrase and symbolization of 'only' statements. In particular, the paraphrase utilizes both 'no' and 'not'. However, the details are importantly different.

Recall that

only A are B

is paraphrased:

no non A are B

Statements involving 'the only' are similarly paraphrased; specifically,

the only AB are CD
CD are the only AB

are paraphrased:

no AB are not CD

So, for example, we have the following paraphrases and symbolizations of (1)-(4).

(1) the only dangerous snakes are poisonous snakes
no dangerous snakes are not poisonous snakes
no DS are not PS
~∃x([Dx & Sx] & ~[Px & Sx])
or: the only dangerous snakes are poisonous
no dangerous snakes are not poisonous
no DS are not P
~∃x([Dx & Sx] & ~Px)

(2) the only dangerous reptiles are poisonous snakes
no dangerous reptiles are not poisonous snakes
no DR are not PS
~∃x([Dx & Rx] & ~[Px & Sx])

(3) the only dangerous animals are poisonous snakes
no dangerous animals are not poisonous snakes
no DS are not PS
~∃x([Dx & Ax] & ~[Px & Sx])

(4) the only dangerous things are poisonous snakes
no dangerous things are not poisonous snakes
no D are not PS
~∃x(Dx & ~[Px & Sx])

Two features of the above should be noted, about (1) and (4). Both involve situations in which only three predicates are involved. In (1), the predicate 'is a snake' is repeated, and is equivalent to the sentence in which the second occurrence is simply dropped. In particular,

the only AB are CB

is equivalent to

the only AB are C,

which is paraphrased and symbolized:

no AB are not C
~∃x([Ax & Bx] & ~Cx)

In (4), the predicate 'is a thing' is vacuous; hence, it is not symbolized. In particular,

the only A things are CD

is equivalent to

the only A are CD,

which is paraphrased and symbolized:

no A are not CD
$\sim\exists x(Ax \,\&\, \sim[Cx \,\&\, Dx])$.

Note: Students who seek the *shortest symbolization* of a given statement may wish to consider the following equivalent symbolization. Recall that

no A are not B $\quad \sim\exists x(Ax \,\&\, \sim Bx)$

is equivalent to

every A is B $\quad \forall x(Ax \rightarrow Bx)$

Accordingly,

the only AB are CD,

which is paraphrased:

no AB are not CD $\quad \sim\exists x([Ax \,\&\, Bx] \,\&\, \sim[Cx \,\&\, Dx])$

may also be paraphrased:

every AB is CD $\forall x([Ax \,\&\, Bx] \rightarrow [Cx \,\&\, Dx])$

Both symbolizations count as correct symbolizations; however, only the double-negative symbolizations will be given in the answers to the exercises.

18. DISJUNCTIVE COMBINATIONS OF PREDICATES

In Section 13, we examined many conjunctive predicate combinations, ones that may be symbolized by conjunctions. The curious thing about the logical structure of English is that often the word 'and', our archetypical word for conjunction, is used in a manner that does not allow it to be mechanically translated as a conjunction.

Consider the following two examples.

all Cats and Dogs are Suitable pets
only Cats and Dogs are Suitable pets

First, notice that 'suitable' does not combine conjunctively; for example, a suitable pet is (usually!) quite different from a suitable meal. We must accordingly treat the predicate combination 'suitable pet' as simple: 'Sx' stands for 'x is a suitable pet'.

Let us concentrate on the first one for a moment. As a first attempt at translation, let us consider the following.

$\forall x([Cx \,\&\, Dx] \rightarrow Sx)$ WRONG!!!

What is wrong with this translation? Well, translating it back into English, piece by piece, yields the following:

for any thing x,
if x is a cat
and x is a dog,
then x is a suitable pet

in other words,

for any thing x,
if x is both a cat and a dog,
then x is a suitable pet

This is surely true, but only because nothing is both a cat and a dog! By contrast, the original sentence is false, since cats and dogs do not *all* make suitable pets; many are not house-trained, many have rabies, etc.

The above translation is quite amusing, but nevertheless wrong. What is the correct translation? In particular, how does the word 'and' operate in the above sentence? One possible way to interpret 'and' as a genuine conjunction is to transform the original sentence into the following equivalent sentence.

all Cats are Suitable pets,
and
all Dogs are Suitable pets

This sentence is a conjunction, which is symbolized as follows.

$$\forall x(Cx \rightarrow Sx) \ \&\ \forall x(Dx \rightarrow Sx)$$

This formula involves two quantifiers; multiply-quantified formulas are the topic of a later section (Section 19). On the other hand, this formula is logically equivalent to the following singly-quantified formula.

$$\forall x([Cx \vee Dx] \rightarrow Sx),$$

which reads:

for any thing x:
if x is a cat
or x is a dog,
then x is a suitable pet

Thus, in some sense, to be explained shortly, the word 'and' is translated as a disjunction in this sentence.

In order to more fully understand what is going on, let us consider the second example.

only Cats and Dogs are Suitable pets

First, let us apply our earlier technique, transforming this sentence into the corresponding conjunction.

only Cats are Suitable pets, **and**
only Dogs are Suitable pets

As you can see, the simple transformation technique has failed, since the latter sentence is certainly not equivalent to the original. For, unlike the original sentence, the latter implies that any suitable pet is both a cat and a dog!

O.K., the first technique doesn't work. What about the second technique, which involves symbolizing the sentence using disjunction rather than conjunction? Let's see if this surprise attack will also work on the second example.

First, the overall form is:

only $\mathcal{A}$ are S,

where '$\mathcal{A}$' stands for 'Cats and Dogs'.

Its overall symbolization is therefore (using the ∀-version on 'only'):

$$\forall x(\sim\mathcal{A}x \rightarrow \sim Sx)$$

Next, we propose the following disjunctive analysis of the pseudo-atomic formula 'Ax':

$$\mathcal{A}x \ :: \ [Cx \vee Dx]$$

Thus, the final proposed symbolization is:

$$\forall x(\sim[Cx \vee Dx] \rightarrow \sim Sx).$$

Recalling that the negation of 'either-or' is 'neither-nor', this formula reads:

for any thing x:
    if x is neither a Cat nor a Dog,
        then x is not a Suitable pet

This is equivalent to:

for any thing x:
    if x is a Suitable pet,
        then x is either a Cat or a Dog

This seems to be a suitable translation of the original sentence.

The disjunction-approach seems to work. But how can one logically say that sometimes 'and' is translated as disjunction, when usually it is translated as conjunction? This does not make sense, unless we can tell when 'and' is conjunction, and when 'and' is disjunction.

As usual in natural language, the underlying logico-grammatical laws/rules are incredibly complex. But let us see if we can make a small amount of sense out of 'and'.

The key may lie in the distinction between singular and plural terms. Whereas predicate logic uses singular terms exclusively, natural English uses plural terms just as frequently as singular terms. The problem is in translating from plural-talk to singular-talk.

For example, the expressions,

cats, dogs, cats and dogs, suitable pets

are all plural terms; each one refers to a *class* or *set*. Let us name these classes as follows.

C: the class of all cats
D: the class of all dogs
E: the class of all cats and dogs
S: the class of all suitable pets

Now, let us consider the associated sentences. First, the sentences

all cats are suitable pets
all dogs are suitable pets
all cats and dogs are suitable pets

may be understood as asserting the following, respectively.

every member of class C (i.e., cats)
is also a member of class S (i.e., suitable pets)

every member of class D (i.e., dogs)
is also a member of class S (i.e., suitable pets)

every member of class E (i.e., cats and dogs)
is also a member of class S (i.e., suitable pets)

The notion of membership in a class is fairly straightforward in most cases. In particular, we have the following equivalences.

x is a member of C	::	x is a cat	::	Cx
x is a member of D	::	x is a dog	::	Dx
x is a member of S	::	x is a suitable pet	::	Sx

But the key equivalence concerns the class E, cats-and-dogs, which is given as follows.

x is a member of E :: x is a cat **or** x is a dog :: $[Cx \vee Dx]$

In other words, to say that x is a member of the class cats-and-dogs is to say that x is a cat *or* x is a dog (it surely is not to say that x is both a cat and a dog!). If x is a cat *or* x is a dog, then x is in the class cats-and-dogs; conversely, if x is in the class cats-and-dogs, then x is a cat *or* x is a dog.

So, when we translate the above sentences, using the above equivalences, we obtain:

for any x, if x is a cat,
then x is a suitable pet;
∀x(Cx → Sx)

for any x, if x is a dog,
then x is a suitable pet;
∀x(Dx → Sx)

for any x, if x is a cat *or* x is a dog,
then x is a suitable pet;
∀x([Cx ∨ Dx] → Sx)

Now let's go back and do the example involving 'only'.

only cats and dogs are suitable pets

which may be paraphrased as:

only members of E are members of S,

which is symbolized as:

∀x(~Ex → ~Sx)

But 'Ex' means 'x is a member of the class cats-and-dogs', which means 'x is a cat *or* x is a dog', so we have as our final symbolization:

∀x(~[Cx ∨ Dx] → ~Sx)

Let us try one last example in this section.

the only mammals that are suitable pets are cats and dogs.

Once again, we have the compound-class expression 'cats and dogs'. The overall form is

the only M that are S are E,

which we know can be symbolized in a number of ways, including the following.

~∃x([Mx & Sx] & ~Ex)
∀x([Mx & Sx] → Ex)

But 'Ex' is short for [Cx ∨ Dx], so substituting back in, we obtain:

~∃x([Mx & Sx] & ~[Cx ∨ Dx])
∀x([Mx & Sx] → [Cx ∨ Dx])

which are read as follows.

it is not true that:
  there is something x such that:
    it is a mammal and it is a suitable pet,
      but it is neither a cat nor a dog

for any thing x,
if x is a mammal and x is a suitable pet,
then x is a cat or x is a dog

19. MULTIPLE QUANTIFICATION IN MONADIC PREDICATE LOGIC

So far, we have concentrated on quantified formulas and negations of quantified formulas. A quantified formula is a formula whose principal connective is either a universal or an existential quantifier.

The grammar of predicate logic includes the grammar of sentential logic. In other words, when one has one or more predicate logic formulas, then one can combine them with sentential connectives in order to form more complex formulas. For example, if one has quantified formulas or negated quantified formulas $\mathcal{A}$ and $\mathcal{B}$, then one can combine them using conjunction (&), disjunction (∨), conditional (→), and biconditional (↔).

Consider the following formulas, together with possible English translations.

(1a)	∀xFx	everyone is friendly
(2a)	∃xFx	someone is friendly
(3a)	∃x~Fx	someone is unfriendly
(4a)	~∃xFx	no one is friendly
(5a)	∀x~Fx	everyone is unfriendly
(6a)	~∀xFx	not everyone is friendly
(1b)	∀xHx	everyone is happy
(2b)	∃xHx	someone is happy
(3b)	∃x~Hx	someone is unhappy
(4b)	~∃xHx	no one is happy
(5b)	∀x~Hx	everyone is unhappy
(6b)	~∀xHx	not everyone is happy

We can take any two of the above formulas (sentences) and combine them with any two-place connective. For example, we can combine them with conjunction. The following are a few examples.

(c1)	∀xFx & ∀xHx	everyone is friendly, and everyone is happy
(c2)	∀xFx & ~∀xHx	everyone is friendly, but not everyone is happy
(c3)	∃xHx & ∃x~Hx	someone is happy, and someone is unhappy
(c4)	~∃xFx & ∀xHx	no one is friendly, but everyone is happy

Similarly, we can combine any pair of the above formulas (sentences) with the conditional connective. The following are a few examples.

(c5)	$\forall xFx \rightarrow \forall xHx$	if everyone is friendly, then everyone is happy
(c6)	$\exists xFx \rightarrow \exists xHx$	if someone is friendly, then someone is happy
(c7)	$\sim\exists xFx \rightarrow \forall x{\sim}Hx$	if no one is friendly, then everyone is unhappy
(c8)	$\exists x{\sim}Fx \rightarrow {\sim}\exists xHx$	if someone is unfriendly, then no one is happy

At this point, probably the most important thing to recognize is the novelty of the above formulas. They are unlike any formula we have discussed so far. In particular, each one involves two quantifier expressions, whereas every previous example has involved at most one quantifier.

Let us pursue the difference for a moment. Consider the following pair of formulas.

(u1) $\forall x(Fx \rightarrow Hx)$
(u2) $\forall xFx \rightarrow \forall xHx$

They read as follows.

(r1) everything is such that:
  if it is F, then it is H

  every F is H

(r2) if everything is such that it is F,
  then everything is such that it is H

  if everything is F,
  then everything is H

What is the logical relation between (r1) and (r2)? Well, they are *not* equivalent; although (r1) implies (r2), (r2) does *not* imply (r1).

To see that (r2) does not imply (r1), consider the following counter-example to the argument form.

if everyone is a Freshman, then everyone is happy
therefore, every Freshman is happy

First, this concrete argument has the right form. Furthermore, the conclusion is false. So, what about the premise? This is a conditional; the antecedent is 'everyone is a Freshman'; this is false; the consequent is 'everyone is happy'; this is also false. Therefore, recalling the truth table for arrow (F→F=T), the conditional is true.

Whereas this argument is invalid, its converse is valid, but not sound. Its validity will be demonstrated in a later chapter.

Let us consider another example of the difference between a singly-quantified formula and a similar-looking multiply-quantified formula. Consider the following pair.

(e1) ∃x(Fx & Hx)
(e2) ∃xFx & ∃xHx

The colloquial readings are given as follows.

(c1) something is both F and H [or: some F is H]
(c2) something is F, and something is H

Once again the formulas are not logically equivalent; however, (c1) does imply (c2). For suppose that something is both F and H; then, it is F, and hence something is F; furthermore, it is H, and hence something is H. Hence, something is F, and something is H. [We will examine this style of reasoning in detail in the chapter on derivations in predicate logic.]

So (c1) implies (c2). In order to see that (c2) does not imply (c1), consider the following counterexample.

someone is female, and someone is male
therefore, someone is both male and female

The premise is surely true, but the conclusion is false. Legally, if not biologically, everyone is *exclusively* male *or* female; no one is both male and female.

Having seen the basic theme (namely, combining quantified formulas with sentential connectives), let us now consider the three most basic variations on this theme.

First, one can combine the simple quantified formulas, listed above, using *non-standard* connectives ('unless', 'only if', etc.) Second, one can combine more complex quantified formulas (every A is B, every AB is C, etc.) using standard connectives. Finally, one can combine complex quantified formulas using non-standard connectives.

The following are examples of these three variations

(1a) everyone is happy, only if everyone is friendly
(1b) no one is happy, unless everyone is friendly
(2a) if every student is happy, then every Freshman is happy
(2b) every Freshman is a student, but not every student is a Freshman
(3a) every Freshman is Happy, only if every student is happy
(3b) no Student is happy, unless every student is friendly

Now, in translating English statements like the above, which involve more than one quantifier, and one or more *explicit* statement connectives, the best strategy is the following.

(1) Identify the overall sentential structure; i.e., identify the explicit sentential connectives;
(2) Identify the various (quantified) parts;
(3) Symbolize the overall sentential structure;
(4) Symbolize each (quantified) part;
(5) Substitute the symbolized parts into the overall sentential form.

This is pretty much the same strategy as for sentential symbolizations. The key difference is that, whereas in sentential logic one combines atomic formulas (capital letters), in predicate logic one combines quantified formulas as well.

With this strategy in mind, let us go back to the above examples.

Example 1

(1a) everyone is happy, only if everyone is friendly

The overall form of this sentence is:

$\mathcal{A}$ only if $\mathcal{B}$,

which is symbolized:

$\sim\mathcal{B} \rightarrow \sim\mathcal{A}$

The parts, and their respective symbolizations, are:

$\mathcal{A}$:	everyone is happy	∀xHx
$\mathcal{B}$:	everyone is friendly	∀xFx

So the final symbolization is:

$\sim\forall xFx \rightarrow \sim\forall xHx$

Example 2

(1b) no one is happy, unless everyone is friendly

The overall form is

$\mathcal{A}$ unless $\mathcal{B}$,

which is symbolized:

$\sim\mathcal{B} \rightarrow \mathcal{A}$

The parts, and their respective symbolizations, are:

$\mathcal{A}$:	no one is happy	~∃xHx
$\mathcal{B}$:	everyone is friendly	∀xFx

So the final symbolization is:

$\sim\forall x Fx \rightarrow \sim\exists x Hx$

Example 3

(2a) if every student is happy, then every Freshman is happy

The overall form of this sentence is:

if $\mathcal{A}$, then $\mathcal{B}$,

which is symbolized

$\mathcal{A} \rightarrow \mathcal{B}$

The parts, and their respective symbolizations, are:

$\mathcal{A}$:	every student is happy	$\forall x(Sx \rightarrow Hx)$
$\mathcal{B}$:	every Freshman is happy	$\forall x(Fx \rightarrow Hx)$

So the final symbolization is:

$\forall x(Sx \rightarrow Hx) \rightarrow \forall x(Fx \rightarrow Hx)$

Example 4

(2b) every Freshman is a student, but not every student is a Freshman.

The overall form of this sentence is:

$\mathcal{A}$ but $\mathcal{B}$ (i.e., $\mathcal{A}$ and $\mathcal{B}$),

which is symbolized

$\mathcal{A}$ & $\mathcal{B}$.

The parts, and their respective symbolizations, are:

$\mathcal{A}$:	every Freshman is a student	$\forall x(Fx \rightarrow Sx)$
$\mathcal{B}$:	not every student is a Freshman	$\sim\forall x(Sx \rightarrow Fx)$

So the final symbolization is:

$\forall x(Fx \rightarrow Sx)$ & $\sim\forall x(Sx \rightarrow Fx)$

Example 5

(3a) every Freshman is Happy, only if every student is happy

The overall form of this sentence is:

$\mathcal{A}$ only if $\mathcal{B}$,

which is symbolized

$\sim\mathcal{B} \rightarrow \sim\mathcal{A}$.

The parts, and their respective symbolizations, are:

$\mathcal{A}$:	every Freshman is happy	$\forall x(Fx \rightarrow Hx)$
$\mathcal{B}$:	every student is happy	$\forall x(Sx \rightarrow Hx)$

So the final symbolization is:

$$\sim\forall x(Sx \rightarrow Hx) \rightarrow \sim\forall x(Fx \rightarrow Hx)$$

Example 6

(3b) no Student is happy, unless every student is friendly

The overall form of this sentence is:

$\mathcal{A}$ unless $\mathcal{B}$,

which is symbolized

$$\sim\mathcal{B} \rightarrow \mathcal{A}$$

The parts, and their respective symbolizations, are:

$\mathcal{A}$:	no student is happy	$\sim\exists x(Sx \,\&\, Hx)$
$\mathcal{B}$:	every student is friendly	$\forall x(Sx \rightarrow Fx)$

So the final symbolization is:

$$\sim\forall x(Sx \rightarrow Fx) \rightarrow \sim\exists x(Sx \,\&\, Hx)$$

These are examples of the basic variations on the basic theme. There are also more complicated variations available. But in attacking a sentence that has a combination of several quantifiers and one or more sentential connectives (perhaps non-standard), the strategy is the same as before.

20. 'ANY' AND OTHER WIDE SCOPE QUANTIFIERS

Some quantifier expressions are occasionally used in ways that lead to confusion in symbolization in predicate logic. The troublesome expressions are:

any, anything, anyone, a, some.

Let us consider 'anyone' first. Clearly, this quantifier expression is sometimes equivalent to 'everyone', as seen in the following examples.

(1a)	anyone can fix your car	$\forall xFx$
(1b)	everyone can fix your car	$\forall xFx$
(2a)	if Jones can fix your car, then anyone can (fix your car)	$Fj \rightarrow \forall xFx$

(2b) if Jones can fix your car,
then everyone can (fix your car) $Fj \rightarrow \forall xFx$

Here, 'j' stands for 'Jones', 'F_' stands for '_ can fix your car', and '$\forall$x' stands for 'every *person* x is such that...'

So far, our working hypothesis is that 'anyone' and 'everyone' are completely interchangeable. However, this hypothesis is quickly refuted when we interchange the roles of antecedent and consequent in (2a) and (2b), in which case we obtain the following statements.

(a) if anyone can fix your car, then Jones can (fix your car)

(e) if everyone can fix your car, then Jones can (fix your car)

Clearly, these are not equivalent! Whereas the former sentence could very well be an ad in the yellow pages, bragging about Jones' mechanical abilities, the latter would be a truly stupid ad, since it merely states a logical truth – that Jones can fix your car supposing *everyone* can.

Now, the symbolization of (e) is straightforward, it is a conditional with 'everyone can fix your car' [symbolized: $\forall xFx$] as antecedent and with 'Jones can fix your car' [symbolized: Fj] as consequent. It is accordingly symbolized as follows.

(e') $\forall xFx \rightarrow Fj$

Notice that the main connective is arrow, and not a universal quantifier; in particular, when we read it literally, it goes as follows.

if everyone is F, *then* j is F

But what happens if we get confused and put in parentheses, so that '$\forall$x' is the main connective, and not '$\rightarrow$'? In that case, we obtain the following formula,

$\forall x(Fx \rightarrow Fj)$,

which says something quite different from (e); but what? Well, the main connective is '$\forall$x', so the literal reading goes as follows.

everyone is such that: if he/she is F, then j is F.

Every universal formula is, in effect, a shorthand expression for a (possibly infinite) list of formulas, one formula for every individual in the universe. For example,

$\forall xFx$

is short for the following list:

Fa
Fb
Fc
etc.

And,

$\forall x(Fx \rightarrow Gx)$

is short for the following list:

Fa → Ga
Fb → Gb
Fc → Gc
etc.

So, following this same pattern, the formula in question,

$\forall x(Fx \rightarrow Fj)$

is short for the following list:

Fa → Fj
Fb → Fj
Fc → Fj
etc.

This list says, using the original scheme of abbreviation:

if a can fix your car, then Jones can
if b can fix your car, then Jones can
if c can fix your car, then Jones can
etc.

In other words,

if *anyone* can fix your car, then Jones can

This sentence, of course, is one of our original sentences, which we now see is symbolized in predicate logic as follows.

$\forall x(Fx \rightarrow Fj)$

In other words, although the English sentence *looks like* a conditional with 'anyone can fix your car' as its antecedent, in actuality, the sentence is a universal conditional. Although 'if...then...' appears to be the main connective, in fact 'anyone' is the main connective.

Consider another pair of examples involving 'any' versus 'every'.

(e) Jones does not know *everyone*

(a) Jones does not know *anyone*

As in the earlier case, 'everyone' and 'anyone' are not interchangeable. Whereas (e) is a negation of a universal, (a) is just the opposite, being a universal of a negation. The following are the respective symbolizations in *monadic* predicate logic, followed by their respective readings.

(e') $\sim\forall xKx$

it is not true that
everyone is such that
Jones knows him/her.

(a') $\forall x\sim Kx$

everyone is such that
it is not true that
Jones knows him/her.

Another way to express the latter is:

(a'') Jones knows no one.

Note: 'Kx' stands for 'Jones knows x', or 'x is known by Jones'. This can be further analyzed using a two place predicate '...knows...'; however, this further analysis is unnecessary to make the point about the difference between 'any' and 'every'.

The moral concerning 'any' versus 'every' seems to be this: On the one hand, the apparent grammatical position of 'every' coincides with its true logical position, in a sentence. On the other hand, the apparent grammatical position of 'any' does not coincide with its true logical position in a sentence. In particular, 'any' appears to be deeper inside the sentence than the affiliated sentential connectives, but its actual logical position is at the outside of the sentence. In short:

The scope of 'any' is wide.
The scope of 'every' is narrow.

Now what is worse is that 'any' is not the only wide-scope universal quantifier used in English; there are others, as witnessed by the following examples.

if a skunk enters, then every person will leave
if a skunk enters, then it won't be welcomed
a number is even if and only if it is divisible by 2
if someone were to enter, he/she would be surprised

We will deal with these particular examples shortly. First, let's consider what the problem might be. Clearly, both 'a' and 'some' are occasionally used as existential quantifiers; for example,

a tree grows in Brooklyn,

and

some tree grows in Brooklyn

both mean

> at least one tree grows in Brooklyn,

which may be paraphrased as

> there is at least one thing such that
> it is a tree
> and it grows in Brooklyn,

which is symbolized (in monadic logic, at least) as follows:

> ∃x(Tx & Gx)

But what if I say

> if a tree grows in Brooklyn, then it is sturdy

This is a *much* harder symbolization problem! The problem is how do the quantifier 'a', the pronoun 'it', and the connective 'if-then' interact logically.

Consider an analogous example. which might be clearer.

> if a number is divisible by 2, then it is even.

Here, we are clearly not talking about some *particular* number, which is even if it is divisible by 2; rather, we are talking about every/any number. In particular, this sentence can be paraphrased as

> any number that is divisible by 2 is even,

or

> every number is such that:
> if it is divisible by 2,
> then it is even.

These are symbolized as follows,

> ∀x(Dx → Ex),

where '∀x' means 'every *number* is such that' or 'for any *number*'.

Going back to the Brooklyn tree example, it is symbolized in a parallel manner,

> ∀x(Gx → Sx),

where, in this case, '∀x' means 'every tree is such that' or 'for any tree'.

> every tree is such that:
> if it grows in Brooklyn,
> then it is sturdy

Now let us symbolize the earlier sentences.

if a skunk enters, then every person will leave

$\forall x(Sx \rightarrow [Ex \rightarrow \forall x(Px \rightarrow Lx)])$

if a skunk enters, then it won't be welcomed

$\forall x(Sx \rightarrow [Ex \rightarrow \sim Wx])$

a number is even if and only if it is divisible by 2

$\forall x(Nx \rightarrow [Ex \leftrightarrow Dx])$

if someone were to enter, he/she would be surprised

$\forall x(Ex \rightarrow Sx)$

By way of concluding this section, we observe that in certain *special* circumstances sentences containing wide-scope universal quantifiers ('a', 'any', etc.) can be translated into corresponding sentences containing narrow-scope existential quantifiers.

Let us go back to the example concerning the mechanic Jones.

if anyone can fix your car, then Jones can (fix your car).

One way to look at this is by way of a round-about paraphrase that goes as follows.

if Jones can*not* fix your car, then no one can (fix your car)

This is, just as it appears, a conditional, which is symbolized as follows.

$\sim Fj \rightarrow \sim\exists xFx$

if j is not F, then no one is F

Now, you will recall the following equivalence of sentential logic:

$\sim\mathcal{A} \rightarrow \sim\mathcal{B} \; :: \; \mathcal{B} \rightarrow \mathcal{A}$

Accordingly, the above formula is equivalent to the following formula.

$\exists xFx \rightarrow Fj$

which translates into colloquial English as follows.

if *someone* can fix your car, then Jones can (fix your car).

This is consistent with our original symbolization of the sentence, since the following is an equivalence of predicate logic (as we will be able to demonstrate in a later chapter!)

$\forall x(Fx \rightarrow Fj) \; :: \; \exists xFx \rightarrow Fj$

This is a special case of a more general scheme given as follows.

$\forall x(F[x] \rightarrow \mathcal{B}) \; :: \; \exists xF[x] \rightarrow \mathcal{B}$

Here, F[x] is any formula in which ‘x’ occurs "free", and $\mathcal{B}$ is any formula in which ‘x’ is does not occur "free" (Consult later appendix concerning freedom and bondage of variables.)

Rather than dwell on the general problem, let us consider a few special cases. First, let us do an example contrasting ‘if every...’ and ‘if any...’.

if everyone fails the exam, then everyone will be sad

if anyone fails the exam, then everyone will be sad

Whereas ‘everyone’ is a narrow-scope universal quantifier, ‘anyone’ is a wide-scope universal quantifier, so the symbolizations go as follows.

$\forall xFx \rightarrow \forall xSx$

$\forall x(Fx \rightarrow \forall xSx)$

Remember, the latter is short for the following (possibly infinite) list.

$Fa \rightarrow \forall xSx$	if a fails, then everyone will be sad
$Fb \rightarrow \forall xSx$	if b fails, then everyone will be sad
$Fc \rightarrow \forall xSx$	if c fails, then everyone will be sad
$Fd \rightarrow \forall xSx$	if d fails, then everyone will be sad
etc.	

Now, in the formula

$\forall x(Fx \rightarrow \forall xSx)$,

‘x’ is free in ‘Fx’, but ‘x’ is not free in ‘$\forall$xSx’, so we can apply the above-mentioned equivalence, to obtain:

$\exists xFx \rightarrow \forall xSx$,

which reads

if someone fails, then everyone will be sad

But what about the following:

if anyone fails the exam, he/she will be sad

This is symbolized the same as any ‘if any...’ statement:

$\forall x(Fx \rightarrow Sx)$,

which is short for the following (infinite) list:

$Fa \rightarrow Sa$	if a fails, then a will be sad
$Fb \rightarrow Sb$	if b fails, then b will be sad
$Fc \rightarrow Sc$	if c fails, then c will be sad
etc.	

This is *not* equivalent to a corresponding conditional with a narrow-scope existential quantifier, for example,

$$\exists xFx \rightarrow Sx,$$

which is equivalent to

$$\exists xFx \rightarrow Sy,$$

which reads:

if someone fails, then this (person) will be sad,

where 'this' points at whomever the person speaking chooses.

21. EXERCISES FOR CHAPTER 6

Directions for every exercise set:

Using the suggested abbreviations (the capitalized words), translate each of the following into the language of predicate logic.

EXERCISE SET A

1. JAY is a FRESHMAN.
2. KAY is a JUNIOR.
3. JAY and KAY are STUDENTS.
4. JAY is TALLER than KAY.
5. JAY is not SMARTER than KAY.
6. FRAN INTRODUCED JAY to KAY.
7. FRAN did not INTRODUCE KAY to JAY.
8. CHRIS is TALLER than both JAY and KAY.
9. JAY and KAY are MARRIED (to each other).
10. Both JAY and KAY are MARRIED.
11. Neither JAY nor KAY is MARRIED.
12. Although JAY and KAY are both MARRIED, they are not MARRIED to each other.
13. Neither JAY nor KAY is a SENIOR.
14. If JAY is a SOPHOMORE, then so is KAY.
15. If JAY and KAY LIVE off-campus, then neither of them is a FRESHMAN.
16. If neither JAY nor KAY is a FRESHMAN, then both of them are SOPHOMORES.
17. JAY and KAY are not ROOMMATES unless they are MARRIED.
18. JAY or KAY is the STUDENT body president, but not both.
19. JAY and KAY are FRIENDS if and only if they are ROOMMATES.
20. JAY and KAY are neither SIBLINGS nor COUSINS.

EXERCISE SET B

21. Everything is POSSIBLE.
22. Something is POSSIBLE.
23. Nothing is POSSIBLE.
24. Something is not POSSIBLE.
25. Not everything is POSSIBLE.
26. Everything is imPOSSIBLE.
27. Nothing is imPOSSIBLE.
28. Something is imPOSSIBLE.
29. Not everything is imPOSSIBLE.
30 Not a thing can be CHANGED.
31. Everyone is PERFECT.
32. Someone is PERFECT.
33. No one is PERFECT.
34. Someone is not PERFECT.
35. Not everyone is PERFECT.
36. Everyone is imPERFECT.
37. No one is imPERFECT.
38. Someone is imPERFECT.
39. Not everyone is imPERFECT
40. Not a single person CAME.

EXERCISE SET C

41. Every STUDENT is HAPPY.
42. Some STUDENT is HAPPY.
43. No STUDENT is HAPPY.
44. Some STUDENT is not HAPPY.
45. Not every STUDENT is HAPPY.
46. Every STUDENT is unHAPPY.
47. Some STUDENT is unHAPPY.
48. No STUDENT is unHAPPY.
49. Not every STUDENT is unHAPPY.
50. Not a single STUDENT is HAPPY.
51. All SNAKES HIBERNATE.
52. Some SENATORS are HONEST.
53. No SCOUNDRELS are HONEST.
54. Some SENATORS are not HONEST.
55. Not all SNAKES are HARMFUL.
56. All SKUNKS are unHAPPY.
57. Some SENATORS are unHAPPY.
58. No SCOUNDRELS are unHAPPY.
59. Not all SNAKES are unHAPPY.
60. Not a single SCOUNDREL is HONEST.

EXERCISE SET D

61. No one who is HONEST is a POLITICIAN.
62. No one who isn't COORDINATED is an ATHLETE.
63. Anyone who is ATHLETIC is WELL-ADJUSTED.
64. Everyone who is SENSITIVE is HEALTHY.
65. At least one ATHLETE is not BOORISH.
66. There is at least one POLITICIAN who is HONEST.
67. Everyone who isn't VACATIONING is WORKING.
68. Everything is either MATERIAL or SPIRITUAL.
69. Nothing is both MATERIAL and SPIRITUAL.
70. At least one thing is neither MATERIAL nor SPIRITUAL.

EXERCISE SET E

71. Every CLEVER STUDENT is AMBITIOUS.
72. Every AMBITIOUS STUDENT is CLEVER.
73. Every STUDENT is both CLEVER and AMBITIOUS.
74. Every STUDENT is either CLEVER or not AMBITIOUS.
75. Every STUDENT who is AMBITIOUS is CLEVER.
76. Every STUDENT who is CLEVER is AMBITIOUS.
77. Some CLEVER STUDENTS are AMBITIOUS.
78. Some CLEVER STUDENTS are not AMBITIOUS.
79. Not every CLEVER STUDENT is AMBITIOUS.
80. Not every AMBITIOUS STUDENT is CLEVER.
81. Some AMBITIOUS STUDENTS are not CLEVER.
82. No AMBITIOUS STUDENT is CLEVER.
83. No CLEVER STUDENT is AMBITIOUS.
84. No STUDENT is either CLEVER or AMBITIOUS.
85. No STUDENT is both CLEVER and AMBITIOUS.
86. Every AMBITIOUS PERSON is a CLEVER STUDENT.
87. No AMBITIOUS PERSON is a CLEVER STUDENT.
88. Some AMBITIOUS PERSONS are not CLEVER STUDENTS.
89. Not every AMBITIOUS PERSON is a CLEVER STUDENT.
90. Not all CLEVER PERSONS are STUDENTS.

EXERCISE SET F

91. Only MEMBERS are ALLOWED to enter.
92. Only CITIZENS who are REGISTERED are ALLOWED to vote.
93. The only non-MEMBERS who are ALLOWED inside are GUESTS.
94. DOGS are the only PETS worth having.
95. DOGS are not the only PETS worth having.
96. The only DANGEROUS SNAKES are the ones that are POISONOUS.
97. The only DANGEROUS things are POISONOUS SNAKES.
98. Only POISONOUS SNAKES are DANGEROUS (snakes).
99. Only POISONOUS SNAKES are DANGEROUS ANIMALS.
100. The only FRESHMEN who PASS intro logic are the ones who WORK.

EXERCISE SET G

101. All HORSES and COWS are FARM animals.
102. All CATS and DOGS make EXCELLENT pets.
103. RAINY days and MONDAYS always get me DOWN.
104. CATS and DOGS are the only SUITABLE pets.
105. The only PERSONS INSIDE are MEMBERS and GUESTS.
106. The only CATS and DOGS that are SUITABLE pets are the ones that have been HOUSE-trained.
107. CATS and DOGS are the only ANIMALS that are SUITABLE pets.
108. No CATS or DOGS are SOLD here.
109. No CATS or DOGS are SOLD, that are not VACCINATED.
110. CATS and DOGS that have RABIES are not SUITABLE pets.

EXERCISE SET H

111. If nothing is sPIRITUAL, then nothing is SACRED.
112. If everything is MATERIAL, then nothing is SACRED.
113. Not everything is MATERIAL, provided that something is SACRED.
114. If everything is SACRED, then all COWS are SACRED.
115. If nothing is SACRED, then no COW is SACRED.
116. If all COWS are SACRED, then everything is SACRED.
117. All FRESHMEN are STUDENTS, but not all STUDENTS are FRESHMEN.
118. If every STUDENT is CLEVER, then every FRESHMAN is CLEVER.
119. If every BIRD can FLY, then every BIRD is DANGEROUS.
120. If some SNAKE is not POISONOUS, then not every SNAKE is DANGEROUS.
121. No PROFESSOR is HAPPY, unless some STUDENTS are CLEVER.
122. All COWS are SACRED, only if no COW is BUTCHERED.
123. Some SNAKES are not DANGEROUS, only if some SNAKES are not POISONOUS.
124. If everything is a COW, and every COW is SACRED, then everything is SACRED.
125. If everything is a COW, and no COW is SACRED, then nothing is SACRED.
126. If every BOSTONIAN CAB driver is a MANIAC, then no BOSTONIAN PEDESTRIAN is SAFE.
127. If everyone is FRIENDLY, then everyone is HAPPY.
128. Unless every PROFESSOR is FRIENDLY, no STUDENT is HAPPY.
129. Every STUDENT is HAPPY, only if every PROFESSOR is FRIENDLY.
130. No STUDENT is unHAPPY, unless every PROFESSOR is unFRIENDLY.

EXERCISE SET I

131. If anyone is FRIENDLY, then everyone is HAPPY.
132. If anyone can FIX your car, then SMITH can.
133. If SMITH can't FIX your car, then no one can.
134. If everyone PASSES the exam, then everyone will be HAPPY.
135. If anyone PASSES the exam, then everyone will be HAPPY.
136. If everyone FAILS the exam, then no one will be HAPPY.
137. If anyone FAILS the exam, then no one will be HAPPY.
138. A SKUNK is DANGEROUS if and only if it is RABID.
139. If a CLOWN ENTERS the room, then every PERSON will be SURPRISED.
140. If a CLOWN ENTERS the room, then it will be DISPLEASED if no PERSON is SURPRISED.

22. ANSWERS TO EXERCISES FOR CHAPTER 6

Note: Only one translation is written down in each case; in most cases, there are alternative translations that are equally correct. Your translation is correct if and only if it is equivalent to the answer given below.

EXERCISE SET A

1. Fj
2. Jk
3. Sj & Sk
4. Tjk
5. ~Sjk
6. Ifjk
7. ~Ifkj
8. Tcj & Tck
9. Mjk
10. Mj & Mk
11. ~Mj & ~Mk
12. (Mj & Mk) & ~Mjk
13. ~Sj & ~Sk
14. Sj → Sk
15. (Lj & Lk) → (~Fj & ~Fk)
16. (~Fj & ~Fk) → (Sj & Sk)
17. ~Mjk → ~Rjk
18. (Sj ∨ Sk) & ~(Sj & Sk)
19. Fjk ↔ Rjk
20. ~Sjk & ~Cjk

EXERCISE SET B

21. ∀xPx
22. ∃xPx
23. ~∃xPx
24. ∃x~Px
25. ~∀xPx
26. ∀x~Px
27. ~∃x~Px
28. ∃x~Px
29. ~∀x~Px
30. ~∃xCx
31. ∀xPx
32. ∃xPx
33. ~∃xPx
34. ∃x~Px
35. ~∀xPx
36. ∀x~Px
37. ~∃x~Px
38. ∃x~Px
39. ~∀x~Px
40. ~∃xCx

EXERCISE SET C

41. $\forall x(Sx \rightarrow Hx)$
42. $\exists x(Sx \,\&\, Hx)$
43. $\sim\exists x(Sx \,\&\, Hx)$
44. $\exists x(Sx \,\&\, \sim Hx)$
45. $\sim\forall x(Sx \rightarrow Hx)$
46. $\forall x(Sx \rightarrow \sim Hx)$
47. $\exists x(Sx \,\&\, \sim Hx)$
48. $\sim\exists x(Sx \,\&\, \sim Hx)$
49. $\sim\forall x(Sx \rightarrow \sim Hx)$
50. $\sim\exists x(Sx \,\&\, Hx)$
51. $\forall x(Sx \rightarrow Hx)$
52. $\exists x(Sx \,\&\, Hx)$
53. $\sim\exists x(Sx \,\&\, Hx)$
54. $\exists x(Sx \,\&\, \sim Hx)$
55. $\sim\forall x(Sx \rightarrow Hx)$
56. $\forall x(Sx \rightarrow \sim Hx)$
57. $\exists x(Sx \,\&\, \sim Hx)$
58. $\sim\exists x(Sx \,\&\, \sim Hx)$
59. $\sim\forall x(Sx \rightarrow \sim Hx)$
60. $\sim\exists x(Sx \,\&\, Hx)$

EXERCISE SET D

61. $\sim\exists x(Hx \,\&\, Px)$
62. $\sim\exists x(\sim Cx \,\&\, Ax)$
63. $\forall x(Ax \rightarrow Wx)$
64. $\forall x(Sx \rightarrow Hx)$
65. $\exists x(Ax \,\&\, \sim Bx)$
66. $\exists x(Px \,\&\, Hx)$
67. $\forall x(\sim Vx \rightarrow Wx)$
68. $\forall x(Mx \vee Sx)$
69. $\sim\exists x(Mx \,\&\, Sx)$
70. $\exists x(\sim Mx \,\&\, \sim Sx)$

EXERCISE SET E

71. $\forall x([Cx \,\&\, Sx] \rightarrow Ax)$
72. $\forall x([Ax \,\&\, Sx] \rightarrow Cx)$
73. $\forall x(Sx \rightarrow [Cx \,\&\, Ax])$
74. $\forall x(Sx \rightarrow [Cx \vee \sim Ax])$
75. $\forall x([Sx \,\&\, Ax] \rightarrow Cx)$
76. $\forall x([Sx \,\&\, Cx] \rightarrow Ax)$
77. $\exists x([Cx \,\&\, Sx] \,\&\, Ax)$
78. $\exists x([Cx \,\&\, Sx] \,\&\, \sim Ax)$
79. $\sim\forall x([Cx \,\&\, Sx] \rightarrow Ax)$
80. $\sim\forall x([Ax \,\&\, Sx] \rightarrow Cx)$
81. $\exists x([Ax \,\&\, Sx] \,\&\, \sim Cx)$
82. $\sim\exists x([Ax \,\&\, Sx] \,\&\, Cx)$
83. $\sim\exists x([Cx \,\&\, Sx] \,\&\, Ax)$
84. $\sim\exists x(Sx \,\&\, [Cx \vee Ax])$
85. $\sim\exists x(Sx \,\&\, [Cx \,\&\, Ax])$
86. $\forall x([Ax \,\&\, Px] \rightarrow [Cx \,\&\, Sx])$
87. $\sim\exists x([Ax \,\&\, Px] \,\&\, [Cx \,\&\, Sx])$
88. $\exists x([Ax \,\&\, Px] \,\&\, \sim[Cx \,\&\, Sx])$
89. $\sim\forall x([Ax \,\&\, Px]) \rightarrow [Cx \,\&\, Sx])$
90. $\sim\forall x([Cx \,\&\, Px] \rightarrow Sx)$

EXERCISE SET F

91. $\sim\exists x(\sim Mx \,\&\, Ax)$
92. $\sim\exists x(\sim[Cx \,\&\, Rx] \,\&\, Ax)$
93. $\sim\exists x([\sim Mx \,\&\, Ax] \,\&\, \sim Gx)$
94. $\sim\exists x(Px \,\&\, \sim Dx)$
95. $\exists x(Px \,\&\, \sim Dx)$
96. $\sim\exists x([Dx \,\&\, Sx] \,\&\, \sim Px)$
97. $\sim\exists x(Dx \,\&\, \sim[Px \,\&\, Sx])$
98. $\sim\exists x(\sim[Px \,\&\, Sx] \,\&\, [Dx \,\&\, Sx])$
99. $\sim\exists x(\sim[Px \,\&\, Sx] \,\&\, [Dx \,\&\, Ax])$
100. $\sim\exists x([Fx \,\&\, Px] \,\&\, \sim Wx)$

EXERCISE SET G

101. $\forall x([Hx \vee Cx] \rightarrow Fx)$
102. $\forall x([Cx \vee Dx] \rightarrow Ex)$
103. $\forall x([Rx \vee Mx] \rightarrow Dx)$
104. $\sim\exists x(Sx \,\&\, \sim[Cx \vee Dx])$
105. $\sim\exists x([Px \,\&\, Ix] \,\&\, \sim[Mx \vee Gx])$
106. $\sim\exists x(\{[Cx \vee Dx] \,\&\, Sx\} \,\&\, \sim Hx)$
107. $\sim\exists x([Ax \,\&\, Sx] \,\&\, \sim[Cx \vee Dx])$
108. $\sim\exists x([Cx \vee Dx] \,\&\, Sx)$
109. $\sim\exists x([(Cx \vee Dx) \,\&\, \sim Vx] \,\&\, Sx)$
110. $\forall x([(Cx \vee Dx) \,\&\, Rx] \rightarrow \sim Sx)$

EXERCISE SET H

111. ~∃xPx → ~∃xSx
112. ∀xMx → ~∃xSx
113. ∃xSx → ~∀xMx
114. ∀xSx → ∀x(Cx → Sx)
115. ~∃xSx → ~∃x(Cx & Sx)
116. ∀x(Cx → Sx) → ∀xSx
117. ∀x(Fx → Sx) & ~∀x(Sx → Fx)
118. ∀x(Sx → Cx) → ∀x(Fx → Cx)
119. ∀x(Bx → Fx) → ∀x(Bx → Dx)
120. ∃x(Sx & ~Px) → ~∀x(Sx → Dx)
121. ~∃x(Sx & Cx) → ~∃x(Px & Hx)
122. ∃x(Cx & Bx) → ~∀x(Cx → Sx)
123. ~∃x(Sx & ~Px) → ~∃x(Sx & ~Dx)
124. [∀xCx & ∀x(Cx → Sx)] → ∀xSx
125. [∀xCx & ~∃x(Cx & Sx)] → ~∃xSx
126. ∀x([Bx&Cx] → Mx) → ~∃x([Bx&Px] & Sx)
127. ∀xFx → ∀xHx
128. ~∀x(Px → Fx) → ~∃x(Sx & Hx)
129. ~∀x(Px → Fx) → ~∀x(Sx → Hx)
130. ~∀x(Px → ~Fx) → ~∃x(Sx & ~Hx)

EXERCISE SET I

131. ∀x(Fx → ∀xHx)
132. ∀x(Fx → Fs)
133. ~Fs → ~∃xFx
134. ∀xPx → ∀xHx
135. ∀x(Px → ∀xHx)
136. ∀xFx → ~∃xHx
137. ∀x(Fx → ~∃xHx)
138. ∀x(Sx → [Dx ↔ Rx])
139. ∀x{[Cx & Ex] → ∀x(Px → Sx)}
140. ∀x{[Cx & Ex] → [~∃y(Py & Sy) → Dx]}

7 TRANSLATIONS IN POLYADIC PREDICATE LOGIC

1. INTRODUCTION

Recall that predicate logic can be conveniently divided into monadic predicate logic, on the one hand, and polyadic predicate logic, on the other. Whereas the former deals exclusively with 1-place (monadic) predicates, the latter deals with all predicates (1-place, 2-place, etc.). In the present chapter, we turn to quantification in the context of polyadic predicate logic.

The reason for being interested in polyadic logic is simple: although monadic predicate logic reveals much more logical structure in English sentences than does sentential logic, monadic logic often does not reveal enough logical structure.

Consider the following argument.

(A) Every Freshman is a student
/Anyone who respects every student respects every Freshman

If we symbolize this in monadic logic, we obtain the following.

$\forall x(Fx \rightarrow Sx)$ [every F is S]
/ $\forall x(Kx \rightarrow Lx)$ [every K is L]

The following is the translation scheme:

Fx:	x is a Freshman
Sx:	x is a student
Kx:	x respects every student
Lx:	x respects every Freshman

The trouble with this analysis, which is the best we can do in monadic predicate logic, is that the resulting argument form is invalid. Yet, the original concrete argument is valid. This means that our analysis of the logical form of (A) is inadequate.

In order to provide an adequate analysis, we need to provide a deeper analysis of the formulas,

Kx:	x respects every student
Lx:	x respects every Freshman

These formulas are logically analyzed into the following items:

student:	Sy:	y is a student
Freshman:	Fy:	y is a Freshman
respects	Rxy:	x respects y
every:	$\forall$y:	for any person y

Thus, the formulas are symbolized as follows

Kx: x respects every student

$\forall y(Sy \rightarrow Rxy)$

for any person y:
if y is a student, then x respects y

Lx: x respects every Freshman;

$\forall y(Fy \rightarrow Rxy)$

for any person y:
if y is a Freshman, then x respects y

Thus, the argument form, according to our new analysis is:

$\forall x(Fx \rightarrow Sx)$
$/ \forall x(\forall y(Sy \rightarrow Rxy) \rightarrow \forall y(Fy \rightarrow Rxy))$

This argument form is valid, as we will be able to demonstrate in a later chapter. It is a fairly complex example, so it may not be entirely clear at the moment. Don't worry just yet! The important point right now is to realize that many sentences and arguments have further logical structure whose proper elucidation requires polyadic predicate logic. The example above is fairly complex. In the next section, we start with more basic examples of polyadic quantification.

2. SIMPLE POLYADIC QUANTIFICATION

In the present section, we examine the simplest class of examples of polyadic quantification – those involving an atomic formula constructed from a two-place predicate. First, recall that a two-place predicate is an expression that forms a formula (open or closed) when combined with two singular terms.

For example, consider the two-place predicate '...respects...', abbreviated R. With this predicate, we can form various formulas, including the following.

(1)	Jay respects Kay	Rjk
(2)	he respects Kay	Rxk
(3)	Jay respects her	Rjy
(4)	he respects her	Rxy
(5)	she respects herself	Rxx

The particular pronouns used above are completely arbitrary (any third person singular pronoun will do).

Now, the grammar of predicate logic has the following feature: if we have a formula, we can prefix it with a quantifier, and the resulting expression is also a formula. This merely restates the idea that quantifiers are one-place connectives.

Occasionally, however, quantifying a formula is trivial or pointless; for example,

∀xRjk everyone is such that Jay respects Kay

says exactly the same thing as

Rjk Jay respects Kay

This is an example of trivial (or vacuous) quantification.

In other cases, quantification is significant. For example, beginning with formulas (2)-(5), we can construct the following formulas, which are accompanied by English paraphrases.

(2a)	∀xRxk	everyone respects Kay
(2b)	∃xRxk	someone respects Kay
(3a)	∀yRjy	Jay respects everyone
(3b)	∃yRjy	Jay respects someone
(4a)	∀xRxy	everyone respects her
(4b)	∃xRxy	someone respects her
(4c)	∀yRxy	he respects everyone
(4d)	∃yRxy	he respects someone
(5a)	∀xRxx	everyone respects him(her)self
(5b)	∃xRxx	someone respects him(her)self

Now, (4a)-(4d) have variables that can be further quantified in a significant way. So prefixing (4a)-(5b) yields the following formulas.

(4a1) ∀y∀xRxy
(4a2) ∃y∀xRxy
(4b1) ∀y∃xRxy
(4b2) ∃y∃xRxy
(4c1) ∀x∀yRxy
(4c2) ∃x∀yRxy
(4d1) ∀x∃yRxy
(4d2) ∃x∃yRxy

How do we translate such formulas into English. As it turns out, there is a handy step-by-step procedure for translating formulas (4a1)-(4d2) into colloquial English – supposing that we are discussing people exclusively, and supposing that the predicate is '...respects...' This procedure is given as follows.

Step 1: Look at the first quantifier, and read it as follows:

(a)	universal (∀)	everyone
(b)	existential (∃)	there is someone who

Step 2: Look to see which variable is quantified (is it 'x' or 'y'?), then check where that variable appears in the quantified formula; does it appear in the first (active) position, or does it appear in the second (passive) position? If it appears in the first (active) position, then read the verb in the active voice as 'respects'. If it appears in the second (passive) position, then read the verb in the passive voice as 'is respected by' (passive voice).

(a)	active	respects
(b)	passive	is respected by

Step 3: Look at the second quantifier, and read it as follows:

(a)	universal (∀)	everyone
(b)	existential (∃)	someone or other

Step 4: String together the components obtained in steps (1)-(3) to produce the colloquial English sentence.

With this procedure in mind, let us do a few examples.

Example 1: ∀x∃yRxy

(1) the first quantifier is universal, so we read it as: **everyone**

(2) the variable x appears in the active position, so we read the verb in the active voice: **respects**

(3) the second quantifier is existential, so we read it as: **someone (or other)**

(4) altogether: **everyone respects someone (or other)**

Example 2: $\exists x \forall y Ryx$

(1) the first quantifier is existential,
so we read it as: **there is someone who**

(2) the variable x appears in the passive position,
so we read the verb in the passive voice: **is respected by**

(3) the second quantifier is universal,
so we read it as: **everyone**

(4) altogether: **there is someone who is respected by everyone**

By following the above procedure, we can translate all the above formulas in colloquial English as follows.

(4a1) $\forall y \forall x Rxy$: everyone is respected by everyone

(4a2) $\exists y \forall x Rxy$: there is someone who is respected by everyone

(4b1) $\forall y \exists x Rxy$: everyone is respected by someone or other

(4b2) $\exists y \exists x Rxy$: there is someone who is respected by someone or other

(4c1) $\forall x \forall y Rxy$: everyone respects everyone

(4c2) $\exists x \forall y Rxy$: there is someone who respects everyone

(4d1) $\forall x \exists y Rxy$: everyone respects someone or other

(4d2) $\exists x \exists y Rxy$: there is someone who respects someone or other

Before continuing, it is important to understand the significance of the expression 'or other'. In Example 1, the final translation is

everyone respects someone or other

Dropping 'or other' yields

everyone respects someone.

This is fine so long as we are completely clear what is meant by the last sentence – namely, that everyone respects someone, not necessarily the same person in each case.

A familiar grammatical transformation converts active sentences into passive ones; for example,

Jay respects Kay

can be transformed into

Kay is respected by Jay.

Both are symbolized the same way.

Rjk

If we perform the same grammatical transformation on

everyone respects someone,

we obtain:

someone is respected by everyone,

which *might* be thought to be equivalent to

there is someone who is respected by everyone.

The following lists the various sentences.

(1) everyone respects someone or other
(2) everyone respects someone
(3) someone is respected by everyone
(4) there is someone who is respected by everyone

The problem we face is simple: (1) and (4) are *not equivalent*; although (4) implies (1), (1) does not imply (4).

In order to see this, consider a very small world with only three persons in it: Adam (a), Eve (e), and Cain (c). For the sake of argument, suppose that Cain respects Adam (but not vice versa), Adam respects Eve (but not vice versa), and Eve respects Cain (but not vice versa). Also, suppose that no one respects him(her)self (although the argument does not depend upon this). Thus, we have the following state of affairs.

Rae	Adam respects Eve
Rec	Eve respects Cain
Rca	Cain respects Adam
~Rea	Eve doesn't respect Adam
~Rce	Cain doesn't respect Eve
~Rac	Adam doesn't respect Cain
~Rcc	Cain doesn't respect himself
~Raa	Adam doesn't respect himself
~Ree	Eve doesn't respect herself

Now, to say that everyone respects someone *or other* is to say everyone respects someone, but not necessarily the same person in each case. In particular, it is to say *all* of the following:

Adam respects someone	∃xRax
Eve respects someone	∃xRex
Cain respects someone	∃xRcx

The first is true, since Adam respects Eve; the second is true, since Eve respects Cain; finally, the third is true, since Cain respects Adam. Thus, in the very small world we are imagining, everyone respects someone *or other*, but not necessarily the same person in each case.

They all respect someone, but there is no single person they all respect. To say that there is someone who is respected by everyone is to say that *at least one* of the following is true.

Adam is respected by everyone	∀xRxa
Eve is respected by everyone	∀xRxe
Cain is respected by everyone	∀xRxc

But the first is false, since Eve doesn't respect Adam; the second is false, since Cain doesn't respect Eve, and the third is false, since Adam doesn't respect Cain. Also, in this world, no one respects him(her)self, but that doesn't make any difference. Thus, in this world, it is not true that there is someone who is respected by everyone, although it is true that everyone respects someone or other.

Thus, sentences (1) and (4) are not equivalent. It follows that the following can't all be true:

(1) is equivalent to (2)
(2) is equivalent to (3)
(3) is equivalent to (4)

For then we would have that (1) and (4) are equivalent, which we have just shown is not the case.

The problem is that (2) and (3) are ambiguous. *Usually*, (2) means the same thing as (1), so that the 'or other' is not necessary. But, *sometimes*, (2) means the same thing as (4), so that the 'or other' is definitely necessary to distinguish (1) and (2). It is best to avoid (2) in favor of (1), if that is what is meant. On the other hand, (3) *usually* means the same thing as (4), but occasionally it is equivalent to (1).

In other words, it is best to avoid (2) and (3) altogether, and say either (1) or (4), depending on what is meant.

3. NEGATIONS OF SIMPLE POLYADIC QUANTIFIERS

What happens when we take the formulas considered in Section 2 and introduce a negation (~) at any of the three possible positions? That is what we consider in the present section.

The quantified formulas obtainable from the atomic formulas 'Rxy' and 'Ryx' are the following.

(1)	∀x∀yRxy	∀y∀xRyx	everyone respects everyone
(2)	∀y∀xRxy	∀x∀yRyx	everyone is respected by everyone
(3)	∃x∃yRxy	∃y∃xRyx	someone respects someone
(4)	∃y∃xRxy	∃x∃yRyx	someone is respected by someone
(5)	∀x∃yRxy	∀y∃xRyx	everyone respects someone
(6)	∃y∀xRxy	∃x∀yRyx	someone is respected by everyone
(7)	∀y∃xRxy	∀x∃yRyx	everyone is respected by someone or other
(8)	∃x∀yRxy	∃y∀xRyx	someone respects everyone

Now, at any stage in the construction of these formulas, we could interpolate a negation connective. That gives us not just 8 formulas but 64 distinct formulas (plus alphabetic variants). The basic form is the following.

SIGN..QUANTIFIER..SIGN..QUANTIFIER..SIGN..FORMULA

Each sign is either negative or positive (i.e., negated or not negated); each quantifier is either universal or existential; finally, the formula has the first quantified variable in active or passive position. All told, there are 64 (2×2×2×2×2×2) combinations!

Let us consider two examples.

(e1) ~∀x~∃y~Rxy

In this formula, all the signs are negative, the first quantifier is universal, the second quantifier is existential, the first quantified variable ('x') is in active position.

(e2) ~∃x∀y~Ryx

In this formula the first and third signs are negative, the second sign is positive, the first quantifier is existential, the second quantifier is universal, the first quantified variable ('x') is in passive position.

There are 54 more combinations! We have seen the latter two combinations, not to mention the original eight, which are the combinations in which every sign is positive.

But how does one translate formulas with negations into colloquial English? This is considerably trickier than before. The problem concerns where to place the negation operator in the colloquial sentence. Consider the following sentences.

(1) j dislikes k;
(2) j doesn't like k;
(3) it is not true that j likes k.

The problem is that sentence (2) is actually ambiguous in meaning between the sentence (1) and sentence (3). Furthermore, this is not a harmless ambiguity, since (1) and (3) are *not* equivalent. In particular, the following is not valid in ordinary English.

> it is not true that Jay likes Kay;
> therefore, Jay dislikes Kay.

The premise may be true simply because Jay doesn't even know Kay, so he can't like her. But he doesn't dislike her either, for the same reason – he doesn't know her.

Now, the problem is that, when someone utters the following,

> I don't like spinach,

he or she *usually* means,

> I *dis*like spinach,

although he/she *might* go on to say,

> but I don't dislike spinach, either (since I've never tried it),

Given that ordinary English seldom provides us with simple negations, we need some scheme for expressing them. Toward this end, let us employ the somewhat awkward expression 'fails to...' to construct simple negations. In particular, let us adopt the following translation.

> x fails to Respect y :: not(x Respects y)

With this in mind, let us proceed. Recall that a simple double-quantified formula has the following form.

SIGN..QUANTIFIER..SIGN..QUANTIFIER..SIGN..FORMULA

Let us further parse this construction as follows.

[SIGN-QUANTIFIER]..[SIGN-QUANTIFIER]..[SIGN-FORMULA]

In particular, let us use the word quantifier to refer to the combination sign-quantifier. In this case, there are four quantifiers (plus alphabetic variants):

$\forall x$, $\sim\forall x$, $\exists x$, $\sim\exists x$

We are now, finally, in a position to offer a systematic translation scheme, given as follows.

Step 1: Look at the first quantifier, and read it as follows:

(a)	universal ($\forall$)	everyone
(b)	existential ($\exists$)	there is someone who
(c)	negation universal ($\sim\forall$)	not everyone
(d)	negation existential ($\sim\exists$)	there is no one who

Step 2: Check the quantified formula, and check whether the first quantified variable occurs in the active or passive position, and read the verb as follows:

(a)	positive active	respects
(b)	positive passive	is respected by
(c)	negative active	fails to respect
(d)	negative passive	fails to be respected by

Step 3: Look at the second quantifier, and read it as follows:

(a)	universal (∀)	everyone
(b)	existential (∃)	someone or other
(c)	negation universal (~∀)	not...everyone*
(d)	negation existential (~∃)	no one

*Here, it is understood that 'not' goes in front of the verb phrase.

Step 4: String together the components obtained in steps (1)-(3) to produce the colloquial English sentence.

With this procedure in mind, let us do a few examples.

Example 1: ∃x~∃yRyx

(1) the first quantifier is existential, so we read it as: **there is someone who**

(2) the quantified formula is positive, and the first quantified variable 'x' is in the passive position, so we read the verb as: **is respected by**

(3) the second quantifier is negation-existential, so we read it as: **no one**

(4) altogether: **there is someone who is respected by no one**

Example 2: ~∀x∃y~Rxy

(1) the first quantifier is negation-universal, so we read it as: **not everyone**

(2) the quantified formula is negative, and the first quantified variable 'x' is in the active position, so we read the verb as: **fails to respect**

(3) the second quantifier is existential,
so we read it as: **someone (or other)**

(4) altogether: **not everyone fails to respect someone (or other)**

Example 3: ∃x~∀yRxy

(1) the first quantifier is existential,
so we read it as: **there is someone who**

(2) the quantified formula is positive,
and the first quantified variable 'x'
is in active position, so we read the verb as: **respects**

(4) the second quantifier is negation-universal,
so we read it as: **not...everyone**

(5) altogether: **there is someone who respects not...everyone**

(5*) or, more properly: **there is someone who does not respect everyone**

4. THE UNIVERSE OF DISCOURSE

The reader has probably noticed a small discrepancy in the manner in which the quantifiers are read. On the one hand, the usual readings are the following.

∀x: every*thing* x is such that...
for any*thing* x...

∃x: some*thing* x is such that...
there is at least one *thing* x such that...

On the other hand, in the previous sections in particular, the following readings are used.

∀x: every *person* x is such that...
for any *person* x...

∃x: some *person* x is such that...
there is at least one *person* x such that...

In other words, depending on the specific example, the various quantifiers are read differently. If we are talking *exclusively* about persons, then it is convenient to read '∀x' as 'everyone' and '∃x' as 'someone', rather than the more general 'everything' and 'something'. If, on the other hand, we are talking *exclusively* about *numbers* (as in arithmetic), then it is equally convenient to read '∀x' as 'every number' and '∃x' as 'some number'.

The reason that this is allowed is that, for any symbolic context (formula or argument), we can agree to specify the associated *universe of discourse*. The universe of discourse is, in any given context, the set of all the possible things that the constants and variables refer to.

Thus, depending upon the particular universe of discourse, U, we read the various quantifiers differently.

In symbolizing English sentences, one must first establish exactly what U is. For sake of simplifying our choices, in the exercises, we allow only two possible choices for U, namely:

U = things (in general)

U = persons

In particular, if the sentence uses 'everyone' or 'someone', then the student is allowed to set U=persons, but if the sentence uses 'every **person**' or 'some **person**', then the student must set U=things.

In some cases (but never in the exercises) both 'every(some)one' and 'every(some)thing' appear in the same sentence. In such cases, one must explicitly supply the predicate '...is a person' in order to symbolize the sentence.

Consider the following example.

there is some*one* who hates every*thing*,

which means

there is some *person* who hates every *thing*.

The following is **not** a correct translation.

∃x∀yHxy WRONG!!!

In translating this back into English, we first must specify the reading of the quantifiers, which is to say we must specify the universe of discourse. In the present context at least, there are only two choices; either U=persons or U=things. So the two possible readings are:

there is some *person* who hates every *person*

there is some *thing* that hates every *thing*

Neither of these corresponds to the original sentence. In particular, the following is *not* an admissible reading of the above formula.

there is some *person* who hates every *thing* WRONG!!!

The principle at work here may be stated as follows.

> One cannot change the universe of discourse
> in the middle of a sentence.
>
> All the quantifiers in a sentence
> must have a uniform reading

5. QUANTIFIER SPECIFICATION

So, how do we symbolize

there is some*one* (some *person*) who hates every*thing*.

First, we must choose a universe of discourse that is large enough to encompass everything that we are talking about. In the context of intro logic, if we are talking about anything whatsoever that is not a person, then we must set U=things. In that case, we have to specify which things in the sentence are persons by employing the predicate '...is a person'. The following paraphrase makes significant headway.

there is something such that
    it is a person who hates everything

Now we have a sentence with uniform quantifiers. Continuing the translation yields the following sequence.

there is something such that	∃x
it is a person and	(Px &
it hates everything	∀yHxy)
	∃x(Px & ∀yHxy)

Let's do another example much like the previous one.

everyone hates something (or other)

This means

every person hates something (or other)

which can be paraphrased pretty much like every other sentence of the form 'every A is B':

everything is such that	∀x
if it is a person,	(Px →
then it hates something	
(or other)	∃yHxy)
	∀x(Px → ∃yHxy)

At this point, let us compare the sentences.

there is something that hates everything	∃x∀yHxy
there is some person who hates everything	∃x(Px & ∀yHxy)
everything hates something (or other)	∀x∃yHxy
every person hates something (or other)	∀x(Px → ∃yHxy)

The general forms of the above may be formulated as follows.

there is something that is K	∃xKx
there is some person who is K	∃x(Px & Kx)
everything is K	∀xKx
every person is K	∀x(Px → Kx)

We have already seen this particular transition – from completely general claims to more specialized claims. This maneuver, which might be called *quantifier specification*, still works.

everything is B:	∀x.........Bx
every A is B:	∀x(Ax → Bx)
something is B:	∃x.........Bx
some A is B:	∃x(Ax & Bx)

Quantifier specification is the process of modifying quantifiers by further specifying (or delimiting) the domain of discussion. The following are simple examples of quantifier specification.

converting 'everything' into 'every physical object'
converting 'everyone' into 'every student'

converting 'something' into 'some physical object'
converting 'someone' into 'some student'

The general process (in the special case of a simple predicate P) is described as follows.

SIMPLE QUANTIFIER SPECIFICATION:

Where **v** is any variable, **P** is any one-place predicate, and **F** is any formula, quantifier specification involves the following substitutions.

substitute ∀**v**(**Pv** → **F**) for ∀**vF**

substitute ∃**v**(**Pv** & **F**) for ∃**vF**

Note carefully the use of '→' in one and '&' in the other.

Examples

something is evil	$\exists xEx$
some physical thing is evil	$\exists x(Px \& Ex)$
everything is evil	$\forall xEx$
every physical thing is evil	$\forall x(Px \rightarrow Ex)$
someone respects everyone	$\exists x\forall yRxy$
some student respects everyone	$\exists x(Sx \& \forall yRxy)$
everyone respects someone	$\forall x\exists yRxy$
every student respects someone	$\forall x(Sx \rightarrow \exists yRxy)$

So far we have dealt exclusively with the outermost quantifier. However, we can apply quantifier specification to any quantifier in a formula. Consider the following example:

everyone respects someone (or other) $\forall x\exists yRxy$

versus

everyone respects some student (or other) ???

In applying quantifier specification, we note the following.

overall formula:	$\exists yRxy$
specified quantifier:	$\exists y$
specifying predicate:	Sy
modified formula:	Rxy

So applying the procedure, we obtain:

resulting formula: $\exists y(Sy \& Rxy)$

So plugging this back into our original formula, we obtain

everyone respects some student (or other)
$\forall x\exists y(Sy \& Rxy)$.

The more or less literal reading of the latter formula is:

for any person x,
there is a person y such that,
y is a student
and x respects y.

More colloquially,

for any person, there is a person such that
the latter is a student and the former respects the latter.

Still more colloquially,

for any person, there is a person such that
the latter is a student whom the former respects.

We can deal with the following in the same way.

there is someone who respects every student

This results from

there is someone who respects everyone
∃x∀yRxy,

by specifying the second quantifier, as follows:

overall formula:	∀yRxy
specified quantifier:	∀y
specifying predicate:	Sy
modified formula:	Rxy

So applying the procedure, we obtain:

resulting formula: ∀y(Sy → Rxy)

So plugging this back into our original formula, we obtain

there is someone who respects every student
∃x∀y(Sy → Rxy)

The more or less literal reading of the latter formula is:

there is a person x such that,
for any person y,
if y is a student,
then x respects y.

More colloquially,

there is a person such that,
for any person,
if the latter is a student
then the former respects the latter.

Still more colloquially,

there is a person such that,
for any student,
the former respects the latter.

So far, we have only done examples in which a single quantifier is specified by a predicate. We can also do examples in which both quantifiers are specified, and by different predicates. The principles remain the same; they are simply applied more generally. Consider the following examples.

(1) there is someone who respects everyone
(1a) there is a student who respects every professor
(1b) there is a professor who respects every student

(2) there is someone who is respected by everyone
(2a) there is a student who is respected by every professor
(2b) there is a professor who is respected by every student

(3) everyone respects someone or other
(3a) every student respects some professor or other
(3b) every professor respects some student or other

(4) everyone is respected by someone or other
(4a) every student is respected by some professor or other
(4b) every professor is respected by some student or other

The following are the corresponding formulas; in each case, the latter two are obtained from the first one by specifying the quantifiers appropriately.

(1) $\exists$x.........$\forall$y........Rxy
(1a) $\exists x(Sx \,\&\, \forall y(Py \to Rxy))$
(1b) $\exists x(Px \,\&\, \forall y(Sy \to Rxy))$

(2) $\exists$x.........$\forall$y........Ryx
(2a) $\exists x(Sx \,\&\, \forall y(Py \to Ryx))$
(2b) $\exists x(Px \,\&\, \forall y(Sy \to Ryx))$

(3) $\forall$x........$\exists$y.........Rxy
(3a) $\forall x(Sx \to \exists y(Py \,\&\, Rxy))$
(3b) $\forall x(Px \to \exists y(Sy \,\&\, Rxy))$

(4) $\forall$x........$\exists$y.........Ryx
(4a) $\forall x(Sx \to \exists y(Py \,\&\, Ryx))$
(4b) $\forall x(Px \to \exists y(Sy \,\&\, Ryx))$

6. COMPLEX PREDICATES

In order to further understand the translations that appear in the previous sections, and in order to be prepared for more complex translations still, we now examine the notion of *complex predicate*.

Roughly, complex predicates stand to simple (ordinary) predicates as complex (molecular) formulas stand to simple (atomic) formulas. Like ordinary predicates, complex predicates have places; there are one-place, two-place, etc., complex predicates. However, we are going to concentrate exclusively on one-place complex predicates.

The notion of a complex one-place predicate depends on the notion of a free occurrence of a variable. This is discussed in detail in an appendix. Briefly, an occurrence of a variable in a formula is *bound* if it falls inside the scope of a quantifier governing that variable; otherwise, the occurrence is *free*.

Examples

(1)	Fx	the one and only occurrence of 'x' is free.
(2)	∀x(Fx → Gx)	all three occurrences of 'x' are bound by '∀x'.
(3)	∀xRxy	every occurrence of 'x' is bound; the one and only occurrence of 'y' is free.

Next, to say that a variable (say, 'x') is *free* in a formula F is to say that at least one occurrence of 'x' is free in F; on the other hand, to say that 'x' is *bound* in F is to say that no occurrence of 'x' is free in F. For example, in the following formulas, 'x' is free, but 'y' is bound.

(f1) ∀yRxy
(f2) ∃yRxy
(f3) ∀yRyx
(f4) ∃yRyx

Any formula with exactly one free variable (perhaps with many occurrences) may be thought of as a complex one-place predicate. To see how this works, let us translate formulas (1)-(4) into nearly colloquial English.

(e1) x (he/she) respects everyone
(e2) x (he/she) respects someone
(e3) x (he/she) is respected by everyone
(e4) x (he/she) is respected by someone

Now, if we say of someone that he(she) respects everyone, then we are attributing a complex predicate to that person. We can abbreviate this complex predicate '$\mathcal{A}$x', which stands for 'x respects everyone'. Similarly with all the other formulas above; each one corresponds to a complex predicate, which can be abbreviated by a single letter. These abbreviations may be summarized by the following schemes.

$\mathcal{A}$x :: ∀yRxy
$\mathcal{B}$x :: ∃yRxy
$\mathcal{C}$x :: ∀yRyx
$\mathcal{D}$x :: ∃yRyx

Here, '::' basically means '...is short for...'.

Now, complex predicates can be used in sentences just like ordinary predicates. For example, we can say the following:

some Freshman is $\mathcal{A}$
every Freshman is $\mathcal{B}$
no Freshman is $\mathcal{C}$
some Freshman is not $\mathcal{D}$

Recalling what '$\mathcal{A}$', '$\mathcal{B}$', '$\mathcal{C}$', and '$\mathcal{D}$' are short for, these are read colloquially as follows.

some Freshman respects everyone
every Freshman respects someone or other
no Freshman is respected by everyone
some Freshman is not respected by someone (or other)

These have the following as overall symbolizations.

$\exists x(Fx \;\&\; \mathcal{A}x)$
$\forall x(Fx \rightarrow \mathcal{B}x)$
$\sim\exists x(Fx \;\&\; \mathcal{C}x)$
$\exists x(Fx \;\&\; \sim\mathcal{D}x)$

But '$\mathcal{A}$x', '$\mathcal{B}$x', '$\mathcal{C}$x', and '$\mathcal{D}$x' are short for more complex formulas, which when substituted yield the following formulas.

$\exists x(Fx \;\&\; \forall yRxy)$
$\forall x(Fx \rightarrow \exists yRxy)$
$\sim\exists x(Fx \;\&\; \forall yRyx)$
$\exists x(Fx \;\&\; \sim\exists yRyx)$

We can also make the following claims.

every $\mathcal{A}$ is $\mathcal{B}$
every $\mathcal{A}$ is $\mathcal{C}$
every $\mathcal{A}$ is $\mathcal{D}$

Given what '$\mathcal{A}$', '$\mathcal{B}$', '$\mathcal{C}$', and '$\mathcal{D}$' are short for, these read colloquially as follows.

every one who respects everyone respects someone
every one who respects everyone is respected by everyone
every one who respects everyone is respected by someone

The overall symbolizations of these sentences are given as follows.

$\forall x(\mathcal{A}x \rightarrow \mathcal{B}x)$
$\forall x(\mathcal{A}x \rightarrow \mathcal{C}x)$
$\forall x(\mathcal{D}x \rightarrow \mathcal{D}x)$

But '$\mathcal{A}$x', '$\mathcal{B}$x', '$\mathcal{C}$x', and '$\mathcal{D}$x' are short for more complex formulas, which when substituted yield the following formulas.

$\forall x(\forall yRxy \rightarrow \exists yRxy)$
$\forall x(\forall yRxy \rightarrow \forall yRyx)$
$\forall x(\forall yRxy \rightarrow \exists yRyx)$

Let's now consider somewhat more complicated complex predicates, given as follows.

$\mathcal{A}$x: x respects every professor
$\mathcal{B}$x: x is respected by every student
$\mathcal{C}$x: x respects at least one professor
$\mathcal{D}$x: x is respected by at least one student

Given the symbolizations of the formulas to the right, we have the following abbreviations.

$\mathcal{A}$x :: $\forall$y(Py → Rxy)
$\mathcal{B}$x :: $\forall$y(Sy → Ryx)
$\mathcal{C}$x :: $\exists$y(Py & Rxy)
$\mathcal{D}$x :: $\exists$y(Sy & Ryx)

We can combine these complex predicates with simple predicates or with each other. The following are examples.

(1) some S is $\mathcal{A}$
(2) some P is $\mathcal{B}$
(3) every S is $\mathcal{C}$
(4) every P is $\mathcal{D}$

The colloquial readings are:

(r1) there is a student who respects every professor
(r2) there is a professor who is respected by every student
(r3) every student respects at least one professor
(some professor or other)
(r4) every professor is respected by at least one student
(some student or other)

And the overall symbolizations are given as follows.

(o1) $\exists$x(Sx & $\mathcal{A}$x)
(o2) $\exists$x(Px & $\mathcal{B}$x)
(o3) $\forall$x(Sx → $\mathcal{C}$x)
(o4) $\forall$x(Px → $\mathcal{D}$x)

But '$\mathcal{A}$x', '$\mathcal{B}$x', '$\mathcal{C}$x', '$\mathcal{D}$x' are short for more complex formulas, which when substituted yield the following formulas.

(f1) $\exists$x(Sx & $\forall$y(Py → Rxy))
(f2) $\exists$x(Px & $\forall$y(Sy → Ryx))
(f3) $\forall$x(Sx → $\exists$y(Py & Rxy))
(f4) $\forall$x(Px → $\exists$y(Sy & Ryx))

These correspond to the formulas obtained by the technique of quantifier specification, presented in the previous section.

The advantage of understanding complex predicates is that it allows us to combine the complex predicates into the same formula. The following are examples.

$\mathcal{A}$x: x respects every professor
$\mathcal{B}$x: x is respected by every student
$\mathcal{C}$x: x is respected by at least one professor

no $\mathcal{A}$ is $\mathcal{B}$

every $\mathcal{B}$ is $\mathcal{C}$

These may be read colloquially as

no one who respects every professor is respected by every student

everyone who is respected by every student is respected by at least one professor

The overall symbolizations are, respectively,

$\sim\exists x(\mathcal{A}x \ \& \ \mathcal{B}x)$
$\forall x(\mathcal{B}x \rightarrow \mathcal{C}x)$

but '$\mathcal{A}$x', '$\mathcal{B}$x', and '$\mathcal{C}$x' stand for more complex formulas, which when substituted yield the following formulas.

$\sim\exists x(\forall y(Py \rightarrow Rxy) \ \& \ \forall y(Sy \rightarrow Ryx))$
$\forall x(\forall y(Sy \rightarrow Ryx) \rightarrow \exists y(Py \ \& \ Ryx))$

7. THREE-PLACE PREDICATES

So far, we have concentrated on two-place predicates. In the present section, we look at examples that involve quantification over formulas based on three-place predicates.

As mentioned in the previous chapter, there are numerous three place predicate expressions in English. The most common, perhaps, are constructed from verbs that take a subject, a direct object, and an indirect object. For example, in the sentence

Kay loaned her car to Jay

may be grammatically analyzed thus:

subject:	Kay
verb:	loaned
direct object:	her car
indirect object:	Jay

The remaining word, 'to', marks 'Jay' as the indirect object of the verb. In general, prepositions such as 'to' and 'from', as well as others, are used to mark indirect objects. The following sentence uses 'from' to mark the indirect object.

Jay borrowed Kay's car (from Kay)

Letting 'c' name the particular individual car in question, the above sentences can be symbolized as follows.

Lkcj
Bjck

The convention is to write subject first, direct object second, and indirect object last.

As usual, variables (pronouns) may replace one or more of the constants (proper nouns) in above formulas, and as usual, the resulting formulas can be quantified, either universally or existentially. The following are examples.

Kay loaned her car to him(her)	Lkcx
Kay loaned her car to someone	∃xLkcx
Kay loaned her car to everyone	∀xLkcx
Jay borrowed it from Kay	Bjxk
Jay borrowed something from Kay	∃xBjxk
Jay borrowed everything from Kay	∀xBjxk

As before, we can also further specify the quantifiers. Rather than saying 'someone' or 'everyone', we can say 'some student' or 'every student'; rather than saying 'something' or 'everything', we can say 'some car' or 'every car'. Quantifier specification works the same as before.

Kay loaned her car to some student	∃x(Sx & Lkcx)
Kay loaned her car to every student	∀x(Sx → Lkcx)
Jay borrowed some car from Kay	∃x(Cx & Bjxk)
Jay borrowed every car from Kay	∀x(Cx → Bjxk)

These are examples of single-quantification; we can quantify over every place in a predicate, so in the predicates we are considering, we can quantify over three places.

Two quantifiers first; let's change our example slightly. First note the following:

x rents y to z ↔ z rents y from x

For example,

Avis rents this car to Jay iff Jay rents this car from Avis.

Letting 'Rxyz' stand for 'x rents y **to** z', consider the following.

Example 1

every student has rented a car from Avis

$\forall x(Sx \rightarrow \exists y(Cy \;\&\; Rayx))$

for any x,
  if x is a student,
    then there is a y such that,
      y is a car
        and Avis has rented y to x

Example 2

there is at least one car that Avis has rented to every student

$\exists x(Cx \;\&\; \forall y(Sy \rightarrow Raxy))$

there is an x such that,
  x is a car
    and for any y,
      if y is a student,
        then Avis has rented x to y

8. 'ANY' REVISITED

Recall that certain quantifier expressions of English are wide-scope universal quantifiers. The most prominent wide-scope quantifier is 'any', whose standard derivatives are 'anything' and 'anyone'. Also recall that other words are also occasionally used as wide-scope universal quantifiers – including 'a' and 'some'; these are discussed in the next section.

To say that 'any' is a wide-scope universal quantifier is to say that, when it is attached to another logical expression, the scope of 'any' is wider than the scope of the attached expression.

In the context of monadic predicate logic, 'any' most frequently attaches to 'if' to produce the 'if any' locution. In particular, statements of the form:

if anything is A, then $\mathcal{B}$

appears to have the form:

if $\mathcal{A}$, then $\mathcal{B}$,

but because of the wide-scope of 'any', the sentence *really* has the form:

for anything (if it is A, then $\mathcal{B}$)

which is symbolized:

$\forall x(Ax \rightarrow \mathcal{B})$

In monadic logic, 'any' usually attaches to 'if'. In polyadic logic, 'any' often attaches to other words as well, most particularly 'no' and 'not', as in the following examples.

no one respects *any* one
Jay does *not* respect *any* one

Let us consider the second example, since it is easier. One way to understand this sentence is to itemize its content, which might go as follows.

Jay does not respect Adams	~Rja
Jay does not respect Brown	~Rjb
Jay does not respect Carter	~Rjc
Jay does not respect Dickens	~Rjd
Jay does not respect Evans	~Rje
Jay does not respect Field	~Rjf
etc.	

in short:

Jay does not respect *anyone*.

Given that 'Jay does not respect anyone' summarizes the list,

~Rja
~Rjb
~Rjc
~Rjd
~Rje
~Rjf
etc.

it is natural to regard 'Jay does not respect anyone' as a universally quantified statement, namely,

$\forall x{\sim}Rjx$.

Notice that the main logical operator is '$\forall x$'; the formula is a universally quantified formula.

Another way to symbolize the above 'any' statement employs the following series of paraphrases.

Jay does not respect anyone
Jay does not respect x, for any x
for any x, Jay does not respect x

$\forall x{\sim}Rjx$

Before considering more complex examples, let us contrast the any-sentence with the corresponding every-sentence.

Jay does not respect *any*one

versus

Jay does not respect *every*one

The latter certainly does not entail the former; 'any' and 'every' are not interchangeable, but we already know that. Also, we already know how to paraphrase and symbolize the latter sentence:

Jay does not respect *every*one
not(Jay does respect everyone)
it is not true that Jay respects everyone
  not everyone is respected by Jay

$\sim\forall xRjx$

Notice carefully that, although both 'any' and 'every' are universal quantifiers, they are quite different in meaning. The difference pertains to their respective scopes, which is summarized as follows, in respect to 'not'.

> 'not' has wider scope than 'every';
>
> 'any' has wider scope than 'not'.

not everyone	$\sim\forall x$
not anyone	$\forall x\sim$

Having considered the basic 'not any' form, let us next consider quantifier specification. For example, consider the following pair.

Jay does not respect *every* Freshman;
Jay does not respect *any* Freshman.

We already know how to paraphrase and symbolize the first one, as follows.

Jay does not respect every Freshman
not(Jay does respect every Freshman)
it is not true that Jay respects every Freshman
not every Freshman is respected by Jay

$\sim\forall x(Fx \rightarrow Rjx)$

The corresponding 'any' statement is more subtle. One approach involves the following series of paraphrases.

Jay does not respect any Freshman
Jay does not respect x, for any Freshman x
for any Freshman x, Jay does not respect x

$\forall x(Fx \rightarrow \sim Rjx)$

Notice that this is obtained from

Jay does not respect anyone
$\forall x \sim Rjx$

by quantifier specification, as described in an earlier section.

9. COMBINATIONS OF 'NO' AND 'ANY'

As mentioned in the previous section, 'any' attaches to 'if', 'not', and 'no' to form special compounds. We have already seen how 'any' interacts with 'if' and 'not'; in the present section, we examine how 'any' interacts with 'no'. Consider the following example.

(a) no Senior respects any Freshman

First we observe that 'any' and 'every' are not interchangeable. In particular, (a) is *not* equivalent to the following formula, which results by replacing 'any' by 'every'.

(e) no Senior respects every Freshman

The latter is equivalent to the following.

(e') there is no Senior who Respects every Freshman

The latter is symbolized, in parts, as follows.

(1) there is no S who R's every F
(2) there is no S who is $\mathcal{A}$
(3) no S is $\mathcal{A}$
(4) $\sim\exists x(Sx \,\&\, \mathcal{A}x)$
$\mathcal{A}x$ :: x R's every F :: $\forall y(Fy \rightarrow Rxy)$
(5) $\sim\exists x(Sx \,\&\, \forall y(Fy \rightarrow Rxy))$

Now let us go back and do the 'any' example (a); if we symbolize it in parts, we might proceed as follows.

(1) no S R's any F
(2) no S is $\mathcal{A}$
(3) $\sim\exists x(Sx \,\&\, Kx)$
$\mathcal{A}x$ :: x R's any F
???

The problem is that the complex predicate '$\mathcal{A}$' involves 'any', which cannot be straightforwardly symbolized in isolation; 'any' requires a *correlative* word to which it attaches.

At this point, it might be useful to recall (previous chapter) that 'no A is B' may be plausibly symbolized in either of the following ways.

(s1) $\sim\exists x(Ax \,\&\, Bx)$
(s2) $\forall x(Ax \rightarrow \sim Bx)$

These are logically equivalent, as we will demonstrate in the following chapter, so either counts as a correct symbolization. Each symbolization has its advantages; the first one shows the relation between 'no A is B' and 'some A is B' – they are negations of one another. The second one shows the relation between 'no A is B' and 'every A is unB' – they are equivalent.

In choosing a standard symbolization for 'no A is B' we settled on (s1) because it uses a single logical operator – namely $\sim\exists x$ – to represent 'no'. However, there are a few sentences of English that are more profitably symbolized using the second scheme, especially sentences involving 'any'.

So let us approach sentence (a) using the alternative symbolization of 'no'.

(a) no Senior respects any Freshman
(1) no S R's any F
(2) no S is $\mathcal{A}$
(3) $\forall x(Sx \rightarrow \sim\mathcal{A}x)$
$\mathcal{A}x$ :: x R's any F
???

Once again, we get stuck, because we can't symbolize '$\mathcal{A}x$' in isolation. However, we can rephrase (3) by treating '$\sim\mathcal{A}x$' as a unit, '$\mathcal{B}x$', in which the negation gets attached to 'any'.

(4) $\forall x(Sx \rightarrow \mathcal{B}x)$
$\mathcal{B}x$ :: x does not R any F
$\forall y(Fy \rightarrow \sim Rxy)$

Substituting the symbolization of '$\mathcal{B}x$' into (4), we obtain the following formula.

(5) $\forall x(Sx \rightarrow \forall y(Fy \rightarrow \sim Rxy))$

The latter formula reads

for any x,
    if x is a Senior,
        then for any y,
            if y is a Freshman,
                then x does not respect y

The latter may be read more colloquially as follows.

for any Senior, for any Freshman,
    the Senior does not respect the Freshman

On the other hand, if we follow the suggested translation scheme from earlier in the chapter, (5) is read colloquially as follows.

every Senior fails to respect every Freshman

The following is a somewhat more complex example.

no woman respects any man who does not respect her

We attack this in parts, but we note that one of the parts is a no-any combination. So the overall form is:

(1) no W R's any $\mathcal{A}$

As we already saw, this may be symbolized:

(2) $\forall x(Wx \rightarrow \forall y(\mathcal{A}y \rightarrow \sim Rxy))$

$\mathcal{A}y$ :: y is a man who does not respect her (x) :: $(My \ \& \sim Ryx)$

Substituting the symbolization of '$\mathcal{A}y$' into (2), we obtain:

(3) $\forall x(Wx \rightarrow \forall y([My \ \& \sim Ryx] \rightarrow \sim Rxy))$

for any x,
    if x is a woman,
        then for any y,
            if y is a man
                and y does not respect x,
                    then x does not respect y

Because of our wish to symbolize 'any' as a wide-scope universal quantifier, our symbolization of 'no A R's any B' is different from our symbolization of 'no A is $\mathcal{B}$'. Specifically, we have the following symbolization.

(1) no A R's any B
$\forall x(Ax \rightarrow \forall y(By \rightarrow \sim Rxy))$

(2) no A is $\mathcal{B}$
$\sim\exists x(Ax \ \& \ \mathcal{B}x)$

We conclude with an alternative symbolization which preserves 'no' but sacrifices the universal quantifier reading of 'any'. We start with (2) and perform two logical transformations, both based on the following equivalence.

(e) $\forall x(\mathcal{A} \rightarrow {\sim}\mathcal{B})$ :: ${\sim}\exists x(\mathcal{A}\ \&\ \mathcal{B})$.

(3) $\forall x(Ax \rightarrow {\sim}\exists y(By\ \&\ Rxy))$
(4) ${\sim}\exists x(Ax\ \&\ \exists y(By\ \&\ Rxy))$

The latter is read:

it is not true that
there is an x such that,
x is A,
and there is a y such that,
y is B,
and x R's y

Following our earlier scheme, we read (4) as

no A R's some B or other
no A R's at least one B

10. MORE WIDE-SCOPE QUANTIFIERS

Recall from the previous chapter the following example.

if anyone can fix your car, then Jones can (fix your car).

for anyone x, if x can fix your car Jones can

$\forall x(Fx \rightarrow Fj)$

Monadic logic does not do full justice to this sentence; in particular, we must symbolize 'can fix your car' as a simple one-place predicate, even though it includes a direct object. This expression is more adequately analyzed as a complex one-place predicate derived from the two-place predicate '...can fix...' and the singular term 'your car'. Then the symbolization goes as follows.

$\forall x(Fxc \rightarrow Fjc)$

This analysis is based on construing the expression 'your car' as referring to a particular car, named c in this context. In this reading, the speaker is speaking to a particular person, about a particular car.

This is not entirely accurate, because we now include cars in our domain of discourse, so we need to specify the quantifier to persons, as follows.

$\forall x[Px \rightarrow (Fxc \rightarrow Fjc)]$

However, there is another equally plausible analysis of the original sentence, which construes the ‘you’ in the sentence, not as a particular person to whom the speaker is speaking, but as a universal quantifier. In this case, the following is a more precise paraphrase.

> if anyone can fix anyone's car, then Jones can fix it.

The use of ‘you’ as a universal quantifier is actually quite common in English. The following is representative.

> you can't win at Las Vegas

can be paraphrased as

> no one can win at Las Vegas.

Another example:

> if you murder someone, then you are guilty of a capital crime,

can be paraphrased as

> if anyone murders someone, then he/she is guilty of a capital crime.

More about this example in a moment. First, let us finish the car example. By saying that ‘your car’ means ‘anyone’s car' we are saying that the formula

$$\forall x[Px \rightarrow (Fxc \rightarrow Fjc)]$$

is true, not just of c, but of every car, which is to say that the following formula holds.

$$\forall y\{Cy \rightarrow \forall x[Px \rightarrow (Fxc \rightarrow Fjc)]\}$$

An alternative symbolization puts all the quantifiers in front.

$$\forall x\forall y([(Px \,\&\, Cy) \,\&\, Fxy] \rightarrow Fjy)$$

Now, let us consider the murder example, which also involves two wide-scope universal quantifiers. First, notice that the word ‘someone’ does not act as an existential quantifier in this sentence. In this sentence, the most plausible reading of ‘someone’ is ‘anyone’.

> if anyone murders anyone,
> then the former is guilty of a capital crime

Let us treat the predicate ‘...is guilty of a capital crime’ as simple, symbolizing it simply as ‘G’. Simple versus complex predicates is not the issue at the moment. The issue is that there are two occurrences of ‘any’. How do we deal with sentences of the form

> if any... any... , then ...

The best way to treat the appearance of two wide-scope quantifiers is to treat them as double-universal quantifiers, thus:

for any x, for any y, if..., then...

So the murder-example is symbolized as follows.

$\forall x \forall y(Mxy \rightarrow Gx)$

Another example that has a similar form is the following.

if someone injures someone, then the latter sues the former

Once again, there are two wide-scope quantifiers, both being occurrences of 'someone'. This can be paraphrased and symbolized as follows.

if someone injures someone, then the latter sues the former

for any two people, if the former injures the latter,
then the latter sues the former

for any x, for any y, if x injures y, then y sues x

$\forall x \forall y(Ixy \rightarrow Syx)$

Next, let us consider the word 'a', which (like 'some' and 'any') is often used as a wide-scope quantifier. Consider the following two examples, which have the same form.

if a solid object is heated, it expands
if a game is rained out, it is rescheduled

These appear, at first glance, to be conditionals, but the occurrence of 'a' with the attached pronoun 'it' indicates that they are actually universal statements. The following is a plausible paraphrase of the first one.

if a solid object is heated, it expands

for any solid object, if it is heated, then it expands

for any S, if it is H, then it is E

for any S (x), if x is H, then x is E

for any x, if x is S, then if x is H, then x is E

$\forall x(Sx \rightarrow [Hx \rightarrow Ex])$

The word 'a' ('an') can also appear twice in the antecedent of a conditional, as in the following example.

if a student misses an exam, then he/she retakes that exam

This may be paraphrased and symbolized as follows.

if a S M's an E, then the S R's the E

for any S, for any E, if the S M's the E, then the S R's the E

for any S x, for any E y, if x M's y, then x R's y

$\forall x(Sx \rightarrow \forall y[Ey \rightarrow (Mxy \rightarrow Rxy)])$

Having seen examples involving various wide-scope quantifiers, including 'any', 'some', and 'a', it is important to recognize how they differ from one another. Compare the following sentences.

if a politician isn't respected by a citizen, then the politician is displeased;
if a politician isn't respected by any citizen, then the politician is displeased.

The difference is between 'a citizen' and 'any citizen'. Curiously, 'any citizen' attaches to 'not' (in the contraction 'isn't'), whereas 'a citizen' attaches to 'if'. In both cases, the quantifier 'a politician' attaches to 'if'. The former is paraphrased and symbolized as follows.

if a politician isn't respected by a citizen, then he/she is displeased
for any P, for any C, if the P isn't R'ed by the C, then the P is D
for any P x, for any C y, if x isn't R'ed by y, then x is D
$\forall x(Px \rightarrow \forall y[Cy \rightarrow (\sim Ryx \rightarrow Dx)])$

The following example further illustrates the difference between 'a' and 'any'.

if no one respects a politician, then the politician isn't re-elected;

If we substitute 'any politician' for 'a politician', we obtain a sentence of dubious grammaticality.

?? if no one respects any politician, then the politician isn't re-elected;

The reason this is grammatically dubious is that 'any' attaches to 'no', which is closer than 'if', and hence 'any' does not attach to the quasi-pronoun 'the politician'. By contrast, 'a' attaches to 'if' and 'the politician'; it does not attach to 'no'.

The rule of thumb that prevails is the following.

'any' attaches to the nearest logical operator from the following list: 'if', 'no', 'not'

'a' attaches to the nearest occurrence of 'if'

By way of concluding this section, we consider how 'a' interacts with 'every', which is a special case of how it interacts with 'if'. Recall that sentences of the form everyone who is *A* is *B*

are given an overall paraphrase/symbolization as follows

for anyone, if he/she is *A*, he/she is *B*
$\forall x(Ax \rightarrow Bx)$

In particular, many sentences involving 'every' are paraphrased using 'if-then'. Consider the following.

every person who likes a movie recommends it

Let us simplify matters by treating 'recommends' as a two-place predicate. Then the sentence is paraphrased and symbolized as follows.

for any person x, if x likes a movie, then x recommends it

for any person x,
for any movie y, if x likes y, then x recommends y
$\forall x(Px \rightarrow \forall y[My \rightarrow (Lxy \rightarrow Rxy)])$

11. EXERCISES FOR CHAPTER 7

Directions: Using the suggested abbreviations (the capitalized words), translate each of the following into the language of predicate logic.

EXERCISE SET A

1. Everyone RESPECTS JAY.
2. JAY RESPECTS everyone.
3. Someone RESPECTS JAY.
4. JAY RESPECTS someone.
5. Someone doesn't RESPECT JAY.
6. There is someone JAY does not RESPECT.
7. No one RESPECTS JAY.
8. JAY RESPECTS no one.
9. JAY doesn't RESPECT everyone.
10. Not everyone RESPECTS JAY.
11. Everyone RESPECTS everyone.
12. Everyone is RESPECTED by everyone.
13. Everyone RESPECTS someone (or other).
14. Everyone is RESPECTED by someone (or other).
15. There is someone who RESPECTS everyone.
16. There is someone who is RESPECTED by everyone.
17. Someone RESPECTS someone.
18. Someone is RESPECTED by someone.
19. Every event is CAUSED by some event or other (U=events).
20. There is some event that CAUSES every event.

EXERCISE B

21. There is no one who RESPECTS everyone.
22. There is no one who is RESPECTED by everyone.
23. There is someone who RESPECTS no one.
24. There is someone whom no one RESPECTS.
25. Not everyone RESPECTS everyone.
26. Not everyone is RESPECTED by everyone.

27. Not everyone RESPECTS someone or other.
28. Not everyone is RESPECTED by someone or other.
29. There is no one who doesn't RESPECT someone or other.
30. There is no one who isn't RESPECTED by someone or other.
31. There is no one who doesn't RESPECT everyone.
32. There is no one who isn't RESPECTED by everyone.
33. There is no one who isn't RESPECTED by at least one person.
34. There is no one who RESPECTS no one.
35. There is no one who is RESPECTED by no one.
36. There is no one who doesn't RESPECT at least one person.
37. For any person there is someone he/she doesn't RESPECT.
38. For any person there is someone who doesn't RESPECT him/her.
39. For any event there is an event that doesn't CAUSE it. (U=events)
40. There is no event that is not CAUSED by some event or other.

EXERCISE SET C

41. Every FRESHMAN RESPECTS someone or other.
42. Every FRESHMAN IS RESPECTED BY someone or other.
43. Everyone RESPECTS some FRESHMAN or other.
44. Everyone is RESPECTED by some FRESHMAN or other.
45. There is some FRESHMAN who RESPECTS everyone.
46. There is some FRESHMAN who is RESPECTED by everyone.
47. There is some one who RESPECTS every FRESHMAN.
48. There is some one who is RESPECTED by every FRESHMAN.
49. There is no FRESHMAN who is RESPECTED by everyone.
50. There is no one who RESPECTS every FRESHMAN.

EXERCISE SET D

51. Every PROFESSOR is RESPECTED by some STUDENT or other.
52. Every PROFESSOR RESPECTS some STUDENT or other.
53. Every STUDENT is RESPECTED by some PROFESSOR or other.
54. Every STUDENT RESPECTS some PROFESSOR or other.

55. For every PROFESSOR, there is a STUDENT who doesn't RESPECT that professor.

56. For every STUDENT, there is a PROFESSOR who doesn't RESPECT that student.

57. For every PROFESSOR, there is a STUDENT whom the professor doesn't RESPECT.

58. For every STUDENT, there is a PROFESSOR whom the student doesn't RESPECT.

59. There is a STUDENT who RESPECTS every PROFESSOR.

60. There is a PROFESSOR who RESPECTS every STUDENT.

61. There is a STUDENT who is RESPECTED by every PROFESSOR.

62. There is a PROFESSOR who is RESPECTED by every STUDENT.

63. There is a STUDENT who RESPECTS no PROFESSOR.

64. There is a PROFESSOR who RESPECTS no STUDENT.

65. There is a STUDENT who is RESPECTED by no PROFESSOR.

66. There is a PROFESSOR who is RESPECTED by no STUDENT.

67. There is no STUDENT who RESPECTS every PROFESSOR.

68. There is no PROFESSOR who RESPECTS every STUDENT.

69. There is no STUDENT who is RESPECTED by every PROFESSOR.

70. There is no STUDENT who RESPECTS no PROFESSOR.

71. There is no PROFESSOR who RESPECTS no STUDENT.

72. There is no STUDENT who is RESPECTED by no PROFESSOR.

73. There is no PROFESSOR who is RESPECTED by no STUDENT.

74. There is a STUDENT who doesn't RESPECT every PROFESSOR.

75. There is a PROFESSOR who doesn't RESPECT every STUDENT.

76. There is a PROFESSOR who isn't RESPECTED by every STUDENT.

77. There is a STUDENT who isn't RESPECTED by every PROFESSOR.

78. There is no STUDENT whom every PROFESSOR RESPECTS.

79. There is no PROFESSOR whom every STUDENT RESPECTS.

80. There is no PROFESSOR who isn't RESPECTED by every STUDENT.

EXERCISE E

81. Everyone who RESPECTS him(her)self RESPECTS everyone.
82. Everyone who RESPECTS him(her)self is RESPECTED by everyone.
83. Everyone who RESPECTS everyone is RESPECTED by everyone.
84. Everyone who RESPECTS every FRESHMAN is RESPECTED by every FRESHMAN.
85. Anyone who is SHORTER than every JOCKEY is SHORTER than JAY.
86. Anyone who is TALLER than KAY is TALLER than every STUDENT.
87. Anyone who is TALLER than every BASKETBALL player is TALLER than every JOCKEY.
88. JAY RESPECTS everyone who RESPECTS KAY.
89. JAY RESPECTS no one who RESPECTS KAY.
90. Everyone who KNOWS JAY RESPECTS at least one person who KNOWS KAY.
91. At least one person RESPECTS no one who RESPECTS JAY.
92. There is a GANGSTER who is FEARED by everyone who KNOWS him.
93. There is a PROFESSOR who is RESPECTED by every STUDENT who KNOWS him(her).
94. There is a STUDENT who is RESPECTED by every PROFESSOR who RESPECTS him(her)self.
95. There is a PROFESSOR who RESPECTS every STUDENT who ENROLLS in every COURSE the professor OFFERS.
96. Every STUDENT who KNOWS JAY RESPECTS every PROFESSOR who RESPECTS JAY.
97. There is a PROFESSOR who RESPECTS no STUDENT who doesn't RESPECT him(her)self.
98. There is a PROFESSOR who RESPECTS no STUDENT who doesn't RESPECT every PROFESSOR.
99. There is no PROFESSOR who doesn't RESPECT every STUDENT who ENROLLS in every COURSE he/she TEACHES.
100. Every STUDENT RESPECTS every PROFESSOR who RESPECTS every STUDENT.
101. Only MISANTHROPES HATE everyone.
102. Only SAINTS LOVE everyone.
103. The only MORTALS who are RESPECTED by everyone are movie STARS.

104. MORONS are the only people who IDOLIZE every movie STAR.

105. Only MORONS RESPECT only POLITICIANS.

EXERCISE F

106. JAY RECOMMENDS every BOOK he LIKES to KAY.

107. JAY LIKES every BOOK RECOMMENDED to him by KAY.

108. Every MAGAZINE that JAY READS is BORROWED from KAY.

109. Every BOOK that KAY LENDS to JAY she STEALS from CHRIS.

110. For every PROFESSOR, there is a STUDENT who LIKES every BOOK the professor RECOMMENDS to the student.

EXERCISE SET G

111. JAY doesn't RESPECT anyone.

112. JAY isn't RESPECTED by anyone.

113. There is someone who doesn't RESPECT anyone.

114. There is no one who isn't RESPECTED by anyone.

115. There is no one who doesn't RESPECT anyone.

116. JAY doesn't RESPECT any POLITICIAN.

117. JAY isn't RESPECTED by any POLITICIAN.

118. There is someone who isn't RESPECTED by any POLITICIAN.

119. There is no one who doesn't RESPECT any POLITICIAN.

120. There is at least one STUDENT who doesn't RESPECT any POLITICIAN.

121. There is no STUDENT who doesn't RESPECT any PROFESSOR.

122. There is no STUDENT who isn't RESPECTED by any PROFESSOR.

123. No STUDENT RESPECTS any POLITICIAN.

124. No STUDENT is RESPECTED by any POLITICIAN.

125. Everyone KNOWS someone who doesn't RESPECT any POLITICIAN.

126. Every STUDENT KNOWS at least one STUDENT who doesn't RESPECT any POLITICIAN.

127. No one who KNOWS JAY RESPECTS anyone who KNOWS KAY.

128. There is someone who doesn't RESPECT anyone who RESPECTS JAY.

129. No STUDENT who KNOWS JAY RESPECTS any PROFESSOR who RESPECTS JAY.

130. There is a PROFESSOR who doesn't RESPECT any STUDENT who doesn't RESPECT him(her).

131. There is a PROFESSOR who doesn't RESPECT any STUDENT who doesn't RESPECT every PROFESSOR.

132. If JAY can CRACK a SAFE, then every PERSON can CRACK it.

133. If KAY can't crack a SAFE, then no PERSON can CRACK it.

134. If a SKUNK ENTERS the room, then every PERSON will NOTICE it.

135. If a CLOWN ENTERS a ROOM, then every PERSON IN the room will NOTICE the clown.

136. If a MAN BITES a DOG, then every WITNESS is SURPRISED at him.

137. If a TRESPASSER is CAUGHT by one of my ALLIGATORS, he/she will be EATEN by that alligator.

138. Any FRIEND of YOURS is a FRIEND of MINE (o=you)

139. Anyone who BEFRIENDS any ENEMY of YOURS is an ENEMY of MINE

140. Any person who LOVES a SLOB is him(her)self a SLOB.

12. ANSWERS TO EXERCISES FOR CHAPTER 7

EXERCISE SET A

1. $\forall xRxj$
2. $\forall xRjx$
3. $\exists xRxj$
4. $\exists xRjx$
5. $\exists x{\sim}Rxj$
6. $\exists x{\sim}Rjx$
7. ${\sim}\exists xRxj$
8. ${\sim}\exists xRjx$
9. ${\sim}\forall xRjx$
10. ${\sim}\forall xRxj$
11. $\forall x\forall yRxy$
12. $\forall x\forall yRyx$
13. $\forall x\exists yRxy$
14. $\forall x\exists yRyx$
15. $\exists x\forall yRxy$
16. $\exists x\forall yRyx$
17. $\exists x\exists yRxy$
18. $\exists x\exists yRyx$
19. $\forall x\exists yCyx$
20. $\exists x\forall yCxy$

EXERCISE SET B

21. ${\sim}\exists x\forall yRxy$
22. ${\sim}\exists x\forall yRyx$
23. $\exists x{\sim}\exists yRxy$
24. $\exists x{\sim}\exists yRyx$
25. ${\sim}\forall x\forall yRxy$
26. ${\sim}\forall x\forall yRyx$
27. ${\sim}\forall x\exists yRxy$
28. ${\sim}\forall x\exists yRyx$
29. ${\sim}\exists x{\sim}\exists yRxy$
30. ${\sim}\exists x{\sim}\exists yRyx$
31. ${\sim}\exists x{\sim}\forall yRxy$
32. ${\sim}\exists x{\sim}\forall yRyx$
33. ${\sim}\exists x{\sim}\exists yRyx$
34. ${\sim}\exists x{\sim}\exists yRxy$
35. ${\sim}\exists x{\sim}\exists yRyx$
36. ${\sim}\exists x{\sim}\exists yRxy$
37. $\forall x\exists y{\sim}Rxy$
38. $\forall x\exists y{\sim}Ryx$
39. $\forall x\exists y{\sim}Cyx$
40. ${\sim}\exists x{\sim}\exists yCyx$

EXERCISE SET C

41. $\forall x(Fx \rightarrow \exists yRxy)$
42. $\forall x(Fx \rightarrow \exists yRyx)$
43. $\forall x\exists y(Fy \,\&\, Rxy)$
44. $\forall x\exists y(Fy \,\&\, Ryx)$
45. $\exists x(Fx \,\&\, \forall yRxy)$
46. $\exists x(Fx \,\&\, \forall yRyx)$
47. $\exists x\forall y(Fy \rightarrow Rxy)$
48. $\exists x\forall y(Fy \rightarrow Ryx)$
49. $\sim\exists x(Fx \,\&\, \forall yRyx)$
50. $\sim\exists x\forall y(Fy \rightarrow Rxy)$

EXERCISE SET D

51. $\forall x(Px \rightarrow \exists y(Sy \,\&\, Ryx))$
52. $\forall x(Px \rightarrow \exists y(Sy \,\&\, Rxy))$
53. $\forall x(Sx \rightarrow \exists y(Py \,\&\, Ryx))$
54. $\forall x(Sx \rightarrow \exists y(Py \,\&\, Rxy))$
55. $\forall x(Px \rightarrow \exists y(Sy \,\&\, \sim Ryx))$
56. $\forall x(Sx \rightarrow \exists y(Py \,\&\, \sim Ryx))$
57. $\forall x(Px \rightarrow \exists y(Sy \,\&\, \sim Rxy))$
58. $\forall x(Sx \rightarrow \exists y(Py \,\&\, \sim Rxy))$
59. $\exists x(Sx \,\&\, \forall y(Py \rightarrow Rxy))$
60. $\exists x(Px \,\&\, \forall y(Sy \rightarrow Rxy))$
61. $\exists x(Sx \,\&\, \forall y(Py \rightarrow Ryx))$
62. $\exists x(Px \,\&\, \forall y(Sy \rightarrow Ryx))$
63. $\exists x(Sx \,\&\, \sim\exists y(Py \,\&\, Rxy))$
64. $\exists x(Px \,\&\, \sim\exists y(Sy \,\&\, Rxy))$
65. $\exists x(Sx \,\&\, \sim\exists y(Py \,\&\, Ryx))$
66. $\exists x(Px \,\&\, \sim\exists y(Sy \,\&\, Ryx))$
67. $\sim\exists x(Sx \,\&\, \forall y(Py \rightarrow Rxy))$
68. $\sim\exists x(Px \,\&\, \forall y(Sy \rightarrow Rxy))$
69. $\sim\exists x(Sx \,\&\, \forall y(Py \rightarrow Ryx))$
70. $\sim\exists x(Sx \,\&\, \sim\exists y(Py \,\&\, Rxy))$
71. $\sim\exists x(Px \,\&\, \sim\exists y(Sy \,\&\, Rxy))$
72. $\sim\exists x(Sx \,\&\, \sim\exists y(Py \,\&\, Ryx))$
73. $\sim\exists x(Px \,\&\, \sim\exists y(Sy \,\&\, Ryx))$
74. $\exists x(Sx \,\&\, \sim\forall y(Py \rightarrow Rxy))$
75. $\exists x(Px \,\&\, \sim\forall y(Sy \rightarrow Rxy))$
76. $\exists x(Px \,\&\, \sim\forall y(Sy \rightarrow Ryx))$
77. $\exists x(Sx \,\&\, \sim\forall y(Py \rightarrow Ryx))$
78. $\sim\exists x(Sx \,\&\, \forall y(Py \rightarrow Rxy))$
79. $\sim\exists x(Px \,\&\, \forall y(Sy \rightarrow Ryx))$
80. $\sim\exists x(Px \,\&\, \sim\forall y(Sy \rightarrow Ryx))$

EXERCISE SET E

81. $\forall x(Rxx \rightarrow \forall yRxy)$
82. $\forall x(Rxx \rightarrow \forall yRyx)$
83. $\forall x(\forall yRxy \rightarrow \forall yRyx)$
84. $\forall x[\forall y(Fy \rightarrow Rxy) \rightarrow \forall y(Fy \rightarrow Ryx)]$
85. $\forall x[\forall y(Jy \rightarrow Sxy) \rightarrow Sxj]$
86. $\forall x[Txk \rightarrow \forall y(Sy \rightarrow Txy)]$
87. $\forall x[\forall y(By \rightarrow Txy) \rightarrow \forall y(Jy \rightarrow Txy)]$
88. $\forall x(Rxk \rightarrow Rjx)$
89. $\sim\exists x(Rxk \,\&\, Rjx)$;
90. $\forall x(Kxj \rightarrow \exists y(Kyk \,\&\, Rxy))$
91. $\exists x\sim\exists y(Ryj \,\&\, Rxy)$
92. $\exists x(Gx \,\&\, \forall y(Kyx \rightarrow Fyx))$
93. $\exists x(Px \,\&\, \forall y([Sy \,\&\, Kyx] \rightarrow Ryx))$
94. $\exists x(Sx \,\&\, \forall y([Py \,\&\, Ryy] \rightarrow Ryx))$
95. $\exists x[Px \,\&\, \forall y(\{Sy \,\&\, \forall z([Cz \,\&\, Oxz] \rightarrow Eyz)\} \rightarrow Rxy)]$
96. $\forall x\{[Sx \,\&\, Kxj] \rightarrow \forall y([Py \,\&\, Ryj] \rightarrow Rxy)\}$
97. $\exists x(Px \,\&\, \sim\exists y([Sy \,\&\, \sim Ryy] \,\&\, Rxy))$
98. $\exists x(Px \,\&\, \sim\exists y([Sy \,\&\, \sim\forall z(Pz \rightarrow Ryz)] \,\&\, Rxy))$
99. $\sim\exists x\{Px \,\&\, \sim\forall y([Sy \,\&\, \forall z([Cz \,\&\, Txz) \rightarrow Eyz)] \rightarrow Rxy)\}$
100. $\forall x\{Sx \rightarrow \forall y([Py \,\&\, \forall z(Sz \rightarrow Ryz)] \rightarrow Rxy)\}$
101. $\sim\exists x(\sim Mx \,\&\, \forall yHxy)$
102. $\sim\exists x(\sim Sx \,\&\, \forall yLxy)$
103. $\sim\exists x([Mx \,\&\, \forall yRyx] \,\&\, \sim Sx)$
104. $\sim\exists x(\sim Mx \,\&\, \forall y(Sy \rightarrow Ixy))$
105. $\sim\exists x(\sim Mx \,\&\, \sim\exists y(\sim Py \,\&\, Rxy))$

EXERCISE SET F

106. $\forall x([Bx \,\&\, Ljx] \rightarrow Rjxk)$
107. $\forall x([Bx \,\&\, Rkxj] \rightarrow Ljx)$
108. $\forall x([Mx \,\&\, Rjx] \rightarrow Bjxk)$
109. $\forall x([Bx \,\&\, Lkxj] \rightarrow Skxc)$
110. $\forall x\{Px \rightarrow \exists y(Sy \,\&\, \forall z([Bz \,\&\, Rxzy] \rightarrow Lyz))\}$

EXERCISE SET G

111. ∀x~Rjx
112. ∀x~Rxj
113. ∃x∀y~Rxy
114. ~∃x∀y~Ryx
115. ~∃x∀y~Rxy
116. ∀x(Px → ~Rjx)
117. ∀x(Px → ~Rxj)
118. ∃x∀y(Py → ~Ryx)
119. ~∃x∀y(Py → ~Rxy)
120. ∃x(Sx & ∀y(Py → ~Rxy))
121. ~∃x(Sx & ∀y(Py → ~Rxy))
122. ~∃x(Sx & ∀y(Py → ~Ryx))
123. ∀x(Px → ~∃y(Sy & Ryx))
124. ∀x(Px → ~∃y(Sy & Rxy))
125. ∀x∃y(Kxy & ∀z(Pz → ~Ryz))
126. ∀x(Sx → ∃y([Sy & Kxy] & ∀z(Pz → ~Ryz))
127. ∀x(Kxk → ~∃y(Kyj & Ryx))
128. ∃x∀y(Ryj → ~Rxy)
129. ∀x([Px & Rxj] → ~∃y([Sy & Kyj] & Ryx))
130. ∃x(Px & ∀y([Sy & ~Ryx] → ~Rxy))
131. ∃x(Px & ∀y([Sy & ~∀z(Pz → Ryz] → ~Rxy))
132. ∀x([Sx & Cjx] → ∀y(Py → Cyx))
133. ∀x([Sx & ~Ckx] → ~∃y(Py & Cyx))
134. ∀x([Sx & Ex] → ∀y(Py → Nyx))
135. ∀x∀y([(Cx & Ry) & Exy] → ∀z([Pz & Izy]→ Nzx))
136. ∀x∀y([(Mx&Dy) & Bxy] → ∀z(Wz → Szx))
137. ∀x∀y([(Tx & Ay) & Cyx] → Eyx)
138. ∀x(Fxo → Fxm)
139. ∀x∀y([Eyo & Bxy] → Exm)
140. ∀x∀y([Sy & Lxy] → Sx]

8 DERIVATIONS IN PREDICATE LOGIC

1. INTRODUCTION

Having discussed the grammar of predicate logic and its relation to English, we now turn to the problem of argument validity in predicate logic.

Recall that, in Chapter 5, we developed the technique of formal derivation in the context of sentential logic – specifically System SL. This is a technique to deduce conclusions from premises in sentential logic. In particular, if an argument is *valid* in sentential logic, then we *can* (in principle) construct a derivation of its conclusion from its premises in System SL, and if it is *invalid*, then we *cannot* construct such a derivation.

In the present chapter, we examine the corresponding deductive system for predicate logic – what will be called System PL (short for 'predicate logic'). As you might expect, since the syntax (grammar) of predicate logic is considerably more complex than the syntax of sentential logic, the method of derivation in System PL is correspondingly more complex than System SL.

On the other hand, anyone who has already mastered sentential logic derivations can also master predicate logic derivations. The transition primarily involves (1) getting used to the new symbols and (2) practicing doing the new derivations (just like in sentential logic!). The practical converse, unfortunately, is also true. Anyone who hasn't already mastered sentential logic derivations will have tremendous difficulty with predicate logic derivations. Of course, it's still not too late to figure out sentential derivations!

2. THE RULES OF SENTENTIAL LOGIC

We begin by stating the first principle of predicate logic derivations. To wit,

> Every rule of System SL (sentential logic) is also a rule of System PL (predicate logic).

The converse is not true; as we shall see in later sections, there are several rules peculiar to predicate logic, i.e., rules that do not arise in sentential logic.

Since predicate logic adopts all the derivation rules of sentential logic, it is a good idea to review the salient features of sentential logic derivations.

First of all, the derivation rules divide into two categories; on the one hand, there are *inference rules*, which are upward-oriented; on the other hand, there are *show rules*, which are downward-oriented.

There are numerous inference rules, but they divide into four basic categories.

(I1)	Introduction Rules (In-Rules): &I, ∨I, ↔I, **X**I
(I2)	Simple Elimination Rules (Out-Rules): &O, ∨O, →O, ↔O, **X**O
(I3)	Negation Elimination Rules (Tilde-Out-Rules): ~&O, ~∨O, ~→O, ~↔O
(I4)	Double Negation, Repetition

In addition, there are four show-rules.

(S1)	Direct Derivation
(S2)	Conditional Derivation
(S3)	Indirect Derivation (First Form)
(S4)	Indirect Derivation (Second Form)

As noted at the beginning of the current section, every rule of sentential logic is still operative in predicate logic. However, when applied to predicate logic, the rules of sentential logic look somewhat different, but only because the syntax of predicate logic is different. In particular, instead of formulas that involve only sentential letters and connectives, we are now faced with formulas that involve predicates and quantifiers. Accordingly, when we apply the sentential logic rules to the new formulas, they look somewhat different.

For example, the following are all instances of the arrow-out rule, applied to predicate logic formulas.

(1) Fa › Ga
Fa
____
Ga

(2) ∀xFx → ∀xGx
∀xFx
____
∀xGx

(3) Fa → Ga
~Ga
____
~Fa

(4) ∀x(Fx → Gx) → ∃xFx
~∃xFx
____
~∀x(Fx → Gx)

Thus, in moving from sentential logic to predicate logic, one must first become accustomed to applying the old inference rules to new formulas, as in examples (1)-(4).

The same thing applies to the show rules of sentential logic, and their associated derivation strategies, which remain operative in predicate logic. Just as before, to show a conditional formula, one uses conditional derivation; similarly, to show a negation, or disjunction, or atomic formula, one uses indirect derivation. The only difference is that one must learn to apply these strategies to predicate logic formulas.

For example, consider the following show lines.

(1) SHOW: Fa → Ga

(2) SHOW: ∀xFx → ∀xGx

(3) SHOW: ~Fa

(4) SHOW: ~∃x(Fx & Gx)

(5) SHOW: Rab

(6) SHOW: ∀xFx ∨ ∀xGx

Every one of these is a formula for which we already have a ready-made derivation strategy. In each case, either the formula is atomic, or its main connective is a sentential logic connective.

The formulas in (1) and (2) are conditionals, so we use conditional derivation, as follows.

(1)	SHOW: Fa → Ga	CD
	Fa	As
	SHOW: Ga	??

(2)	SHOW: ∀xFx → ∀xGx	CD
	∀xFx	As
	SHOW: ∀xGx	??

The formulas in (3) and (4) are negations, so we use indirect derivation of the first form, as follows.

(3)	SHOW: ~Fa	ID
	Fa	As
	SHOW: ×	??

(4)	SHOW: ~∃x(Fx & Gx)	ID
	∃x(Fx & Gx)	As
	SHOW: ×	??

The formula in (5) is atomic, so we use indirect derivation, supposing that a direct derivation doesn't look promising.

(5)	SHOW: Rab	ID
	~Rab	As
	SHOW: ×	??

Finally, the formula in (6) is a disjunction, so we use indirect derivation, along with tilde-wedge-out, as follows.

(6)	SHOW: ∀xFx ∨ ∀xGx	ID
	~(∀xFx ∨ ∀xGx)	As
	SHOW: ×	??
	~∀xFx	~∨O
	~∀xGx	~∨O

In conclusion, since predicate logic subsumes sentential logic, all the derivation techniques we have developed for the latter can be transferred to predicate logic. On the other hand, given the additional logical apparatus of predicate logic, in the form of quantifiers, we need additional derivation techniques to deal successfully with predicate logic arguments.

3. THE RULES OF PREDICATE LOGIC: AN OVERVIEW

If we confined ourselves to the rules of sentential logic, we would be unable to derive any interesting conclusions from our premises. All we could derive would be conclusions that follow purely in virtue of sentential logic. On the other hand, as noted at the beginning of Chapter 6, there are valid arguments that can't be shown to be valid using only the resources of sentential logic.

Consider the following (valid) arguments.

∀x(Fx → Hx)	every Freshman is Happy
Fc	Chris is a Freshman
———————	———————
Hc	Chris is Happy

∀x(Sx → Px)	every Snake is Poisonous
∀x([Sx & Px] → Dx)	every Poisonous Snake is Dangerous
Sm	Max is a Snake
———————	———————————
Dm	Max is Dangerous

In either example, if we try to derive the conclusion from the premises, we are stuck very quickly, for we have no means of dealing with those premises that are universal formulas. They are not conditionals, so we can't use arrow-out; they are not conjunctions, so we can't use ampersand-out, etc., etc.

Sentential logic does not provide a rule for dealing with such formulas, so we need special rules for the added logical structure of predicate logic.

In choosing a set of rules for predicate logic, one goal is to follow the general pattern established in sentential logic. In particular, according to this pattern, for each connective, we have a rule for introducing that connective, and a rule for eliminating that connective. Also, for each two-place connective, we have a rule for eliminating negations of formulas with that connective. In sentential logic, with the exception of the conditional for which there is no introduction rule, every connective has both an in-rule and an out-rule, and every connective has a tilde-out-rule. There is no arrow-in inference rule; rather, there is an arrow show-rule, namely, conditional derivation.

In regard to derivations, moving from sentential logic to predicate logic basically involves adding two sets of one-place connectives; on the one hand, there are the universal quantifiers – ∀x, ∀y, ∀z; on the other hand, there are the existential quantifiers – ∃x, ∃y, ∃z. So, following the general pattern for rules, just as we have three rules for each sentential connective, we correspondingly have three rules for universals, and three rules for existentials, which are summarized as follows.

Universal Rules

(1) Universal Derivation (UD)
(2) Universal-Out (∀O)
(3) Tilde-Universal-Out (~∀O)

Existential Rules

(1) Existential-In (∃I)
(2) Existential-Out (∃O)
(3) Tilde-Existential-Out (~∃O)

Thus, predicate logic employs six rules, in addition to all of the rules of sentential logic. Notice carefully, that five of the rules are inference rules (upward-oriented rules), but one of them (universal derivation) is a show-rule (downward-oriented rule), much like conditional derivation. Indeed, universal derivation plays a role in predicate logic very similar to the role of conditional derivation in sentential logic.

[Note: Technically speaking, Existential-Out (∃O) is an *assumption rule*, rather than a true inference rule. See Section 10 for an explanation.]

In the next section, we examine in detail the easiest of the six rules of predicate logic – universal-out.

4. UNIVERSAL OUT

The first, and easiest, rule we examine is universal-elimination (universal-out, for short). As its name suggests, it is a rule designed to decompose any formula whose main connective is a universal quantifier (i.e., ∀x, ∀y, or ∀z).

The official statement of the rule goes as follows.

> **Universal-Out** (∀O)
>
> If one has an available line that is a universal formula, which is to say that it has the form ∀vF[v], where v is any variable, and F[v] is any formula in which v occurs free, then one is entitled to infer any substitution instance of F[v].

In symbols, this may be pictorially summarized as follows.

$$\forall O: \quad \frac{\forall v F[v]}{F[n]}$$

Here,

(1) v is any variable (x, y, z);

(2) n is any name (a-w);

(3) F[v] is any formula, and F[n] is the formula that results when n is substituted for every occurrence of v that is *free in* F[v].

In order to understand this rule, it is best to look at a few examples.

Example 1: ∀xFx

This is by far the easiest example. In this v is x, and F[v] is Fx. To obtain a substitution instance of Fx one simply replaces x by a name, any name. Thus, all of the following follow by ∀O:

Fa, Fb, Fc, Fd, etc.

Example 2: ∀yRyk

This is almost as easy. In this v is y, and F[v] is Ryk. To obtain a substitution instance of Ryk one simply replaces y by a name, any name. Thus, all of the following follow by ∀O:

Rak, Rbk, Rck, Rdk, etc.

In both of these examples, the intuition behind the rule is quite straightforward. In Example 1, the premise says that everything is an F; but if everything is an F, then any particular thing we care to mention is an F, so a is an F, b is an F, c is an F, etc. Similarly, in Example 2, the premise says that everything bears relation R to k (for example, everyone respects Kay); but if everything bears R to k, then any particular thing we care to mention bears R to k, so a bears R to k, b bears R to k, etc.

In examples 1 and 2, the formula F[v] is atomic. In the remaining examples, F[v] is molecular.

Example 3: ∀x(Fx → Gx)

In this v is x, and F[v] is Fx→Gx. To obtain a substitution instance, we replace both occurrences of x by a name, the same name for both occurrences. Thus, all of the following follow by ∀O.

Fa → Ga, Fb → Gb, Fc → Gc, etc.

In this example, the intuition underlying the rule may be less clear than in the first two examples. The premise may be read in many ways in English, some more colloquial than others.

(r1) every F is G
(r2) everything is G if it's F
(r3) everything is such that: if it is F, then it is G.

The last reading (r3) says that everything has a certain property, namely, that if it is F then it is G. But if everything has this property, then any particular thing we care to mention has the property. So a has the property, b has the property, etc. But to say that a has the property is simply to say that if a is F then a is G; to say that b has the property is to say that if b is F then b is G. Both of these are applications of universal-out.

Example 4: ∀x∃yRxy

Here, v is x, and F[v] is ∃yRxy. To obtain a substitution instance of ∃yRxy, one replaces the one and only occurrence of x by a name, any name. Thus, the following all follow by ∀O.

∃yRay, ∃yRby, ∃yRcy, ∃yRdy, etc.

The premise says that everything bears relation R to something or other. For example, it translates the English sentence ‘everyone respects someone (or other)’. But if everyone respects someone (or other), then anyone you care to mention respects someone, so a respects someone, b respects someone, etc.

Example 5: $\forall x(Fx \rightarrow \forall xGx)$

Here, v is x, and F[v] is $Fx \rightarrow \forall xGx$. To obtain a substitution instance, one replaces every *free* occurrence of x in $Fx \rightarrow \forall xGx$ by a name. In this example, the first occurrence is free, but the remaining two are not, so we only replace the first occurrence. Thus, the following all follow by ∀O.

$$Fa \rightarrow \forall xGx,\ \ Fb \rightarrow \forall xGx,\ \ Fc \rightarrow \forall xGx,\ \ \text{etc.}$$

This example is complicated by the presence of a second quantifier governing the same variable, so we have to be especially careful in applying ∀O. Nevertheless, one's intuitions are not violated. The premise says that if anyone is an F then everyone is a G (recall the distinction between 'if any' and 'if every'). From this it follows that if a is an F then everyone is a G, and if b is an F then everyone is a G, etc. But that is precisely what we get when we apply ∀O to the premise.

5. POTENTIAL ERRORS IN APPLYING UNIVERSAL-OUT

There are basically two ways in which one can misapply the rule universal-out: (1) improper substitution; (2) improper application.

In the case of improper substitution, the rule is applied to an appropriate formula, namely, a universal, but an error is made in performing the substitution. Refer to the Appendix concerning correct and incorrect substitution instances. The following are a few examples of improper substitution.

(1) $\forall xRxx$; to infer Rax, Rab, Rba WRONG!!!

(2) $\forall x(Fx \rightarrow Gx)$; to infer $Fa \rightarrow Gb$, $Fb \rightarrow Gc$ WRONG!!!

(3) $\forall x(Fx \rightarrow \forall xGx)$; to infer $Fa \rightarrow \forall aGa$, $Fa \rightarrow \forall xGa$ WRONG!!!

In the case of improper application, one attempts to apply the rule to a line that does not have the appropriate form. Universal-out, as its name is intended to suggest, applies to universal formulas, not to atomic formulas, or existentials, or negations, or conditionals, or biconditional, or conjunctions, or disjunctions.

Recall, in this connection, a very important principle.

INFERENCE RULES APPLY
EXCLUSIVELY TO WHOLE LINES,
NOT TO PIECES OF LINES.

The following are examples of improper application of universal-out.

(4) $\forall xFx \rightarrow \forall xGx$

to infer $Fa \rightarrow \forall xGx$	WRONG!!!
to infer $\forall xFx \rightarrow Ga$	WRONG!!!
to infer $Fa \rightarrow Gb$	WRONG!!!

In each case, the error is the same – specifically, applying universal-out to a formula that does not have the appropriate form. Now, the formula in question is not a universal, but is rather a conditional; so the appropriate elimination rule is not universal-out, but rather arrow-out (which, of course, requires an additional premise).

(5) $\sim\forall xFx$

to infer $\sim Fa$, or $\sim Fb$, or $\sim Fc$ WRONG!!!

Once again, the error involves applying universal-out to a formula that is not a universal. In this case, the formula is a negation. Later, we will have a rule – tilde-universal-out – designed specifically for formulas of this form.

The moral is that you must be able to recognize the major connective of a formula; is it an atomic formula, a conjunction, a disjunction, a conditional, a biconditional, a negation, a universal, or an existential? Otherwise, you can't apply the rules successfully, and hence you can't construct proper derivations.

Of course, sometimes misapplying a rule produces a valid conclusion. Take the following example.

(6) $\forall xFx \rightarrow \forall xGx$

to infer $\forall xFx \rightarrow Ga$
to infer $\forall xFx \rightarrow Gb$
etc.

All of these inferences correspond to valid arguments. But many arguments are valid! The question, at the moment, is whether the inference is an instance of universal out. These inferences are not. In order to show that $\forall xFx \rightarrow Ga$ follows from $\forall xFx \rightarrow \forall xGx$, one must construct a derivation of the conclusion from the premise.

In the next section, we examine this particular derivation, as well as a number of others that employ our new tool, universal-out.

6. EXAMPLES OF DERIVATIONS USING UNIVERSAL-OUT

Having figured out the universal-out rule, we next look at examples of derivations in which this rule is used. We start with the arguments at the beginning of Section 3.

Example 1

(1)	$\forall x(Fx \rightarrow Hx)$	Pr
(2)	Fc	Pr
(3)	~~SHOW~~: Hc	DD
(4)	| $Fc \rightarrow Hc$	1,∀O
(5)	| Hc	2,4,→O

Example 2

(1)	$\forall x(Sx \rightarrow Px)$	Pr
(2)	$\forall x([Sx \,\&\, Px] \rightarrow Dx)$	Pr
(3)	Sm	Pr
(4)	~~SHOW~~: Dm	DD
(5)	| $Sm \rightarrow Pm$	1,∀O
(6)	| $(Sm \,\&\, Pm) \rightarrow Dm$	2,∀O
(7)	| Pm	3,5,→O
(8)	| Sm & Pm	3,7,&I
(9)	| Dm	6,8,→O

The above two examples are quite simple, but they illustrate an important strategic principle for doing derivations in predicate logic.

> REDUCE THE PROBLEM TO A POINT WHERE YOU CAN APPLY RULES OF SENTENTIAL LOGIC.

In each of the above examples, we reduce the problem to the point where we can finish it by applying arrow-out.

Notice in the two derivations above that the tool – namely, universal-out – is specialized to the job at hand. According to universal-out, if we have a line of the form $\forall v F[v]$, we are entitled to write down any instance of the formula F[v]. So, for example, in line (4) of the first example, we are entitled to write down $Fa \rightarrow Ha$, $Fb \rightarrow Hb$, as well as a host of other formulas. But, of all the formulas we are entitled to write down, only one of them is of any use – namely, $Fc \rightarrow Hc$.

Similarly, in the second example, we are entitled by universal-out to instantiate lines (1) and (2) respectively to any name we choose. But of all the permitted instantiations, only those that involve the name m are of any use.

To say that one is *permitted* to do something is quite different from saying that one *must* do it, or even that one *should* do it. At any given point in a game (say, chess), one is *permitted* to make any number of moves, but most of them are stupid (supposing one's goal is to win). A good chess player chooses *good moves* from among the *legal moves*. Similarly, a good derivation builder chooses good moves from among the legal moves. In the first example, it is certainly true that $Fa \rightarrow Ga$ is a permitted step at line (4); but it is pointless because it makes no contribution whatsoever to completing the derivation.

By analogy, standing on your head until you have a splitting headache and are sick to your stomach is not against the law; it's just stupid.

In the examples above, the choice of one particular letter over any other letter as the letter of instantiation is natural and obvious. Other times, as you will later see, there are several names floating around in a derivation, and it may not be obvious which one to use at any given place. Under these circumstances, one must primarily use trial-and-error.

Let us look at some more examples. In the previous section, we looked at an argument that was obtained by a misapplication of universal-out. As noted there, the argument is valid, although it is not an instance of universal-out. Let us now show that it is indeed valid by deriving the conclusion from the premises.

Example 3

(1)	$\forall xFx \rightarrow \forall xGx$	Pr
(2)	~~SHOW~~: $\forall xFx \rightarrow Ga$	CD
(3)		$\forall xFx$	As
(4)		~~SHOW~~: Ga	DD
(5)			$\forall xGx$	1,3,→O
(6)			Ga	5,∀O

Notice, in particular, that the formula in (2) is a conditional, and is accordingly shown by conditional derivation. You are, of course, already very familiar with conditional derivations; to show a conditional, you assume the antecedent and show the consequent.

The following is another example in which a sentential derivation strategy is employed.

Example 4

(1)	$\forall x(Fx \rightarrow Hx)$	Pr
(2)	$\sim Hb$	Pr
(3)	~~SHOW~~: $\sim\forall xFx$	ID
(4)		$\forall xFx$	As
(5)		~~SHOW~~: **×**	DD
(6)			$Fb \rightarrow Hb$	1,∀O
(7)			Fb	4,∀O
(8)			Hb	6,7,→O
(9)			**×**	2,8,**×**I

In line (3), we have to show $\sim\forall xFx$; this is a negation, so we use a tried-and-true strategy for showing negations, namely indirect derivation. To show the negation of a formula, one assumes the formula negated and one shows the generic contradiction, ✕.

We conclude this section by looking at a considerably more complex example, but still an example that requires only one special predicate logic rule, universal-out.

Example 5

(1)	$\forall x(Fx \rightarrow \forall yRxy)$	Pr
(2)	$\forall x\forall y(Rxy \rightarrow \forall zGz)$	Pr
(3)	$\sim Gb$	Pr
(4)	~~SHOW~~: $\sim Fa$	ID
(5)	Fa	As
(6)	~~SHOW~~: ✕	DD
(7)	$Fa \rightarrow \forall yRay$	1,∀O
(8)	$\forall yRay$	5,7,→O
(9)	Rab	8,∀O
(10)	$\forall y(Ray \rightarrow \forall zGz)$	2,∀O
(11)	$Rab \rightarrow \forall zGz$	10,∀O
(12)	$\forall zGz$	9,11,→O
(13)	Gb	12,∀O
(14)	✕	3,13,✕I

If you can figure out this derivation, better yet if you can reproduce it yourself, then you have truly mastered the universal-out rule!

7. EXISTENTIAL IN

Of the six rules of predicate logic that we are eventually going to have, we have now examined only one – universal-out. In the present section, we add one more to the list.

The new rule, existential introduction (existential-in, ∃I) is officially stated as follows.

> **Existential-In** (∃I)
>
> If formula F[n] is an available line, where F[n] is a substitution instance of formula F[v], then one is entitled to infer the existential formula ∃vF[v].

In symbols, this may be pictorially summarized as follows.

∃I:	F[n]
	————
	∃vF[v]

Here,

(1) v is any variable (x, y, z);

(2) n is any name (a-w);

(3) F[v] is any formula, and F[n] is the formula that results when n is substituted for every occurrence of v that is *free in* F[v].

Existential-In is very much like an upside-down version of Universal-Out. However, turning ∀O upside down to produce ∃I brings a small complication. In ∀O, one begins with the formula F[v] with variable v, and one substitutes a name n for the variable v. The only possible complication pertains to free and bound occurrences of v. By contrast, in ∃I, one works backwards; one begins with the substitution instance F[n] with name n, and one "de-substitutes" a variable v for n. Unfortunately, in many cases, de-substitution is radically different from substitution. See examples below.

As with all rules of derivation, the best way to understand ∃I is to look at a few examples.

Example 1

have: Fb	b is F
infer: ∃xFx; ∃yFy; ∃zFz	at least one thing is F

Here, n is 'b', and F[n] is Fb, which is a substitution instance of three different formulas – Fx, Fy, and Fz. So the inferred formulas (which are alphabetic variants of one another; see Appendix) can all be inferred in accordance with ∃I.

In Example 1, the intuition underlying the rule's application is quite straightforward. The premise says that b is F. But if b is F, then at least one thing is F, which is what all three conclusions assert. One might understand this rule as saying that, if *a particular thing* has a property, then *at least one thing* has that property.

Example 2

have: Rjk	j R's k
infer: ∃xRxk, ∃yRyk, ∃zRzk	something R's k
infer: ∃xRjx, ∃yRjy, ∃zRjz	j R's something

Here, we have two choices for n – 'j' and 'k'. Treating 'j' as n, Rjk is a substitution instance of three different formulas – Rxk, Ryk, and Rzk, which are alphabetic variants of one another. Treating 'k' as 'n', Rjk is a substitution instance of three different formulas – Rjx, Rjy, and Rjz, which are alphabetic variants of one another. Thus, two different sets of formulas can be inferred in accordance with ∃I.

In Example 2, letting 'R' be '...respects...' and 'j' be 'Jay' and 'k' be 'Kay', the premise says that Jay respects Kay. The conclusions are basically two (discounting alphabetic variants) – someone respects Kay, and Jay respects someone.

Example 3

have: Fb & Hb

Here, n is 'b', and F[n] is Fb&Hb, which is a substitution instance of nine different formulas:

(f1) Fx & Hx, Fy & Hy, Fz & Hz
(f2) Fb & Hx, Fb & Hy, Fb & Hz
(f3) Fx & Hb, Fy & Hb, Fz & Hb

So the following are all inferences that are in accord with ∃I:

infer: ∃x(Fx & Hx), ∃y(Fy & Hy), ∃z(Fz & Hz)
infer: ∃x(Fb & Hx), ∃y(Fb & Hy), ∃z(Fb & Hz)
infer: ∃x(Fx & Hb), ∃y(Fy & Hb), ∃z(Fz & Hb)

In Example 3, three groups of formulas can be inferred by ∃I. In the case of the first group, the underlying intuition is fairly clear. The premise says that b is F and b is H (i.e., b is both F and H), and the conclusions variously say that at least one thing is both F and H. In the case of the remaining two groups, the intuition is less clear. These are permitted inferences, but they are seldom, if ever, used in actual derivations, so we will not dwell on them here.

In Example 3, there are two groups of conclusions that are somehow extraneous, although they are certainly permitted. The following example is quite similar, insofar as it involves two occurrences of the same name. However, the difference is that the two extra groups of valid conclusions are not only legitimate but also useful.

Example 4

have: Rkk;	k R's itself
infer: ∃xRxx, ∃yRyy, ∃zRzz	something R's itself
infer: ∃xRxk, ∃yRyk, ∃zRzk	something R's k
infer: ∃xRkx, ∃yRky, ∃zRkz	k R's something

Here, n is 'k', and F[n] is Rkk, which is a substitution instance of nine different formulas – Rxx, Rkx, Rxk, as well as the alphabetic variants involving 'y' and 'z'. So the above inferences are all in accord with ∃I.

In Example 4, although the various inferences at first look a bit complicated, they are actually not too hard to understand. Letting 'R' be '...respects...' and 'k' be 'Kay', then the premise says that Kay respects Kay, or more colloquially Kay respects herself. But if Kay respects herself, then we can validly draw all of the following conclusions.

(c1)	someone respects her(him)self	∃xRxx
(c2)	someone respects Kay	∃xRxk
(c3)	Kay respects someone	∃xRkx

All of these follow from the premise 'Kay respects herself', and moreover they are all in accord with ∃I.

In all the previous examples, no premise involves a quantifier. The following is the first such example, which introduces a further complication, as well.

Example 5

have:	∃xRkx	k R's something
infer:	∃y∃xRyx, ∃z∃xRzx	something R's something

Here, n is 'k', and F[n] is ∃xRkx, which is a substitution instance of two different formulas – ∃xRyx, and ∃xRzx, which are alphabetic variants of one another. However, in this example, there is no alphabetic variant involving the variable x'; in other words, ∃xRkx is not a substitution instance of ∃xRxx, because the latter formula doesn't have *any* substitution instances, since it has no free variables!

In Example 5, letting 'R' be '...respects...', and letting 'k' be 'Kay', the premise says that someone (we are not told who in particular) respects Kay. The conclusion says that someone respects someone. If at least one person respects Kay, then it follows that at least one person respects at least one person.

Let us now look at a few examples of derivations that employ ∃I, as well as our earlier rule, ∀O.

Example 1

(1)	∀x(Fx → Hx)	Pr
(2)	Fa	Pr
(3)	SHOW: ∃xHx	DD
(4)		Fa → Ha	1,∀O
(5)		Ha	2,4,→O
(6)		∃xHx	5,∃I

Example 2

(1)	∀x(Gx → Hx)	Pr
(2)	Gb	Pr
(3)	~~SHOW~~: ∃x(Gx & Hx)	DD
(4)	│ Gb → Hb	1,∀O
(5)	│ Hb	2,5,→O
(6)	│ Gb & Hb	2,5,&I
(7)	│ ∃x(Gx & Hx)	6,∃I

Example 3

(1)	∃x~Rxa → ~∃xRax	Pr
(2)	~Raa	Pr
(3)	~~SHOW~~: ~Rab	ID
(4)	│ Rab	As
(5)	│ ~~SHOW~~: ✕	DD
(6)	│ │ ∃x~Rxa	2,∃I
(7)	│ │ ~∃xRax	1,6,→O
(8)	│ │ ∃xRax	4,∃I
(9)	│ │ ✕	7,8,✕I

Example 4

(1)	∀x(∃yRxy → ∀yRxy)	Pr
(2)	Raa	Pr
(3)	~~SHOW~~: Rab	DD
(4)	│ ∃yRay → ∀yRay	1,∀O
(5)	│ ∃yRay	2,∃I
(6)	│ ∀yRay	4,5,→O
(7)	│ Rab	6,∀O

8. UNIVERSAL DERIVATION

We have now studied two rules, universal-out and existential-in. As stated earlier, every connective (other than tilde) has associated with it three rules, an introduction rule, an elimination rule, and a negation-elimination rule. In the present section, we examine the introduction rule for the universal quantifier.

The first important point to observe is that, whereas the introduction rule for the existential quantifier is an *inference rule*, the introduction rule for the universal quantifier is a *show rule*, called universal derivation (UD), compare this with conditional derivation. In other words, the rule is for dealing with lines of the form 'SHOW: ∀v...'.

Suppose one is faced with a derivation problem like the following.

(1)	∀x(Fx → Gx)	Pr
(2)	∀xFx	Pr
(3)	SHOW: ∀xGx	??

How do to go about completing the derivation? At the present, given its form, the only derivation strategies available are direct derivation and indirect derivation (second form). However, in either approach, one quickly gets stuck. This is because, as it stands, our derivation system is inadequate; we cannot derive ∀xFx' with the machinery currently at our disposal. So, we need a new rule.

Now what does the conclusion say? Well, 'for any x, Gx' says that everything is G. This amounts to asserting every item in the following very long list.

(c1) Ga
(c2) Gb
(c3) Gc
(c4) Gd
etc.

This is a very long list, one in which every particular thing in the universe is (eventually) mentioned. [Of course, we run out of ordinary names long before we run out of things to mention; so, in this situation, we have to suppose that we have a truly huge collection of names available.]

Still another way to think about ∀xGx is that it is equivalent to a corresponding infinite conjunction:

(c) Ga & Gb & Gc & Gd & Ge & . . . . .

where every particular thing in the universe is (eventually) mentioned.

Nothing really hinges on the difference between the infinitely long list and the infinite conjunction. After all, in order to show the conjunction, we would have to show every conjunct, which is to say that we would have to show every item in the infinite list.

So our task is to show Ga, Gb, Gc, etc. This is a daunting task, to say the least. Well, let's get started anyway and see what develops.

	(1)	∀x(Fx → Gx)	Pr
	(2)	∀xFx	Pr
a:	(3)	~~SHOW~~: Ga	DD
	(4)	│ Fa → Ga	1,∀O
	(5)	│ Fa	2,∀O
	(6)	│ Ga	4,5,→O
b:	(3)	~~SHOW~~: Gb	DD
	(4)	│ Fb → Gb	1,∀O
	(5)	│ Fb	2,∀O
	(6)	│ Gb	4,5,→O

c:	(3)	~~SHOW~~: Gc	DD
	(4)	| Fc → Gc	1,∀O
	(5)	| Fc	2,∀O
	(6)	| Gc	4,5,→O

d:	(3)	~~SHOW~~: Gd	DD
	(4)	| Fd → Gd	1,∀O
	(5)	| Fd	2,∀O
	(6)	| Gd	4,5,→O

.
.
.

We are making steady progress, but we have a very long way to go! Fortunately, however, having done a few, we can see a distinctive pattern emerging; except for particular names used, the above derivations all look the same. This is a pattern we can use to construct as many derivations of this sort as we care to; for any particular thing we care to mention, we can show that it is G. So we can (eventually!) show that every *particular* thing is G (Ga, Gb, Gc, Gd, etc.), and hence that everything is G (∀xGx).

We have the pattern for all the derivations, but we certainly don't want to (indeed, we can't) construct all of them. How many do we have to do in order to be finished? 5? 25? 100? Well, the answer is that, once we have done *just one* derivation, we already have the pattern (model, mould) for every other derivation, so we can stop after doing just one! The rest look the same, and are redundant, in effect.

This leads to the first (but not final) formulation of the principle of universal derivation.

> **Universal Derivation** (First Approximation)
>
> In order to show a universal formula, which is to say a formula of the form ∀vF[v], it is sufficient to show a substitution instance F[n] of F[v].

This is *not* the whole story, as we will see shortly. However, before facing the complication, let's see what universal derivation, so stated, allows us to do. First, we offer two equivalent solutions to the original problem using universal derivation.

Example 1

a:	(1)	∀x(Fx → Gx)	Pr
	(2)	∀xFx	Pr
	(3)	~~SHOW~~: ∀xGx	UD
	(4)	| ~~SHOW~~: Ga	DD
	(5)	| | Fa → Ga	1,∀O
	(6)	| | Fa	2,∀O
	(7)	| | Ga	5,6,→O

```
b:  (1)  ∀x(Fx → Gx)                    Pr
    (2)  ∀xFx                           Pr
    (3)  SHOW: ∀xGx                     UD
    (4)  | SHOW: Gb                     DD
    (5)  | | Fb → Gb                    1,∀O
    (6)  | | Fb                         2,∀O
    (7)  | | Gb                         5,6,→O
```

Each example above uses universal derivation to show ∀xGx. In each case, the overall technique is the same: one shows a universal formula ∀vF[v] by showing a substitution instance F[n] of F[v].

In order to solidify this idea, let's look at two more examples.

Example 2

```
(1)  ∀x(Fx → Gx)                  Pr
(2)  SHOW: ∀xFx → ∀xGx            CD
(3)  | ∀xFx                       As
(4)  | SHOW: ∀xGx                 UD
(5)  | | SHOW: Ga                 DD
(6)  | | | Fa → Ga                1,∀O
(7)  | | | Fa                     3,∀O
(8)  | | | Ga                     6,7,→O
```

In this example, line (2) asks us to show ∀xFx→∀xGx. One might be tempted to use universal derivation to show this, but this would be completely wrong. Why? Because ∀xFx→∀xGx is not a universal formula, but rather a conditional. Well, we already have a derivation technique for showing conditionals – conditional derivation. That gives us the next two lines; we assume the antecedent, and we show the consequent. So that gets us to line (4), which is to show ∀xGx'. Now, this formula is indeed a universal, so we use universal derivation; this means we immediately write down a further show-line 'SHOW: Ga' (we could also write 'SHOW: Gb', or 'SHOW: Gc', etc.). This is shown by direct derivation.

Example 3

```
(1)   ∀x(Fx → Gx)                 Pr
(2)   ∀x(Gx → Hx)                 Pr
(3)   SHOW: ∀x(Fx → Hx)           UD
(4)   | SHOW: Fa → Ha             CD
(5)   | | Fa                      As
(6)   | | SHOW: Ha                DD
(7)   | | | Fa → Ga               1,∀O
(8)   | | | Ga → Ha               2,∀O
(9)   | | | Ga                    5,7,→O
(10)  | | | Ha                    8,9,→O
```

In this example, we are asked to show ∀x(Fx→Gx), which is a universal formula, so we show it using universal derivation. This means that we immediately write down a new show line, in this case 'SHOW: Fa→Ha'; notice that Fa→Ha is a substitution instance of Fx→Hx. Remember, to show ∀vF[v], one shows F[n], where F[n] is a substitution instance of F[v]. Now the problem is to show Fa→Ha; this is a conditional, so we use conditional derivation.

Having seen three successful uses of universal derivation, let us now examine an illegitimate use. Consider the following "proof" of a clearly invalid argument.

Example 4 (Invalid Argument!!)

(1)	Fa & Ga	Pr	
(2)	SHOW: ∀xGx	UD	
(3)	SHOW: Ga	DD	WRONG!!!
(4)	Ga	1,&O	

First of all, the fact that a is F and a is G does not logically imply that everything (or everyone) is G. From the fact that Adams is a Freshman who is Gloomy it does not follow that everyone is Gloomy. Then what went wrong with our technique? We showed ∀xGx by showing an instance of Gx, namely Ga.

An important clue is forthcoming as soon as we try to generalize the above erroneous derivation to any other name. In the Examples 1-3, the fact that we use 'a' is completely inconsequential; we could just as easily use *any* name, and the derivation goes through with equal success. But with the last example, we can indeed show Ga, but that is all; we cannot show Gb or Gc or Gd. But in order to demonstrate that everything is G, we have to show (in effect) that a is G, b is G, c is G, etc. In the last example, we have actually only shown that a is G.

In Examples 1-3, doing the derivation with 'a' was enough because this one derivation serves as a model for every other derivation. Not so in Example 4. But what is the difference? When is a derivation a model derivation, and when is it *not* a model derivation?

Well, there is at least one conspicuous difference between the good derivations and the bad derivation above. In every good derivation above, no name appears in the derivation before the universal derivation, whereas in the bad derivation above the name 'a' appears in the premises.

This can't be the whole story, however. For consider the following perfectly good derivation.

Example 5

(1)	Fa & Ga	Pr
(2)	∀x(Fx → ∀yGy)	Pr
(3)	∀x(Gx → Fx)	Pr
(4)	~~SHOW~~: ∀xFx	UD
(5)	│ ~~SHOW~~: Fb	DD
(6)	│ │ Fa	1,&O
(7)	│ │ Fa → ∀yGy	2,∀O
(8)	│ │ ∀yGy	6,7,→O
(9)	│ │ Gb	8,∀O
(10)	│ │ Gb → Fb	3,∀O
(11)	│ │ Fb	9,10,→O

In this derivation, which can be generalized to every name, a name occurs earlier, but we refrain from using it as our instance at line (5). We elect to show, not just any instance, but an instance with a letter that is not previously being used in the derivation. We are trying to show that everything is F; we already know that a is F, so it would be no good merely to show that; we show instead that b is F. This is better because we don't know anything about b; so whatever we show about b will hold for everything.

We have seen that universal derivation is not as simple as it might have looked at first glance. The first approximation, which seemed to work for the first three examples, is that to show ∀vF[v] one merely shows F[n], where F[n] is any substitution instance. But this is not right! If the name we choose is already in the derivation, then it can lead to problems, so we must restrict universal derivation accordingly. As it turns out, this adjustment allows Examples 1,2,3,5, but blocks Example 4.

Having seen the adjustment required to make universal derivation work, we now formally present the correct and final version of the universal-elimination rule. The crucial modification is marked with an '▶'.

> **Universal Derivation** (Intuitive Formulation)
>
> In order to show a universal formula,
> which is to say a formula of the form ∀vF[v],
> it is sufficient to show a substitution instance
> F[n] of F[v],
>
> ▶ where n is any ***new*** name,
> which is to say that n does *not* appear
> anywhere earlier in the derivation.

As usual, the official formulation of the rule is more complex.

> **Universal Derivation** (Official Formulation)
>
> If one has a show-line of the form 'SHOW: ∀vF[v]', then if one has '~~SHOW~~: F[n]' as a later available line, where F[n] is a substitution instance of F[v], and n is a new name, and there are no intervening uncancelled show-line, then one may box and cancel 'SHOW: ∀vF[v]'. The annotation is 'UD'

In pictorial terms, similar to the presentations of the other derivation rules (DD, CD, ID), universal derivation (UD) may be presented as follows.

~~SHOW~~: ∀vF[v]	UD
~~SHOW~~: F[n]	n must be **new**; i.e., it cannot occur in any previous line, including the line 'SHOW: ∀vF[v]'.

We conclude this section by examining an argument that involves relational quantification. This example is quite complex, but it illustrates a number of important points.

Example 6

(1)	Raa	Pr
(2)	∀x∀y[Rxy → ∀x∀yRxy]	Pr
(3)	~~SHOW~~: ∀x∀yRyx	UD
(4)	~~SHOW~~: ∀yRyb	UD
(5)	~~SHOW~~: Rcb	DD
(6)	∀y[Ray → ∀x∀yRxy]	2,∀O
(7)	Raa → ∀x∀yRxy	6,∀O
(8)	∀x∀yRxy	1,7,→O
(9)	∀yRcy	8,∀O
(10)	Rcb	9,∀O

Analysis

(3) SHOW: ∀x∀yRyx
this is a universal ∀x...∀yRyx,
so we show it by UD, which is to say that we show an instance of ∀yRyx, where the name must be new. Only 'a' is used so far, so we use the next letter 'b', yielding:

(4) SHOW: ∀yRyb
this is also a universal ∀y...Ryb
so we show it by UD, which is to say that we show an instance of 'Ryb', where the name must be new. Now, both 'a' and 'b' are already in the derivation, so we can't use either of them. So we use the next letter 'c', yielding:

(5) SHOW: Rcb
This is atomic. We use either DD or ID. DD happens to work.

(6) Line (1) is ∀x∀y(Rxy → ∀x∀yRxy),
which is a universal ∀x...∀y(Rxy → ∀x∀yRxy),
so we apply ∀O. The choice of letter is completely free, so we choose 'a', replacing every free occurrence of 'x' by 'a', yielding:

∀y(Ray → ∀x∀yRxy)
This is a universal ∀y...(Ray → ∀x∀yRxy),
so we apply ∀O. The choice of letter is completely free, so we choose 'a', replacing every free occurrence of 'x' by 'a', yielding:

(7) Raa → ∀x∀yRxy
This is a conditional, so we apply →O, in conjunction with line 1, which yields:

(8) ∀x∀yRxy
This is a universal ∀x...∀yRxy,
so we apply ∀O, instantiating 'x' to 'c', yielding:

(9) ∀yRcy
This is a universal ∀y...Rcy,
so we apply ∀O, instantiating 'y' to 'b', yielding:

(10) Rcb
This is what we wanted to show!

By way of concluding this section, let us review the following points.

> Having ∀vF[v] as an *available* line is very different from having 'SHOW: ∀vF[v]' as a line.
>
> In one case you **have** ∀vF[v];
>
> in the other case, you **don't have** ∀vF[v]; rather, you are trying to show it.
>
> ∀O applies when you **have** a universal; you can use **any** name whatsoever.

> UD applies when you **want** a universal;
> you *must* use a **new** name.

9. EXISTENTIAL OUT

We now have three rules; we have both an elimination (out) and an introduction (in) rule for ∀, and we have an introduction rule for ∃. At the moment, however, we do not have an elimination rule for ∃. That is the topic of the current section.

Consider the following derivation problem.

(1)	∀x(Fx → Hx)	Pr
(2)	∃xFx	Pr
(3)	SHOW: ∃xHx	??

One possible English translation of this argument form goes as follows.

(1) every Freshman is happy
(2) at least one person is a Freshman
(3) therefore, at least one person is happy

This is indeed a valid argument. But how do we complete the corresponding derivation? The problem is the second premise, which is an existential formula. At present, we do not have a rule specifically designed to decompose existential formulas.

How should such a rule look? Well, the second premise is ∃xFx, which says that some thing (at least one thing) is F; however, it is not very specific; it doesn't say which particular thing is F. We know that at least one item in the following infinite list is true, but we don't know which one it is.

(1) Fa
(2) Fb
(3) Fc
(4) Fd
etc.

Equivalently, we know that the following infinite disjunction is true.

(d) Fa ∨ Fb ∨ Fc ∨ Fd ∨ ... ∨ ...

[Once again, we pretend that we have sufficiently many names to cover every single thing in the universe.]

The second premise ∃xFx says that at least one thing is F (some thing is F), but it provides no further information as to which thing *in particular* is F. Is it a? Is it b? We don't know given only the information conveyed by ∃xFx. So, what happens if we simply *assume* that a is F. Adding this assumption yields the following substitute problem.

(1)	$\forall x(Fx \to Hx)$	Pr
(2)	$\exists xFx$	Pr
(3)	SHOW: $\exists xHx$	DD
(4)	Fa	???

I write '???' because the status of this line is not obvious at the moment. Let us proceed anyway.

Well, now the problem is much easier! The following is the completed derivation.

a:	(1)	$\forall x(Fx \to Hx)$	Pr
	(2)	$\exists xFx$	Pr
	(3)	~~SHOW~~: $\exists xHx$	DD
	(4)	| Fa	???
	(5)	| $Fa \to Ha$	1,$\forall$O
	(6)	| Ha	4,5,$\to$O
	(7)	| $\exists xHx$	6,$\exists$I

In other words, if we assume that the something that is F is in fact a, then we can complete the derivation.

The problem is that we don't actually know that a is F, but only that something is F. Well, then maybe the something that is F is in fact b. So let us instead assume that b is F. Then we have the following derivation.

b:	(1)	$\forall x(Fx \to Hx)$	Pr
	(2)	$\exists xFx$	Pr
	(3)	~~SHOW~~: $\exists xHx$	DD
	(4)	| Fb	???
	(5)	| $Fb \to Hb$	1,$\forall$O
	(6)	| Hb	4,5,$\to$O
	(7)	| $\exists xHx$	6,$\exists$I

Or perhaps the something that is F is actually c, so let us assume that c is F, in which case we have the following derivation.

c:	(1)	$\forall x(Fx \to Hx)$	Pr
	(2)	$\exists xFx$	Pr
	(3)	~~SHOW~~: $\exists xHx$	DD
	(4)	| Fc	???
	(5)	| $Fc \to Hc$	1,$\forall$O
	(6)	| Hc	4,5,$\to$O
	(7)	| $\exists xHx$	6,$\exists$I

A definite pattern of reasoning begins to appear. We can keep going on and on. It seems that whatever it is that is actually an F (and we know that something is), we can show that something is H. For any particular name, we can construct a derivation using that name. All the resulting derivations would look (virtually) the same, the only difference being the particular letter introduced at line (4).

The generality of the above derivation is reminiscent of universal derivation. Recall that a universal derivation substitutes a single model derivation for infinitely many derivations all of which look virtually the same. The above pattern looks very similar: the first derivation serves as a model of all the rest.

Indeed, we can recast the above derivations in the form of UD by inserting an extra show-line as follows. Remember that one is entitled to write down any show-line at any point in a derivation.

u:			
	(1)	∀x(Fx → Hx)	Pr
	(2)	∃xFx	Pr
	(3)	~~SHOW~~: ∃xHx	DD
	(4)	| ~~SHOW~~: ∀x(Fx → ∃xHx)	UD
	(5)	| | ~~SHOW~~: Fa → ∃xHx	CD
	(6)	| | | Fa	As
	(7)	| | | ~~SHOW~~: ∃xHx	DD
	(8)	| | | Fa → Ha	1,∀O
	(9)	| | | Ha	6,8,→O
	(10)	| | | ∃xHx	9,∃I
	(11)	| ∃xHx	2,4,???

The above derivation is clear until the very last line, since we don't have a rule that deals with lines 2 and 4. In English, the reasoning goes as follows.

(2) at least one thing is F
(4) if anything is F then at least one thing is H
(10) (therefore) at least one thing is H

Without further ado, let us look at the existential-elimination rule.

> **Existential-Out** (∃O)
>
> If a line of the form ∃vF[v] is available, then one can *assume* any substitution instance F[n] of F[v], so long as n is a name that is ***new*** to the derivation. The annotation cites the line number, plus ∃O.

The following is the cartoon version.

∃O:	∃vF[v]	n must be **new**;
	———	i.e., it cannot occur in any previous line,
	F[n]	including the line ∃vF[v].

Note on annotation: When applying ∃O, the annotation appeals to the line number of the existential formula ∃vF[v] and the rule ∃O. In other words, even though ∃O is an *assumption rule*, and not a true inference rule, we annotate derivations as if it were a true inference rule; see below.

Before worrying about the proviso 'so long as n is ...', let us go back now and do our earlier example, now using the rule ∃O. The crucial line is marked by '▶'.

Example 1

	(1)	∀x(Fx → Hx)	Pr
	(2)	∃xFx	Pr
	(3)	~~SHOW~~: ∃xHx	DD
▶	(4)	| Fa	2,∃O
	(5)	| Fa → Ha	1,∀O
	(6)	| Ha	4,5,→O
	(7)	| ∃xHx	6,∃I

In line (4), we apply ∃O to line (2), instantiating 'x' to 'a'; note that 'a' is a *new name*.

The following are two more examples of ∃O.

Example 2

(1)	∀x(Fx → Gx)	Pr
(2)	∃x(Fx & Hx)	Pr
(3)	~~SHOW~~: ∃x(Gx & Hx)	DD
(4)	| Fa & Ha	2,∃O
(5)	| Fa	4,&O
(6)	| Ha	4,&O
(7)	| Fa → Ga	1,∀O
(8)	| Ga	5,7,→O
(9)	| Ga & Ha	6,8,&I
(10)	| ∃x(Gx & Hx)	9,∃I

Example 3

(1)	∀x(Fx → Gx)	Pr
(2)	∀x(Gx → ~Hx)	Pr
(3)	~~SHOW~~: ~∃x(Fx & Hx)	ID
(4)	| ∃x(Fx & Hx)	As
(5)	| ~~SHOW~~: ✕	DD
(6)	| | Fa & Ha	4,∃O
(7)	| | Fa	6,&O
(8)	| | Ha	6,&O
(9)	| | Fa → Ga	1,∀O
(10)	| | Ga	7,9,→O
(11)	| | Ga → ~Ha	2,∀O
(12)	| | ~Ha	10,11,→O
(13)	| | ✕	8,12,✕I

Examples 2 and 3 illustrate an important strategic principle in constructing derivations in predicate logic. In Example 3, when we get to line (6), we have many rules we can apply, including ∀O and ∃O. Which should we apply first? The following are two rules of thumb for dealing with this problem. [Remember, a rule of thumb is just that; it does not work 100% of the time.]

Rule of Thumb 1
Don't apply ∀O unless (until) you have a name in the derivation to which to apply it.

Rule of Thumb 2
If you have a choice between applying ∀O and applying ∃O, apply ∃O first.

The second rule is, in some sense, an application of the first rule. If one has no name to apply ∀O to, then one way to produce a name is to apply ∃O. Thus, one first applies ∃O, thus producing a name, and then applies ∀O.

What happens if you violate the above rules of thumb? Well, nothing very bad; you just end up with extraneous lines in the derivation. Consider the following derivation, which contains a violation of Rules 1 and 2.

Example 2 (revisited):

	(1)	∀x(Fx → Gx)	Pr	
	(2)	∃x(Fx & Hx)	Pr	
	(3)	SHOW: ∃x(Gx & Hx)	DD	
▶	(*)	| Fa → Ga	1,∀O	
	(4)	| Fb & Hb	2,∃O	'b' is new; 'a' isn't.
	(5)	| Fb	4,&O	
	(6)	| Hb	4,&O	
	(7)	| Fb → Gb	1,∀O	
	(8)	| Gb	5,7,→O	
	(9)	| Gb & Hb	6,8,&I	
	(10)	| ∃x(Gx & Hx)	9,∃I	

The line marked '▶' is completely useless; it just gets in the way, as can be seen immediately in line (4). This derivation is *not* incorrect; it would receive full credit on an exam (supposing it was assigned!); rather, it is somewhat disfigured.

In Examples 1-3, there are no names in the derivation except those introduced by ∃O. At the point we apply ∃O, there aren't any names in the derivation, so any name will do! Thus, the requirement that the name be new is easy to satisfy. However, in other problems, additional names are involved, and the requirement is not trivially satisfied.

Nonetheless, the requirement that the name be new is important, because it blocks erroneous derivations (and in particular, erroneous derivations of invalid arguments). Consider the following.

Invalid argument

(A) ∃xFx
∃xGx
/ ∃x(Fx & Gx)

at least one thing is F
at least one thing is G
/ at least one thing is both F and G

There are many counterexamples to this argument; consider two of them.

Counterexamples

at least one number is even
at least one number is odd
/ at least one number is both even and odd

at least one person is female
at least one person is male
/ at least one person is both male and female

Argument (A) is clearly invalid. However, consider the following erroneous derivation.

Example 4 (erroneous derivation)

(1)	∃xFx	Pr	
(2)	∃xGx	Pr	
(3)	SHOW: ∃x(Fx & Gx)	DD	
(4)	Fa	1,∃O	
(5)	Ga	2,∃O	WRONG!!!
(6)	Fa & Ga	4,5,&I	
(7)	∃x(Fx & Gx)	6,∃I	

The reason line (5) is wrong concerns the use of the name 'a', which is definitely not new, since it appears in line (4). To be a proper application of ∃O, the name must be new, so we would have to instantiate Gx to Gb or Gc, anything but Ga. When we correct line (5), the derivation looks like the following.

(1)	∃xFx	Pr	
(2)	∃xGx	Pr	
(3)	SHOW: ∃x(Fx & Gx)	DD	
(4)	Fa	1,∃O	
(5)	Gb	2,∃O	RIGHT!!!
(6)	??????	???	but we can't finish

Now, the derivation cannot be completed, but that is good, because the argument in question is, after all, *invalid*!

The previous examples do not involve multiply quantified formulas, so it is probably a good idea to consider some of those.

Example 5

(1)	$\forall x(Fx \rightarrow \exists yHy)$	Pr
(2)	SHOW: $\exists xFx \rightarrow \exists yHy$	CD
(3)	│ $\exists xFx$	As
(4)	│ SHOW: $\exists yHy$	DD
(5)	│ │ Fa	3,∃O
(6)	│ │ $Fa \rightarrow \exists yHy$	1,∀O
(7)	│ │ $\exists yHy$	5,6,→O

As noted in the previous chapter, the premise may be read

if anything is F, then something is H,

whereas the conclusion may be read

if something is F, then something is H.

Under very special circumstances, ‘if any...’ is equivalent to ‘if some...’; this is one of the circumstances. These two are equivalent. We have shown that the latter follows from the former. To balance things, we now show the converse as well.

Example 6

(1)	$\exists xFx \rightarrow \exists yHy$	Pr
(2)	SHOW: $\forall x(Fx \rightarrow \exists yHy)$	UD
(3)	│ SHOW: $Fa \rightarrow \exists yHy$	CD
(4)	│ │ Fa	As
(5)	│ │ SHOW: $\exists yHy$	DD
(6)	│ │ │ $\exists xFx$	4,∃I
(7)	│ │ │ $\exists yHy$	1,6,→O

Before turning to examples involving relational quantification, we do one more example involving multiple quantification.

Example 7

(1)	∃xFx → ∀x~Gx	Pr
(2)	~~SHOW~~: ∀x[Fx → ~∃yGy]	UD
(3)	│ ~~SHOW~~: Fa → ~∃yGy	CD
(4)	│ │ Fa	As
(5)	│ │ ~~SHOW~~: ~∃yGy	ID
(6)	│ │ │ ∃yGy	As
(7)	│ │ │ ~~SHOW~~: ✕	DD
(8)	│ │ │ │ Gb	6,∃O
(9)	│ │ │ │ ∃xFx	4,∃I
(10)	│ │ │ │ ∀x~Gx	1,9,→O
(11)	│ │ │ │ ~Gb	10, ∀O
(12)	│ │ │ │ ✕	8,11,✕I

As in many previous sections, we conclude this section with some examples that involve relational quantification.

Example 8

(1)	∀x∀y(Kxy → Rxy)	Pr
(2)	∃x∃yKxy	Pr
(3)	~~SHOW~~: ∃x∃yRxy	DD
(4)	│ ∃yKay	2,∃O
(5)	│ Kab	4,∃O
(6)	│ ∀y(Kay → Ray)	1,∀O
(7)	│ Kab → Rab	6,∀O
(8)	│ Rab	5,7,→O
(9)	│ ∃yRay	8,∃I
(10)	│ ∃x∃yRxy	9,∃I

Example 9

(1)	∀x∃yRxy	Pr
(2)	∀x∀y[Rxy → Rxx]	Pr
(3)	∀x[Rxx → ∀yRyx]	Pr
(4)	~~SHOW~~: ∀x∀yRxy	UD
(5)	│ ~~SHOW~~: ∀yRay	UD
(6)	│ │ ~~SHOW~~: Rab	DD
(7)	│ │ │ ∃yRby	1,∀O
(8)	│ │ │ Rbc	7,∃O
(9)	│ │ │ ∀y[Rby → Rbb]	2,∀O
(10)	│ │ │ Rbc → Rbb	9,∀O
(11)	│ │ │ Rbb	8,9,→O
(12)	│ │ │ Rbb → ∀yRyb	3,∀O
(13)	│ │ │ ∀yRyb	11,12,→O
(14)	│ │ │ Rab	13,∀O

10. HOW EXISTENTIAL-OUT DIFFERS FROM THE OTHER RULES

As stated in the previous section, although we annotate existential-out just like other elimination rules (like $\rightarrow$O, $\vee$O, $\forall$O, etc.), it is not a *true inference rule*, but is rather an *assumption rule*. In the present section, we show exactly how $\exists$O is different from the other rules in predicate and sentential logic.

First consider a simple application of the rule $\forall$O.

$$\frac{\forall xFx}{Fa}$$

This is a valid argument of predicate logic, and the corresponding derivation is trivial.

(1)	$\forall xFx$	Pr
(2)	~~SHOW~~: Fa	DD
(3)	| Fa	2,$\forall$O

Next, consider a simple application of the rule $\exists$I.

$$\frac{Fa}{\exists xFx}$$

Again, the argument is valid, and the derivation is trivial.

(1)	Fa	Pr
(2)	~~SHOW~~: $\exists xFx$	DD
(3)	| $\exists xFx$	1,$\exists$I

The same can be said for every inference rule of predicate logic and sentential logic. Specifically, every inference rule corresponds to a valid argument. In each case we derive the conclusion simply by appealing to the rule in question.

But what about $\exists$O? Does it correspond to a valid argument? Earlier, I mentioned that, although the notation makes it look like $\forall$O, it is not really an inference rule, but is rather an *assumption rule*, much like the assumption rules associated with CD and ID

Why is it not a true inference rule? The answer is that it does not correspond to a valid argument in predicate logic! The argument form is the following.

$$\frac{\exists xFx}{Fa}$$

In English, this reads as follows.

something is F
therefore, a is F

That this argument form is invalid is seen by observing the following counterexample.

(1) someone is a pacifist
(2) therefore, Adolf Hitler is a pacifist

If one has ∃xFx, one is entitled to *assume* Fa so long as 'a' is new. So, we can *assume* (for the sake of argument) that Hitler is a pacifist, but we surely cannot deduce the false conclusion that Hitler is/was a pacifist from the true premise that at least one person is a pacifist.

The argument is invalid, but one might still wonder whether we can nonetheless construct a derivation "proving" it is in fact valid. If we could do that, then our derivation system would be inconsistent and useless, so let's hope we cannot!

Well, can we derive Fa from ∃xFx? If we follow the pattern used above, first we write down the problem, then we solve it simply by applying the appropriate rule of inference. Following this pattern, the derivation goes as follows.

(1)	∃xFx	Pr	
(2)	SHOW: Fa	DD	
(3)	| Fa	1,∃O	WRONG!!!

This derivation is erroneous, because in line (3) 'a' is not a permitted substitution according to the ∃O rule, because the letter used is not new, since 'a' already appears in line (2)! We are permitted to write down Fb, Fc, Fd, or a host of other formulas, but none of these makes one bit of progress toward showing Fa. That is good, because Fa does not follow from the premise!

Thus, in spite of the notation, ∃O is quite different from the other rules. When we apply ∃O to an existential formula (say, ∃xFx) to obtain a formula (say, Fc), we are *not* inferring or deducing Fc from ∃xFx. After all, this is not a valid inference. Rather, we are writing down an *assumption*. Some assumptions are permitted and some are not; this is an example of a permitted assumption (provided, of course, the name is new) just like assuming the antecedent in conditional derivation.

11. NEGATION QUANTIFIER ELIMINATION RULES

Earlier in the chapter, I promised six rules, and now we have four of them. The remaining two are tilde-existential-out and tilde-universal-out. As their names are intended to suggest, the former is a rule for eliminating any formula that is a negation of an existential formula, and the latter is a rule for eliminating any formula that is a negation of a universal formulas. These rules are officially given as follows.

Tilde-Existential-Out (~∃O)

If a line of the form ~∃vF[v] is available,
then one can infer the formula ∀v~F[v].

Tilde-Universal-Out (~∀O)

If a line of the form ~∀vF[v] is available,
then one can infer the formula ∃v~F[v].

Schematically, these rules may be presented as follows.

~∃O: ~∃vF[v]
———————
∀v~F[v]

~∀O: ~∀vF[v]
———————
∃v~F[v]

Before continuing, we observe is that both of these rules are *derived rules*, which is to say that they can be derived from the previous rules. In other words, these rules are completely *dispensable*: any conclusion that can be derived using either rule can be derived without using it. They are added for the sake of convenience.

First, let us consider ~∃O, and let us consider its simplest instance (where F[v] is Fx). Then ~∃O amounts to the following argument.

Argument 1

~∃xFx	it is not true that there is at least one thing such that it is F;
———	therefore,
∀x~Fx	everything is such that it is not F.

Recall from the previous chapters that the colloquial translation of the premise is 'nothing is F', and the colloquial translation of the conclusion is 'everything is unF'

The following derivation demonstrates that Argument 1 is valid, by deducing the conclusion from the premise.

(1)	~∃xFx	Pr
(2)	SHOW: ∀x~Fx	UD
(3)	│ SHOW: ~Fa	ID
(4)	│ │ Fa	As
(5)	│ │ SHOW: ✕	DD
(6)	│ │ │ ∃xFx	4,∃I
(7)	│ │ │ ✕	1,6,✕I

Next, let us consider ~∀O, and let us consider the simplest instance.

Argument 2

~∀xFx	it is not true that everything is such that it is F
———	therefore,
∃x~Fx	there is at least one thing such that it is not F

Recall from the previous chapter that the colloquial translation of the premise is 'not everything is F' and the colloquial translation of the conclusion is 'something is not F'.

The following derivation demonstrates that Argument 2 is valid. It employs (lines 1, 5, 11) a seldom-used sentential logic strategy.

▶	(1)	~∀xFx	Pr
	(2)	SHOW: ∃x~Fx	ID
	(3)	│ ~∃x~Fx	As
	(4)	│ SHOW: ✕	DD
▶	(5)	│ │ SHOW: ∀xFx	UD
	(6)	│ │ │ SHOW: Fa	ID
	(7)	│ │ │ │ ~Fa	As
	(8)	│ │ │ │ SHOW: ✕	DD
	(9)	│ │ │ │ │ ∃x~Fx	7,∃I
	(10)	│ │ │ │ │ ✕	3,9,✕I
▶	(11)	│ │ ✕	1,5,✕I

In each derivation, we have only shown the simplest instance of the rule, where F[v] is Fx. However, the complicated instances are shown in precisely the same manner. We can in principle show for any formula F[v] and variable v that ∀v~F[v] follows from ~∃vF[v], and that ∃v~F[v] follows from ~∀vF[v].

Note that the converse arguments are also valid, as demonstrated by the following derivations.

(1)	∀x~Fx	Pr
(2)	SHOW: ~∃xFx	ID
(3)	│ ∃xFx	As
(4)	│ SHOW: ✕	DD
(5)	│ │ Fa	3,∃O
(6)	│ │ ~Fa	1,∀O
(7)	│ │ ✕	5,6,✕I

(1)	∃x~Fx	Pr
(2)	~~SHOW~~: ~∀xFx	ID
(3)		∀xFx	As
(4)		~~SHOW~~: ✕	DD
(5)			~Fa	1,∃O
(6)			Fa	3,∀O
(7)			✕	5,6,✕I

Note carefully, however, that neither of the converse arguments corresponds to any rule in our system. In particular,

> THERE IS NO RULE TILDE-EXISTENTIAL-IN.

> THERE IS NO RULE TILDE-UNIVERSAL-IN.

The corresponding arguments are valid, and accordingly can be demonstrated in our system. However, they are not inference rules. As usual, not every valid argument form corresponds to an inference rule. This is simply a choice we make – we only have negation-connective elimination rules, and no negation-connective introduction rules.

Before proceeding, let us look at several applications of ~∃O and ~∀O to specific formulas, in order to get an idea of what the syntactic possibilities are.

(1) ~∃xFx
―――――
∀x~Fx

(2) ~∃x(Fx & Gx)
―――――
∀x~(Fx & Gx)

(3) ~∃x(Fx & ∀y(Gy → Rxy))
―――――
∀x~(Fx & ∀y(Gy → Rxy))

(4) ~∀xFx
―――――
∃x~Fx

(5) ~∀x(Fx → Gx)
―――――
∃x~(Fx → Gx)

(6) ~∀x(Fx → ∃y(Gy & Rxy))
―――――
∃x~(Fx → ∃y(Gy & Rxy))

Having seen several examples of *proper* applications of $\sim\exists\mathrm{O}$ or $\sim\forall\mathrm{O}$, it is probably a good idea to see examples of *improper* applications.

(7) $\dfrac{\sim(\exists xFx \vee \exists yGy)}{(\forall x{\sim}Fx \vee \exists yGy)}$ WRONG!!!

(8) $\dfrac{\sim\exists xFx \rightarrow \forall xGx}{\forall x{\sim}Fx \rightarrow \forall xGx}$ WRONG!!!

In each example, the error is that the premise does not have the correct form. In (7), the premise is a negation of a disjunction, not a negation of an existential. The appropriate rule is $\sim\vee\mathrm{O}$, not $\sim\exists\mathrm{O}$. In (8), the premise is a conditional, so the appropriate rule is $\rightarrow\mathrm{O}$.

Of course, sometimes an improper application of a rule produces a valid conclusion, and sometimes it does not. (8) is a valid argument, but so are a lot of arguments. The question here is not whether the argument is valid, but whether it is an application of a rule. Some valid arguments correspond to rules, and hence do not have to be explicitly shown; other valid arguments do not correspond to particular rules, and hence must be shown to be valid by constructing a derivation. Recall, as usual:

> INFERENCE RULES APPLY
> EXCLUSIVELY TO WHOLE LINES,
> NOT TO PIECES OF LINES.

(8) is valid, so we can derive its conclusion from its premise. The following is one such derivation. It also illustrates a further point about our new rules.

Example 1

	(1)	$\sim\exists xFx \rightarrow \forall xGx$	Pr
	(2)	~~SHOW~~: $\forall x{\sim}Fx \rightarrow \forall xGx$	CD
	(3)		$\forall x{\sim}Fx$	As
▶	(4)		~~SHOW~~: $\forall xGx$	ID
	(5)			$\sim\forall xGx$	As
	(6)			~~SHOW~~: ✕	DD
	(7)				$\sim\sim\exists xFx$	1,5,$\rightarrow$O
	(8)				$\exists xFx$	7,DN
	(9)				Fa	8,$\exists$O
	(10)				$\sim Fa$	3,$\forall$O
	(11)				✕	9,10,✕I

This derivation is curious in the following way: line (4) is shown by indirect derivation, rather than universal derivation. But this is permissible, since ID is suitable for any kind of formula.

Indeed, once we have the rule $\sim\forall$O, we can show any universal formula by ID. By way of illustration, consider Example 2 from Section 7, first done using UD, then done using ID.

Example 2 (done using UD)

	(1)	$\forall x(Fx \rightarrow Gx)$	Pr
	(2)	~~SHOW~~: $\forall xFx \rightarrow \forall xGx$	CD
	(3)	$\forall xFx$	As
▶	(4)	~~SHOW~~: $\forall xGx$	UD
	(5)	~~SHOW~~: Ga	DD
	(6)	$Fa \rightarrow Ga$	1,$\forall$O
	(7)	Fa	3,$\forall$O
	(8)	Ga	6,7,$\rightarrow$O

Example 2 (done using ID)

	(1)	$\forall x(Fx \rightarrow Gx)$	Pr
	(2)	~~SHOW~~: $\forall xFx \rightarrow \forall xGx$	CD
	(3)	$\forall xFx$	As
▶	(4)	~~SHOW~~: $\forall xGx$	ID
	(5)	$\sim\forall xGx$	As
	(6)	~~SHOW~~: ✕	DD
	(7)	$\exists x\sim Gx$	5,$\sim\forall$O
	(8)	$\sim Ga$	7,$\exists$O
	(9)	$Fa \rightarrow Ga$	1,$\forall$O
	(10)	Fa	3,$\forall$O
	(11)	Ga	9,10,$\rightarrow$O
	(12)	✕	8,11,✕I

Now that we have $\sim\forall$O, it is always possible to show a universal by indirect derivation. However, the resulting derivation is usually longer than the derivation using universal derivation. On rare occasions, the indirect derivation is easier; for example go back and try to do Example 1 using universal derivation.

We conclude this section with a derivation that uses $\sim\forall$O in a straightforward way; it also involves relational quantification.

Example 3

	(1)	$\forall x(\forall yRxy \rightarrow \sim\forall yRyx)$	Pr
	(2)	$\exists x\forall yRxy$	Pr
	(3)	~~SHOW~~: $\exists x\exists y\sim Rxy$	DD
	(4)	$\forall yRay$	2,$\exists$O
	(5)	$\forall yRay \rightarrow \sim\forall yRya$	1,$\forall$O
	(6)	$\sim\forall yRya$	4,5,$\rightarrow$O
▶	(7)	$\exists y\sim Rya$	6,$\sim\forall$O
	(8)	$\sim Rba$	7,$\exists$O
	(9)	$\exists y\sim Rby$	8,$\exists$I
	(10)	$\exists x\exists y\sim Rxy$	9,$\exists$I

12. DIRECT VERSUS INDIRECT DERIVATION OF EXISTENTIALS

Adding ~∀O to our list of rules enables us to show universals using indirect derivation. This particular use of ~∀O is really no big deal, since we already have a derivation technique (i.e., universal derivation) that is perfect for universals.

Whereas we have a derivation scheme (show-rule) specially designed for universal formulas, we do not have such a rule for existential formulas. You may have noticed that, in every previous example involving 'SHOW: ∃vF[v]', we have used direct derivation. This corresponds to a derivation strategy, which is schematically presented as follows.

Direct Derivation Strategy for Existentials

```
~~SHOW~~: ∃vF[v]              DD
|  .
|  .
|  .
|  .
|  F[n]
|  ∃vF[v]                     ∃I
```

But now we have an additional rule, ~∃O, so we can show any existential formula using indirect derivation. This gives rise to a new strategy, which is schematically presented as follows.

Indirect Derivation Strategy for Existentials

```
~~SHOW~~: ∃vF[v]              ID
|  ~∃vF[v]                    As
|  ~~SHOW~~: ×                DD
|  |  ∀v~F[v]                 ~∃O
|  |  .
|  |  .
|  |  ×
```

Many derivation problems can be solved using either strategy. For example, recall Example 1 from Section 8.

Example 1d (DD strategy):

(1)	∀x(Fx → Hx)	Pr
(2)	∃xFx	Pr
(3)	~~SHOW~~: ∃xHx	DD
(4)	| Fa	2,∃O
(5)	| Fa → Ha	1,∀O
(6)	| Ha	4,5,→O
(7)	| ∃xHx	6,∃I

Example 1i (ID strategy)

(1)	∀x(Fx → Hx)	Pr
(2)	∃xFx	Pr
(3)	~~SHOW~~: ∃xHx	ID
(4)	| ~∃xHx	As
(5)	| ~~SHOW~~: **×**	DD
(6)	| | ∀x~Hx	4,~∃O
(7)	| | Fa	2,∃O
(8)	| | Fa → Ha	1,∀O
(9)	| | Ha	7,8,→O
(10)	| | ~Ha	6,∀O
(11)	| | **×**	9,10,**×**I

Comparing these two derivations illustrates an important point. Even though we can use the ID strategy, it may end up producing a longer derivation than if we use the DD strategy instead.

On the other hand, there are derivation problems in which the DD strategy will not work in a straightforward way [recall that every indirect derivation can be converted into a "trick" derivation that does not use ID]; in these problems, it is best to use the ID strategy. Consider the following example; besides illustrating the ID strategy for existentials, it also recalls an important sentential derivation strategy.

Example 2

▶	(1)	$\exists$xFx $\vee$ $\exists$xGx	Pr
▶▶	(2)	~~SHOW~~: $\exists$x(Fx $\vee$ Gx)	ID
	(3)	| ~$\exists$x(Fx $\vee$ Gx)	As
	(4)	| ~~SHOW~~: ✕	DD
	(5)	| | $\forall$x~(Fx $\vee$ Gx)	3,~$\exists$O
▶	(6)	| | ~~SHOW~~: ~$\exists$xFx	ID
	(7)	| | | $\exists$xFx	As
	(8)	| | | ~~SHOW~~: ✕	DD
	(9)	| | | | Fa	7,$\exists$O
	(10)	| | | | ~(Fa $\vee$ Ga)	5,$\forall$O
	(11)	| | | | ~Fa	10,~$\vee$O
	(12)	| | | | ✕	9,11,✕I
▶	(13)	| | $\exists$xGx	1,6,$\vee$O
	(14)	| | Gb	13,$\exists$O
	(15)	| | ~(Fb $\vee$ Gb)	5,$\forall$O
	(16)	| | ~Gb	15,~$\vee$O
	(17)	| | ✕	14,16,✕I

Recall the *wedge-out strategy* from sentential logic:

> **Wedge-Out Strategy**
>
> If you *have* a disjunction (for example, it is a premise), then you try to *find* (or *show*) the negation of one of the disjuncts.

We are following the wedge-out strategy in line (6).

While we are on the topic of sentential derivation strategies, let us recall two other strategies, the first being the wedge-derivation strategy, which is schematically presented as follows.

> **Wedge-Derivation Strategy**

~~SHOW~~: $\mathcal{A} \vee \mathcal{B}$	ID
| ~($\mathcal{A} \vee \mathcal{B}$)	As
| ~~SHOW~~: ✕	DD
| | ~$\mathcal{A}$	~$\vee$O
| | ~$\mathcal{B}$	~$\vee$O
| | .	
| | .	
| | .	
| | ✕	✕I

This strategy is employed in the following example, which is the converse of 2.

Example 2c

(1)	∃x(Fx ∨ Gx)	Pr
(2)	~~SHOW~~: ∃xFx ∨ ∃xGx	ID
(3)	~(∃xFx ∨ ∃xGx)	As
(4)	~~SHOW~~: ✕	DD
(5)	~∃xFx	3,~∨O
(6)	~∃xGx	3,~∨O
(7)	∀x~Fx	5,~∃O
(8)	∀x~Gx	6,~∃O
(9)	Fa ∨ Ga	1,∃O
(10)	~Fa	7,∀O
(11)	~Ga	8,∀O
(12)	Ga	9,10,∨O
(13)	✕	11,12,✕I

Another sentential strategy is the arrow-out strategy, which is given as follows.

Arrow-Out Strategy

If you *have* a conditional (for example, it is a premise), then you try to *find* (or *show*) either the antecedent or the negation of the consequent.

The following example illustrates the arrow-out strategy; it also reiterates a point made in Chapter 6 – namely, that an existential-conditional formula, e.g., ∃x(Fx → Gx), does not say much, and certainly does not say that some F is G.

Example 3

```
►   (1)  ∀xFx → ∃xGx                     Pr
►►  (2)  SHOW: ∃x(Fx → Gx)               ID
    (3)  | ~∃x(Fx → Gx)                  As
    (4)  | SHOW: ×                       DD
    (5)  | | ∀x~(Fx → Gx)                3,~∃O
►   (6)  | | SHOW: ∀xFx                  UD
    (7)  | | | SHOW: Fa                  DD
    (8)  | | | | | ~(Fa → Ga)            5,∀O
    (9)  | | | | | Fa & ~Ga              8,~→O
    (10) | | | | | Fa                    9,&O
►   (11) | | ∃xGx                        1,6,→O
    (12) | | Gb                          11,∃O
    (13) | | ~(Fb → Gb)                  5,∀O
    (14) | | Fb & ~Gb                    13,~→O
    (15) | | ~Gb                         14,&O
    (16) | | ×                           12,15,×I
```

In line (6) above, we apply the arrow-out strategy, electing in particular to show the antecedent.

The converse of the above argument can also be shown, as follows, which demonstrates that ∃x(Fx→Gx) is equivalent to ∀xFx→∃xGx, which says that something is G if everything is F.

Example 3c

```
(1)  ∃x(Fx → Gx)                 Pr
(2)  SHOW: ∀xFx → ∃xGx           CD
(3)  | ∀xFx                      As
(4)  | SHOW: ∃xGx                ID
(5)  | | ~∃xGx                   As
(6)  | | SHOW: ×                 DD
(7)  | | | ∀x~Gx                 5,~∃O
(8)  | | | Fa → Ga               1,∃O
(9)  | | | Fa                    3,∀O
(10) | | | ~Ga                   7,∀O
(11) | | | Ga                    8,9,&I
(12) | | | ×                     10,11,×I
```

Note carefully that the ID strategy is used at line (4), but only for the sake of illustrating this strategy. If one uses the DD strategy, then the resulting derivation is much shorter! This is left as an exercise for the student.

The last several examples of the section involve relational quantification. Many of the problems are done both with and without ID

Example 4

(1) there is a Freshman who respects every Senior
(2) therefore, for every Senior, there is a Freshman who respects him/her

Example 4d (DD strategy)

	(1)	∃x(Fx & ∀y(Sy → Rxy))	Pr
	(2)	~~SHOW~~: ∀x(Sx → ∃y(Fy & Ryx))	UD
	(3)	│ ~~SHOW~~: Sa → ∃y(Fy & Rya)	CD
	(4)	│ │ Sa	As
▶	(5)	│ │ ~~SHOW~~: ∃y(Fy & Rya)	DD
	(6)	│ │ │ Fb & ∀y(Sy → Rby)	1,∃O
	(7)	│ │ │ Fb	6,&O
	(8)	│ │ │ ∀y(Sy → Rby)	6,&O
	(8)	│ │ │ Sa → Rba	8,∀O
	(9)	│ │ │ Rba	4,8,→O
	(10)	│ │ │ Fb & Rba	7,9,&I
	(11)	│ │ │ ∃y(Fy & Rya)	10,∃I

Example 4i (ID strategy)

	(1)	∃x(Fx & ∀y(Sy → Rxy))	Pr
	(2)	~~SHOW~~: ∀x(Sx → ∃y(Fy & Ryx))	UD
	(3)	│ ~~SHOW~~: Sa → ∃y(Fy & Rya)	CD
	(4)	│ │ Sa	As
▶	(5)	│ │ ~~SHOW~~: ∃y(Fy & Rya)	ID
	(6)	│ │ │ ~∃y(Fy & Rya)	As
	(7)	│ │ │ ~~SHOW~~: ✕	DD
	(8)	│ │ │ │ ∀y~(Fy & Rya)	6,~∃O
	(9)	│ │ │ │ Fb & ∀y(Sy → Rby)	1,∃O
	(10)	│ │ │ │ Fb	9,&O
	(11)	│ │ │ │ ∀y(Sy → Rby)	9,&O
	(12)	│ │ │ │ ~(Fb & Rba)	8,∀O
	(13)	│ │ │ │ Sa → Rba	11,∀O
	(14)	│ │ │ │ Rba	4,13,→O
	(15)	│ │ │ │ Fb → ~Rba	12,~&O
	(16)	│ │ │ │ ~Rba	10,15,→O
	(17)	│ │ │ │ ✕	14,16,✕I

Note that this derivation can be shortened by two lines at the end (exercise for the student!)

The previous problem was solved using both ID and DD. The next problem is done both ways as well.

Example 5

(1) there is someone who doesn't respect any Freshman
(2) therefore, for every Freshman, there is someone who doesn't respect him/her.

Example 5d (DD strategy)

	(1)	∃x~∃y(Fy & Ryx)	Pr
	(2)	~~SHOW~~: ∀x(Fx → ∃y~Rxy)	UD
	(3)	│ ~~SHOW~~: Fa → ∃y~Ray	CD
	(4)	│ │ Fa	As
▶	(5)	│ │ ~~SHOW~~: ∃y~Ray	DD
	(6)	│ │ │ ~∃y(Fy & Ryb)	1,∃O
	(7)	│ │ │ ∀y~(Fy & Ryb)	6,~∃O
	(8)	│ │ │ ~(Fa & Rab)	7,∀O
	(9)	│ │ │ Fa → ~Rab	8,~&O
	(10)	│ │ │ ~Rab	4,9,→O
	(11)	│ │ │ ∃y~Ray	10,∃I

Example 5i (ID strategy)

	(1)	∃x~∃y(Fy & Ryx)	Pr
	(2)	~~SHOW~~: ∀x(Fx → ∃y~Rxy)	UD
	(3)	│ ~~SHOW~~: Fa → ∃y~Ray	CD
	(4)	│ │ Fa	As
▶	(5)	│ │ ~~SHOW~~: ∃y~Ray	ID
	(6)	│ │ │ ~∃y~Ray	As
	(7)	│ │ │ ~~SHOW~~: ✕	DD
	(8)	│ │ │ │ ∀y~~Ray	6,~∃O
	(9)	│ │ │ │ ~∃y(Fy & Ryb)	1,∃O
	(10)	│ │ │ │ ∀y~(Fy & Ryb)	9,~∃O
	(11)	│ │ │ │ ~(Fa & Rab)	10,∀O
	(12)	│ │ │ │ Fa → ~Rab	11,~&O
	(12)	│ │ │ │ ~~Rab	8,∀O
	(14)	│ │ │ │ ~Fa	12,13,→O
	(15)	│ │ │ │ ✕	4,14,✕I

The final example of this section is considerably more complex than the previous ones. It is done only once, using ID. Using the ID strategy is hard enough; using the DD strategy is also hard; try it and see!

Example 6

(1) every Freshman respects Adams
(2) there is a Senior who doesn't respect any one who respects Adams
(3) therefore, there is a Senior who doesn't respect any Freshman

	(1)	$\forall x(Fx \rightarrow Rxa)$	Pr
	(2)	$\exists x(Sx \;\&\; {\sim}\exists y(Rya \;\&\; Rxy))$	Pr
▶	(3)	SHOW: $\exists x(Sx \;\&\; {\sim}\exists y(Fy \;\&\; Rxy))$	ID
	(4)	$\quad {\sim}\exists x(Sx \;\&\; {\sim}\exists y(Fy \;\&\; Rxy))$	As
	(5)	$\quad$ SHOW: ✕	DD
	(6)	$\qquad \forall x{\sim}(Sx \;\&\; {\sim}\exists y(Fy \;\&\; Rxy))$	4,$\sim\exists$O
	(7)	$\qquad Sb \;\&\; {\sim}\exists y(Rya \;\&\; Rby)$	2,$\exists$O
	(8)	$\qquad Sb$	7,&O
	(9)	$\qquad {\sim}\exists y(Rya \;\&\; Rby)$	7,&O
	(10)	$\qquad \forall y{\sim}(Rya \;\&\; Rby)$	9,$\sim\exists$O
	(11)	$\qquad {\sim}(Sb \;\&\; {\sim}\exists y(Fy \;\&\; Rby))$	6,$\forall$O
	(12)	$\qquad Sb \rightarrow {\sim}{\sim}\exists y(Fy \;\&\; Rby)$	11,$\sim$&O
	(13)	$\qquad {\sim}{\sim}\exists y(Fy \;\&\; Rby)$	8,12,$\rightarrow$O
	(14)	$\qquad \exists y(Fy \;\&\; Rby)$	13,DN
	(15)	$\qquad Fc \;\&\; Rbc$	14,$\exists$O
	(16)	$\qquad Fc$	15,&O
	(17)	$\qquad Rbc$	15,&O
	(18)	$\qquad Fc \rightarrow Rca$	1,$\forall$O
	(19)	$\qquad Rca$	16,18,$\rightarrow$O
	(20)	$\qquad {\sim}(Rca \;\&\; Rbc)$	10,$\forall$O
	(21)	$\qquad Rca \rightarrow {\sim}Rbc$	20,$\sim$&O
	(22)	$\qquad {\sim}Rbc$	19,21,$\rightarrow$O
	(23)	$\qquad$ ✕	17,22,✕I

What strategy should one employ in showing existential formulas? The following principles might be useful in deciding between the two strategies.

1. If any strategy will work, the ID strategy will. The worst that can happen is that the derivation is longer than it needs to be.
2. If there are no names available, and if there are no existential formulas to instantiate in order to obtain names, then the ID strategy is advisable, although a "trick" derivation is still possible.
3. When it works in a straightforward way (and it usually does), the DD strategy produces a prettier derivation. The worst that can happen is that one has to start over, and use ID
4. If names are obtainable by applying ∃O, then the DD strategy will probably work; however, it might be harder than the ID strategy.

I conclude with the following principle, based on 1-4.

If you want a risk-free technique, use the ID strategy.

If you want more of a challenge, use the DD strategy.

13. APPENDIX 1: THE SYNTAX OF PREDICATE LOGIC

In this appendix, we review the syntactic features of predicate logic that are crucial to understanding derivations in predicate logic. These include the following notions.

(1) principal (major) connective
(2) free occurrence of a variable
(3) substitution instance
(4) alphabetic variant

1. OFFICIAL PRESENTATION OF THE SYNTAX OF PREDICATE LOGIC

A. Singular Terms.

1. Variables: x, y, z;
2. Constants: a, b, c, ..., w;
X. Nothing else is an singular term.

B. Predicate Letters.

0. 0-place predicate letters: A, B, ..., Z;
1. 1-place predicate letters: the same;
2. 2-place predicate letters: the same;
3. 3-place predicate letters: the same;
and so forth...
X. Nothing else is a predicate letter.

C. Quantifiers.

1. Universal Quantifiers: ∀x, ∀y, ∀z.
2. Existential Quantifiers: ∃x, ∃y, ∃z.
X. Nothing else is a quantifier.

D. Atomic Formulas.

1. If P is an n-place predicate letter, and $t_1,...,t_n$ are singular terms, then $Pt_1...t_2$ is an atomic formula.

X. Nothing else is an atomic formula.

E. Formulas.

1. Every atomic formula is a formula.
2. If $\mathcal{A}$ is a formula, then so is $\sim\mathcal{A}$.
3. If $\mathcal{A}$ and $\mathcal{B}$ are formulas, then so are:

(a) $(\mathcal{A} \,\&\, \mathcal{B})$
(b) $(\mathcal{A} \vee \mathcal{B})$
(c) $(\mathcal{A} \rightarrow \mathcal{B})$
(d) $(\mathcal{A} \leftrightarrow \mathcal{B})$.

4. If $\mathcal{A}$ is a formula, then so are:
   $\forall x\mathcal{A}$, $\forall y\mathcal{A}$, $\forall z\mathcal{A}$,
   $\exists x\mathcal{A}$, $\exists y\mathcal{A}$, $\exists z\mathcal{A}$.

X. Nothing else is a formula.

Given the above characterization of the syntax of predicate logic, we see that every formula is exactly one of the following.

1. An atomic formula; there are no connectives:

   Fa, Fx, Rab, Rax, Rxb, etc.

2. A negation; the major connective is negation:

   ~Fa, ~Rxy, ~(Fx & Gx), ~∀xFx, ~∃x∀yRxy, ~∀x(Fx → Gx), etc.

3. A universal; the major connective is a universal quantifier:

   ∀xFx, ∀yRay, ∀x(Fx → Gx), ∀x∃yRxy, ∀x(Fx → ∃yRxy), etc.

4. An existential; the major connective is an existential quantifier:

   ∃zFz, ∃xRax, ∃x(Fx & Gx), ∃y∀xRxy, ∃x(Fx & ∀yRyx), etc.

5. A conjunction; the major connective is ampersand:

   Fx & Gy, ∀xFx & ∃yGy, ∀x(Fx → Gx) & ~∀x(Gx → Fx), etc.
   Fx ∨ Gy, ∀xFx ∨ ∃yGy, ∀x(Fx → Gx) ∨ ~∀x(Gx → Fx), etc.

6. A conditional; the major connective is arrow:

   Fx → Gx, ∀xFx → ∀xGx, ∀x(Fx → Gx) → ∀x(Fx → Hx), etc.

7. A biconditional; the major connective is double-arrow:

   Fx ↔ Gy, ∀xFx ↔ ∃yGy, ∀x(Fx → Gx) ↔ ~∀xGx, etc.

Now, just as in sentential logic, whether a rule of predicate logic applies to a given formula is primarily determined by what the formula's major connective is. (In the case of negations, the immediately subordinate formula must also be considered.) So it is important to be able to recognize the major connective of a formula of predicate logic.

2. FREEDOM AND BONDAGE

A. Variables versus Occurrences of Variables.

How many words are there in this paragraph? Well, it depends on what you mean. This question is actually ambiguous between the following two different questions. (1) How many different (unique) words are used in this paragraph? (2) How long is this paragraph in words, or how many *word occurrences* are there in this paragraph? The answer to the first question is: 46. On the other hand, the answer to the second question is: 93. For example, the word 'the' appears 10 times; which is to say that there are 10 *occurrences* of the word 'the' in this paragraph.

Just as a given word of English (e.g., 'the') can occur many times in a given sentence (or paragraph) of English, a given logic symbol can occur many times in a given formula. And in particular, a given variable can occur many times in a formula. Consider the following examples of occurrences of variables.

(1) Fx

'x' occurs once [or: there is one occurrence of 'x'.]

(2) Rxy

'x' occurs once; 'y' occurs once.

(3) Fx → Hx

'x' occurs twice.

(4) ∀x(Fx → Hx)

'x' occurs three times.

(5) ∀y(Fx → Hy)

'x' occurs once; 'y' occurs twice.

(6) ∀x(Fx → ∀xHx)

'x' occurs four times.

(7) ∀x∀y(Rxy → Ryx)

'x' occurs three times; 'y' occurs three times.

We also speak the same way about occurrences of other symbols and combinations of symbols. So, for example, we can speak of occurrences of '~', or occurrences of '∀x'.

B. Quantifier Scope.

> **Definition**
>
> The ***scope*** of an occurrence of a quantifier is, by definition, the smallest formula containing that occurrence.

The scope of a quantifier is exactly analogous to the scope of a negation sign in a formula of sentential logic. Consider the analogous definition.

> **Definition**
>
> The ***scope*** of an occurrence of '$\sim$' is, by definition, the smallest formula containing that occurrence.

Examples

(1) $\sim P \rightarrow Q$; the scope of $\sim$ is: $\sim P$;
(2) $\sim(P \rightarrow Q)$;the scope of $\sim$ is: $\sim(P \rightarrow Q)$;
(3) $P \rightarrow \sim(R \rightarrow S)$; the scope of $\sim$ is: $\sim(R \rightarrow S)$.

By analogy, consider the following involving universal quantifiers.

(1) $\forall xFx \rightarrow Fa$ the scope of $\forall x$ is: $\forall xFx$
(2) $\forall x(Fx \rightarrow Gx)$ the scope of $\forall x$ is: $\forall x(Fx \rightarrow Gx)$
(3) $Fa \rightarrow \forall x(Gx \rightarrow Hx)$ the scope of $\forall x$ is: $\forall x(Gx \rightarrow Hx)$

As a somewhat more complicated example, consider the following.

(4) $\forall x(\forall yRxy \rightarrow \forall zRzx)$

the scope of $\forall x$ is $\forall x(\forall yRxy \rightarrow \forall zRzx)$
the scope of $\forall y$ is $\forall yRxy$
the scope of $\forall z$ is $\forall zRzx$

As a still more complicated example, consider the following.

(5) $\forall x[\forall xFx \rightarrow \forall y(\forall yGy \rightarrow \forall zRxyz)]$;

the scope of the first $\forall x$ is the whole formula;
the scope of the second $\forall x$ is $\forall xFx$;
the scope of the first $\forall y$ is $\forall y(\forall yGy \rightarrow \forall zRxyz)$;
the scope of the second $\forall y$ is $\forall yGy$;
the scope of the only $\forall z$ is $\forall zRxyz$.

C. Government and Binding

Definition

'∀x' and '∃x' govern the variable 'x';
'∀y' and '∃y' govern the variable 'y';
'∀z' and '∃z' govern the variable 'z';
etc.

Definition

An occurrence of a quantifier *binds* an occurrence of a variable iff:

(1) the quantifier governs the variable,

and

(2) the occurrence of the variable is contained within the scope of the occurrence of the quantifier.

Definition

An occurrence of a quantifier *truly binds* an occurrence of a variable iff:

(1) the occurrence of the quantifier binds the occurrence of the variable,

and

(2) the occurrence of the quantifier is inside the scope of every occurrence of that quantifier that binds the occurrence of the variable.

Example

∀x(Fx → ∀xGx);

In this formula the first '∀x' binds every occurrence of 'x', but it only truly binds the first two occurrences; on the other hand, the second '∀x' truly binds the last two occurrences of 'x'.

D. Free versus Bound Occurrences of Variables

Every given occurrence of a given variable is either *free* or *bound*.

Definition

An occurrence of a variable in a formula F is *bound* in F if and only if that occurrence is bound by some quantifier occurrence in F.

Definition

An occurrence of a variable in a formula F is *free* in F if and only if that occurrence is not bound in F.

Examples

(1) Fx:

the one and only occurrence of 'x' is free in this formula;

(2) ∀x(Fx → Gx):

all three occurrences of 'x' are bound by '∀x';

(3) Fx → ∀xGx:

the first occurrence of 'x' is free; the remaining two occurrences are bound.

(4) ∀x(Fx → ∀xGx):

the first two occurrences of 'x' are bound by the first '∀x'; the second two are bound by the second '∀x'.

(5) ∀x(∀yRxy → ∀zRzx):

every occurrence of every variable is bound.

Notice in example (4) that the variable 'x' occurs within the scope of two different occurrences of '∀x'. It is only the innermost occurrence of '∀x' that truly binds the variable, however. The other occurrence of '∀x' binds the first occurrence of 'x' but none of the remaining ones.

3. SUBSTITUTION INSTANCES

Having described the difference between free and bound occurrences of variables, we turn to the topic of *substitution instance*, which is officially defined as follows.

> **Definition**
>
> Let v be any variable, let F[v] be any formula containing v, and let n be any name. Then a *substitution instance* of the formula F[v] is any formula F[n] obtained from F[v] by substituting occurrences of the name n for each and every occurrence of the variable v that is free in F[v].

Let us look at a few examples; in each example, I give examples of *correct* substitution instances, and then I give examples of *incorrect* substitution instances.

(1) Fx:

Correct: Fa; Fb; Fc; etc.;
Incorrect: Fx; Fy, Fz.

(2) Fx → Gx:

Correct: Fa → Ga; Fb → Gb; Fc → Gc; etc.;
Incorrect: Fa → Gb; Fb → Ga; Fy → Gy.

(3) Rxx:

Correct: Raa; Rbb; Rcc; etc.
Incorrect: Rab, Rba, Rxx.

(4) Fx → ∀xGx:

Correct: Fa → ∀xGx; Fb → ∀xGx; Fc → ∀xGx; etc.
Incorrect: Fy → ∀xGx; Fa → ∀aGa; Fb → ∀bGb.

(5) ∀yRxy:

Correct: ∀yRay; ∀yRby; ∀yRcy; etc.
Incorrect: ∀yRzy; ∀aRaa.

(6) ∀yRxy → ∀zRzx:

Correct: ∀yRay → ∀zRza; ∀yRby → ∀zRzb; ∀yRcy → ∀zRzc;
Incorrect: ∀yRzy → ∀zRza; ∀yRay → ∀zRzb.

In each case, you should convince yourself why the given formula is, or is not, a correct substitution instance.

4. ALPHABETIC VARIANTS

As you will recall, one can symbolize 'everything is F' in one of three ways:

(1) $\forall xFx$
(2) $\forall yFy$
(3) $\forall zFz$

Although these formulas are distinct, they are clearly equivalent. Yet, they are equivalent in a more intimate way than (say) the following formulas.

(4) $\forall x(Fx \rightarrow \forall yHy)$
(5) $\exists xFx \rightarrow \forall yHy$
(6) $\forall x\forall y(Fx \rightarrow Hy)$

(4)-(6) are mutually equivalent in a weaker sense than (1)-(3). If we translate (4)-(6) into English, they might read respectively as follows.

(r4) if anything is F, then everything is H;
(r5) if at least one thing is F, then everything is H;
(r6) for any two things, if the first is F, then the second is H.

These definitely don't sound the same; yet, we can prove that they are logically equivalent.

By contrast, if we translate (1)-(3) into English, they all read exactly the same.

(r1-3) everything is F.

We describe the relation between the various (1)-(3) by saying that they are *alphabetic variants* of one another. They are slightly different symbolic ways of saying exactly the same thing.

The formal definition of alphabetic variants is difficult to give in the general case of unlimited variables. But if we restrict ourselves to just three variables, then the definition is merely complicated.

Definition

A formula F is *closed* iff: no variable occurs free in F.

Definition

Let F_1 and F_2 be *closed* formulas. Then F_1 is an *alphabetic variant* of F_2 iff: F_1 is obtained from F_2 by *permuting* the variables 'x', 'y', 'z', which is to say applying one of the following procedures:

(1) replacing every occurrence of 'x' by 'y' and every occurrence of 'y' by 'x'.

(2) replacing every occurrence of 'x' by 'z' and every occurrence of 'z' by 'x'.

(3) replacing every occurrence of 'y' by 'z' and every occurrence of 'z' by 'y'.

(4) replacing every occurrence of 'x' by 'y' and every occurrence of 'y' by 'z' and every occurrence of 'z' by 'x'.

(5) replacing every occurrence of 'x' by 'z' and every occurrence of 'z' by 'y' and every occurrence of 'y' by 'x'.

Examples

(1) $\forall xFx$; $\forall yFy$; $\forall zFz$;
everyone is F.

(2) $\forall x(Fx \rightarrow Gx)$; $\forall y(Fy \rightarrow Gy)$; $\forall z(Fz \rightarrow Gz)$;
every F is G.

(3) $\forall x\exists yRxy$; $\forall x\exists zRxz$; $\forall y\exists zRyz$; $\forall y\exists xRyx$;
everyone respects someone (or other).

(4) $\forall x(Fx \rightarrow \exists y[Gy \,\&\, \forall z(Rxz \rightarrow Ryz)])$
$\forall x(Fx \rightarrow \exists z[Gz \,\&\, \forall y(Rxy \rightarrow Rzy)])$
$\forall y(Fy \rightarrow \exists z[Gz \,\&\, \forall x(Ryx \rightarrow Rzx)])$
$\forall y(Fy \rightarrow \exists x[Gx \,\&\, \forall z(Ryz \rightarrow Rxz)])$
$\forall z(Fz \rightarrow \exists x[Gx \,\&\, \forall y(Ryz \rightarrow Rxy)])$
$\forall z(Fz \rightarrow \exists y[Gy \,\&\, \forall x(Rzx \rightarrow Ryx)])$
for every F there is a G who respects everyone the F respects.

14. APPENDIX 2: SUMMARY OF RULES FOR SYSTEM PL (PREDICATE LOGIC)

A. Sentential Logic Rules

Every rule of SL (sentential logic) is also a rule of PL (predicate logic).

B. Rules that don't require a new name

In the following, **v** is any variable, **a** and **n** are names, **F[v]** is a formula. Furthermore, **F[a]** is the formula that results when **a** is substituted for **v** at all its *free* occurrences, and similarly, **F[n]** is the formula that results when **n** is so substituted.

Universal-Out (∀O)

∀vF[v]	
F[a]	**a** can be **any** name

Existential-In (∃I)

F[a]	**a** can be **any** name
∃vF[v]	

C. Rules that do require a new name

In the following two rules, **n** must be a **new** name, that is, a name that has not occurred in any previous line of the derivation.

Existential-Out (∃O)

∃v**F**[v]

F[n] — **n** must be a **new** name

Universal Derivation (UD)

~~SHOW~~: ∀v**F**[v]
| ~~SHOW~~: **F**[n] — **n** must be a **new** name
| |
| |

D. Negation Quantifier Elimination Rules

Tilde-Universal-Out (~∀O)

~∀v**F**[v]

∃v~**F**[v]

Tilde-Existential-Out (~∃O)

~∃v**F**[v]

∀v~**F**[v]

15. EXERCISES FOR CHAPTER 8

General Directions: For each of the following, construct a formal derivation of the conclusion, (indicated by '/') from the premises.

EXERCISE SET A (Universal-Out)

(1) ∀x(Fx → Gx) ; ~Gb / ~Fb

(2) ∀x(Fx → Gx) ; ~Gb / ~∀xFx

(3) ∀x(Fx → Gx) ; ~(Fc & Gc) / ~Fc

(4) ∀x[(Fx ∨ Gx) → Hx] ; ∀x[Hx → (Jx & Kx)] / Fa → Ka

(5) ∀x[(Fx & Gx) → Hx] ; Fa & ~Ha / ~Ga

(6) ∀x[~Fx → (Gx ∨ Hx)] ; ∀x(Hx → Gx) / Fa ∨ Ga

(7) ∀x(Fx → ~Gx) ; Fa / ~∀x(Fx → Gx)

(8) ∀x(Fx → Rxx) ; ∀x~Rax / ~Fa

(9) ∀x[Fx → ∀yRxy] ; Fa / Raa

(10) ∀x(Rxx → Fx) ; ∀x∀y(Rxy → Rxx) ; ~Fa / ~Rab

EXERCISE SET B (Existential-In)

(11) ∀x(Fx → Gx) ; Fa / ∃xGx

(12) ∀x(Fx → Gx) ; ∀x(Gx → Hx) ; Fa / ∃x(Gx & Hx)

(13) ~∃x(Fx & Gx) ; Fa / ~Ga

(14) ∃xFx → ∀xGx ; Fa / Gb

(15) ∀x[(Fx ∨ Gx) → Hx] ; ~(Ga ∨ Ha) / ∃x~Fx

(16) ∀x(Rxa → ~Rxb) ; Raa / ∃x~Rxb

(17) ∃xRax → ∀xRxa ; ~Rba / ~Raa

(18) ∀x(Fx → Rxx) ; Fa / ∃xRxa

(19) ∃xRax → ∀xRxa ; ~Raa / ~Rab

(20) ∀x[∃yRxy → ∀yRyx] ; Raa / Rba

EXERCISE SET C (Universal Derivation)

(21) $\forall x(Fx \to Gx)$; $\forall x(Gx \to Hx)$ / $\forall x(Fx \to Hx)$

(22) $\forall x(Fx \to Gx)$; $\forall x[(Fx \,\&\, Gx) \to Hx]$ / $\forall x(Fx \to Hx)$

(23) $\forall x(Fx \to Gx)$; $\forall x([Gx \vee Hx] \to Kx)$ / $\forall x(Fx \to Kx)$

(24) $\forall xFx \,\&\, \forall xGx$ / $\forall x(Fx \,\&\, Gx)$

(25) $\forall xFx \vee \forall xGx$ / $\forall x(Fx \vee Gx)$

(26) $\sim\exists xFx$ / $\forall x(Fx \to Gx)$

(27) $\sim\exists x(Fx \,\&\, Gx)$ / $\forall x(Fx \to \sim Gx)$

(28) $\forall x(Fx \to Gx)$; $\sim\exists x(Gx \,\&\, Hx)$ / $\forall x(Fx \to \sim Hx)$

(29) $\forall x(Fx \to Gx)$ / $\forall xFx \to \forall xGx$

(30) $\forall x((Fx \,\&\, Gx) \to Hx)$ / $\forall x(Fx \to Gx) \to \forall x(Fx \to Hx)$

EXERCISE SET D (Existential-Out)

(31) $\forall x(Fx \to Gx)$; $\exists x(Fx \,\&\, Hx)$ / $\exists x(Gx \,\&\, Hx)$

(32) $\exists x(Fx \,\&\, Gx)$; $\forall x(Hx \to \sim Gx)$ / $\exists x(Fx \,\&\, \sim Hx)$

(33) $\forall x(Fx \to Gx)$; $\forall x(Gx \to Hx)$; $\exists x\sim Hx$ / $\exists x\sim Fx$

(34) $\forall x(Fx \to \sim Gx)$ / $\sim\exists x(Fx \,\&\, Gx)$

(35) $\exists x(Fx \,\&\, \sim Gx)$ / $\sim\forall x(Fx \to Gx)$

(36) $\forall x(Fx \to Gx)$; $\forall x(Gx \to \sim Hx)$ / $\sim\exists x(Fx \,\&\, Hx)$

(37) $\forall x(Gx \to Hx)$; $\exists x(Ix \,\&\, \sim Hx)$; $\forall x(\sim Fx \vee Gx)$ / $\exists x(Ix \,\&\, \sim Fx)$

(38) $\exists xFx \vee \exists xGx$; $\forall x\sim Fx$ / $\exists xGx$

(39) $\forall x(Fx \to Gx)$ / $\exists xFx \to \exists xGx$

(40) $\forall x(Fx \to (Gx \to Hx))$ / $\exists x(Fx \,\&\, Gx) \to \exists x(Fx \,\&\, Hx)$

EXERCISE SET E (Negation Quantifier Elimination)

(41) $\sim\forall x(Fx \to Gx)$ / $\exists x(Fx \& \sim Gx)$

(42) $\sim\forall xFx$ / $\exists x(Fx \to Gx)$

(43) $\forall x(Gx \to Hx)$; $\forall x(Fx \to Gx)$ / $\sim\forall xHx \to \exists x\sim Fx$

(44) $\exists x(Fx \vee Gx)$ / $\exists xFx \vee \exists xGx$

(45) $\exists x(Fx \to Gx)$ / $\exists x\sim Fx \vee \exists xGx$

(46) $\exists xFx \to \forall xFx$ / $\forall xFx \vee \forall x\sim Fx$

(47) $\forall x(Fx \to Gx)$; $\sim\exists x(Gx \& Hx)$ / $\sim\exists x(Fx \& Hx)$

(48) $\exists xFx \vee \exists xGx$ / $\exists x(Fx \vee Gx)$

(49) $\exists x\sim Fx \vee \exists xGx$ / $\exists x(Fx \to Gx)$

(50) $\forall x(Fx \to Gx)$; $\forall x[(Fx \& Gx) \to \sim Hx]$; $\exists xHx$ / $\exists x(Hx \& \sim Fx)$

EXERCISE SET F (Multiple Quantification)

(51) $\forall x(Fx \to Gx)$ / $\forall x(Fx \to \exists yGy)$

(52) $\forall x[Fx \to \forall yGy]$ / $\exists xFx \to \forall xGx$

(53) $\exists xFx \to \forall xGx$ / $\forall x[Fx \to \forall yGy]$

(54) $\exists xFx \to \forall xGx$ / $\forall x\forall y[Fx \to Gy]$

(55) $\forall x\forall y[Fx \to Gy]$ / $\sim\forall xGx \to \sim\exists xFx$

(56) $\exists xFx \to \exists x\sim Gx$ / $\forall x[Fx \to \sim\forall yGy]$

(57) $\exists xFx \to \forall x\sim Gx$ / $\forall x[Fx \to \sim\exists yGy]$

(58) $\forall x[Fx \to \sim\exists yGy]$ / $\exists xFx \to \forall x\sim Gx$

(59) $\forall x[\exists yFy \to Gx]$ / $\forall x\forall y(Fx \to Gy)$

(60) $\exists xFx \to \forall xFx$ / $\forall x\forall y[Fx \leftrightarrow Fy]$

EXERCISE SET G (Relational Quantification)

(61) ∀x∀yRxy / ∀x∀yRyx

(62) ∃xRxx / ∃x∃yRxy

(63) ∃x∃yRxy / ∃x∃yRyx

(64) ∃x∀yRxy / ∀x∃yRyx

(65) ∃x~∃yRxy / ∀x∃y~Ryx

(66) ∃x~∃y(Fy & Rxy) / ∀x(Fx → ∃y~Ryx)

(67) ∀x[Fx → ∃y~Kxy] ; ∃x(Gx & ∀yKxy) / ∃x(Gx & ~Fx)

(68) ∃x[Fx & ~∃y(Gy & Rxy)] / ∀x[Gx → ∃y(Fy & ~Ryx)]

(69) ∃x[Fx & ∀y(Gy → Rxy)] / ∀x[Gx → ∃y(Fy & Ryx)]

(70) ~∃x(Kxa & Lxb) ; ∀x[Kxa → (~Fx → Lxb)] / Kba → Fb

EXERCISE SET H (More Relational Quantification)

(71) ∀x∃yRxy ; ∀x[∃yRxy → Rxx] ; ∀x[Rxx → ∀yRyx] / ∀x∀yRxy

(72) ∀x∃yRxy ; ∀x∀y[Rxy → ∃zRzx] ; ∀x∀y[Ryx → ∀zRxz] / ∀x∀yRxy

(73) ∀x∃yRxy ; ∀x∀y[Rxy → Ryx] ; ∀x[∃yRyx → ∀yRyx] / ∀x∀yRxy

(74) ∃x∃yRxy ; ∀x∀y[Rxy → ∀zRxz] ; ∀x[∀zRxz → ∀yRyx] / ∀x∀yRxy

(75) ∃x∃yRxy ; ∀x[∃yRxy → ∀yRyx] / ∀x∀yRxy

(76) ∀x[Kxa → ∀y(Kyb → Rxy)] ; ∀x(Fx → Kxb) ; ∃x[Kxa & ∃y(Fy & ~Rxy)]
/ ∃xGx

(77) ∃xFx; ∀x[Fx → ∃y(Fy & Ryx)] ; ∀x∀y(Rxy → Ryx) / ∃x∃y(Rxy & Ryx)

(78) ∃x(Fx & Kxa) ; ∃x[Fx & ∀y(Kya → ~Rxy)] / ∃x[Fx & ∃y(Fy & ~Ryx)]

(79) ∃x[Fx & ∀y(Gy → Rxy)] ; ~∃x[Fx & ∃y(Hy & Rxy)] / ~∃x(Gx & Hx)

(80) ∀x(Fx → Kxa) ; ∃x[Gx & ~∃y(Kya & Rxy)] / ∃x[Gx & ~∃y(Fy & Rxy)]

16. ANSWERS TO EXERCISES FOR CHAPTER 8

#1:

(1)	∀x(Fx → Gx)	Pr
(2)	~Gb	Pr
(3)	SHOW: ~Fb	DD
(4)	Fb → Gb	1,∀O
(5)	~Fb	2,4,→O

#2:

(1)	∀x(Fx → Gx)	Pr
(2)	~Gb	Pr
(3)	SHOW: ~∀xFx	ID
(4)	∀xFx	As
(5)	SHOW: ✕	DD
(6)	Fb	4,∀O
(7)	Fb → Gb	1,∀O
(8)	Gb	6,7,→O
(9)	✕	2,8,✕I

#3:

(1)	∀x(Fx → Gx)	Pr
(2)	~(Fc & Gc)	Pr
(3)	SHOW: ~Fc	ID
(4)	Fc	As
(5)	SHOW: ✕	DD
(6)	Fc → Gc	1,∀O
(7)	Fc → ~Gc	2,~&O
(8)	Gc	4,6,→O
(9)	~Gc	4,7,→O
(10)	✕	8,9,✕I

#4:

(1)	∀x[(Fx ∨ Gx) → Hx]	Pr
(2)	∀x[Hx → (Jx & Kx)]	Pr
(3)	SHOW: Fa → Ka	CD
(4)	Fa	As
(5)	SHOW: Ka	DD
(6)	(Fa ∨ Ga) → Ha	1,∀O
(7)	Ha → (Ja & Ka)	2,∀O
(8)	Fa ∨ Ga	4,∨I
(9)	Ha	6,8, →O
(10)	Ja & Ka	7,9,→O
(11)	Ka	10,&O

#5:

(1)	∀x[(Fx & Gx) → Hx]	Pr
(2)	Fa & ~Ha	Pr
(3)	SHOW: ~Ga	ID
(4)	Ga	As
(5)	SHOW: ✕	DD
(6)	(Fa & Ga) → Ha	1,∀O
(7)	Fa	2,&O
(8)	Fa & Ga	4,7,&I
(9)	Ha	6,8,→O
(10)	~Ha	2,&O
(11)	✕	9,10,✕I

#6:

(1)	∀x[~Fx → (Gx ∨ Hx)]	Pr
(2)	∀x(Hx → Gx)	Pr
(3)	SHOW: Fa ∨ Ga	ID
(4)	~(Fa ∨ Ga)	As
(5)	SHOW: ✕	DD
(6)	~Fa	4,~∨O
(7)	~Fa → (Ga ∨ Ha)	1,∀O
(8)	Ga ∨ Ha	6,7,→O
(9)	~Ga	4,~∨O
(10)	Ha	8,9,∨O
(11)	Ha → Ga	2,∀O
(12)	Ga	10,11,→O
(13)	✕	9,12,✕I

#7:

(1)	∀x(Fx → ~Gx)	Pr
(2)	Fa	Pr
(3)	SHOW: ~∀x(Fx → Gx)	ID
(4)	∀x(Fx → Gx)	As
(5)	SHOW: ✕	DD
(6)	Fa → ~Ga	1,∀O
(7)	Fa → Ga	4,∀O
(8)	~Ga	2,6,→O
(9)	Ga	2,7,→O
(10)	✕	8,9,✕I

#8:

(1)	∀x(Fx → Rxx)	Pr
(2)	∀x~Rax	Pr
(3)	SHOW: ~Fa	DD
(4)	Fa → Raa	1,∀O
(5)	~Raa	2,∀O
(6)	~Fa	4,5,→O

#9:

(1)	∀x(Fx → ∀yRxy)	Pr
(2)	Fa	Pr
(3)	SHOW: Raa	DD
(4)	Fa → ∀yRay	1,∀O
(5)	∀yRay	2,4,→O
(6)	Raa	5,∀O

#10:

(1)	∀x(Rxx → Fx)	Pr
(2)	∀x∀y(Rxy → Rxx)	Pr
(3)	~Fa	Pr
(4)	SHOW: ~Rab	DD
(5)	Raa → Fa	1,∀O
(6)	~Raa	3,5,→O
(7)	∀y(Ray → Raa)	2,∀O
(8)	Rab → Raa	7,∀O
(9)	~Rab	6,8,→O

#11:

(1)	∀x(Fx → Gx)	Pr
(2)	Fa	Pr
(3)	SHOW: ∃xGx	DD
(4)	Fa → Ga	1,∀O
(5)	Ga	2,4,→O
(6)	∃xGx	5,∃I

#12:

(1)	∀x(Fx → Gx)	Pr
(2)	∀x(Gx → Hx)	Pr
(3)	Fa	Pr
(4)	SHOW: ∃x(Gx & Hx)	DD
(5)	Fa → Ga	1,∀O
(6)	Ga → Ha	2,∀O
(7)	Ga	3,5,→O
(8)	Ha	6,7,→O
(9)	Ga & Ha	7,8,&I
(10)	∃x(Gx & Hx)	9,∃I

#13:

(1)	~∃x(Fx & Gx)	Pr
(2)	Fa	Pr
(3)	SHOW: ~Ga	DD
(4)	∀x~(Fx & Gx)	1,~∃O
(5)	~(Fa & Ga)	4,∀O
(6)	Fa → ~Ga	5,~&O
(7)	~Ga	2,6,→O

#14:

(1)	∃xFx → ∀xGx	Pr
(2)	Fa	Pr
(3)	SHOW: Gb	DD
(4)	∃xFx	2,∃I
(5)	∀xGx	1,4,→O
(6)	Gb	5,∀O

#15:

(1)	∀x[(Fx ∨ Gx) → Hx]	Pr
(2)	~(Ga ∨ Ha)	Pr
(3)	SHOW: ∃x~Fx	DD
(4)	~Ha	2,~∨O
(5)	(Fa ∨ Ga) → Ha	1,∀O
(6)	~(Fa ∨ Ga)	4,5,→O
(7)	~Fa	6,~∨O
(8)	∃x~Fx	7,∃I

#16:

(1)	∀x(Rxa → ~Rxb)	Pr
(2)	Raa	Pr
(3)	SHOW: ∃x~Rxb	DD
(4)	Raa → ~Rab	1,∀O
(5)	~Rab	2,4,→O
(6)	∃x~Rxb	5,∃I

#17:

(1)	∃xRax → ∀xRxa	Pr
(2)	~Rba	Pr
(3)	SHOW: ~Raa	ID
(4)	Raa	As
(5)	SHOW: ✕	DD
(6)	∃xRax	4,∃I
(7)	∀xRxa	1,6,→O
(8)	Rba	7,∀O
(9)	✕	2,8,✕I

#18:

(1)	∀x(Fx → Rxx)	Pr
(2)	Fa	Pr
(3)	SHOW: ∃xRxa	DD
(4)	Fa → Raa	1,∀O
(5)	Raa	2,4,→O
(6)	∃xRxa	5,∃I

#19:

(1)	∃xRax → ∀xRxa	Pr
(2)	~Raa	Pr
(3)	SHOW: ~Rab	ID
(4)	Rab	As
(5)	SHOW: ✕	DD
(6)	∃xRax	4,∃I
(7)	∀xRxa	1,6,→O
(8)	Raa	7,∀O
(9)	✕	2,8,✕I

#20:

(1)	∀x[∃yRxy → ∀yRyx]	Pr
(2)	Raa	Pr
(3)	SHOW: Rba	DD
(4)	∃yRay → ∀yRya	1,∀O
(5)	∃yRay	2,∃I
(6)	∀yRya	4,5,→O
(7)	Rba	6,∀O

#21:

(1)	∀x(Fx → Gx)	Pr
(2)	∀x(Gx → Hx)	Pr
(3)	~~SHOW~~: ∀x(Fx → Hx)	UD
(4)	~~SHOW~~: Fa → Ha	CD
(5)	Fa	As
(6)	~~SHOW~~: Ha	DD
(7)	Fa → Ga	1,∀O
(8)	Ga → Ha	2,∀O
(9)	Ga	5,7,→O
(10)	Ha	8,9,→O

#22:

(1)	∀x(Fx → Gx)	Pr
(2)	∀x[(Fx & Gx) → Hx]	Pr
(3)	~~SHOW~~: ∀x(Fx → Hx)	UD
(4)	~~SHOW~~: Fa → Ha	CD
(5)	Fa	AS
(6)	~~SHOW~~: Ha	DD
(7)	Fa → Ga	1,∀O
(8)	Ga	5,7,→O
(9)	Fa & Ga	5,8,&I
(10)	(Fa & Ga) → Ha	2,∀O
(11)	Ha	9,10→O

#23:

(1)	∀x(Fx → Gx)	Pr
(2)	∀x[(Gx ∨ Hx) → Kx]	Pr
(3)	~~SHOW~~: ∀x(Fx → Kx)	UD
(4)	~~SHOW~~: Fa → Ka	CD
(5)	Fa	As
(6)	~~SHOW~~: Ka	DD
(7)	Fa → Ga	1,∀O
(8)	Ga	5,7,→O
(9)	Ga ∨ Ha	8,∨I
(10)	(Ga ∨ Ha) → Ka	2,∀O
(11)	Ka	9,10,→O

#24:

(1)	∀xFx & ∀xGx	Pr
(2)	~~SHOW~~: ∀x(Fx & Gx)	UD
(3)	~~SHOW~~: Fa & Ga	DD
(4)	∀xFx	1,&O
(5)	∀xGx	1,&O
(6)	Fa	4,∀O
(7)	Ga	5,∀O
(8)	Fa & Ga	6,7,&I

#25:

(1)	∀xFx ∨ ∀xGx	Pr
(2)	~~SHOW~~: ∀x(Fx ∨ Gx)	UD
(3)	~~SHOW~~: Fa ∨ Ga	ID
(4)	~(Fa ∨ Ga)	As
(5)	~~SHOW~~: ✕	DD
(6)	~Fa	4,~∨O
(7)	~Ga	4,~∨O
(8)	~~SHOW~~: ~∀xFx	ID
(9)	∀xFx	As
(10)	~~SHOW~~: ✕	DD
(11	Fa	9,∀O
(12)	✕	6,11,✕I
(13)	∀xGx	1,8,∨O
(14)	Ga	13,∀O
(15)	✕	7,14,✕I

#26:

(1)	~∃xFx	Pr
(2)	~~SHOW~~: ∀x(Fx → Gx)	UD
(3)	~~SHOW~~: Fa → Ga	CD
(4)	Fa	As
(5)	~~SHOW~~: Ga	ID
(6)	~Ga	As
(7)	~~SHOW~~: ✕	DD
(8)	∀x~Fx	1,~∃O
(9)	~Fa	8,∀O
(10)	✕	4,9,✕I

#27:

(1)	~∃x(Fx & Gx)	Pr
(2)	~~SHOW~~: ∀x(Fx → ~Gx)	UD
(3)	~~SHOW~~: Fa → ~Ga	CD
(4)	Fa	As
(5)	~~SHOW~~: ~Ga	ID
(6)	Ga	As
(7)	~~SHOW~~: ✕	DD
(8)	∀x~(Fx & Gx)	1,~∃O
(9)	~(Fa & Ga)	8,∀O
(10)	Fa & Ga	4,6,&I
(11)	✕	9,10,✕I

#28:

(1)	∀x(Fx → Gx)	Pr
(2)	~∃x(Gx & Hx)	Pr
(3)	~~SHOW~~: ∀x(Fx → ~Hx)	UD
(4)	~~SHOW~~: Fa → ~Ha	CD
(5)	Fa	As
(6)	~~SHOW~~: ~Ha	ID
(7)	Ha	As
(8)	~~SHOW~~: ✕	DD
(9)	Fa → Ga	1,∀O
(10)	Ga	5,9,→O
(11)	Ga & Ha	7,10,&I
(12)	∃x(Gx & Hx)	11,∃I
(13)	✕	2,12,✕I

#29:

(1)	∀x(Fx → Gx)	Pr
(2)	~~SHOW~~: ∀xFx → ∀xGx	CD
(3)	∀xFx	As
(4)	~~SHOW~~: ∀xGx	UD
(5)	~~SHOW~~: Ga	DD
(6)	Fa → Ga	1,∀O
(7)	Fa	3,∀O
(8)	Ga	6,7,→O

#30:

(1)	∀x(Fx & Gx) → Hx)	Pr
(2)	~~SHOW~~: ∀x(Fx→Gx)→∀x(Fx→Hx)	CD
(3)	∀x(Fx → Gx)	As
(4)	~~SHOW~~: ∀x(Fx → Hx)	UD
(5)	~~SHOW~~: Fa → Ha	CD
(6)	Fa	As
(7)	~~SHOW~~: Ha	DD
(8)	Fa → Ga	3,∀O
(9)	Ga	6,8,→O
(10)	Fa & Ga	6,9,&I
(11)	(Fa & Ga) → Ha	1,∀O
(12)	Ha	10,11,→O

#31:

(1)	∀x(Fx → Gx)	Pr
(2)	∃x(Fx & Hx)	Pr
(3)	~~SHOW~~: ∃x(Gx & Hx)	DD
(4)	Fa & Ha	2,∃O
(5)	Fa	4,&O
(6)	Fa → Ga	1,∀O
(7)	Ga	5,6,→O
(8)	Ha	4,&O
(9)	Ga & Ha	7,8,&I
(10)	∃x(Gx & Hx)	9,∃I

#32:

(1)	∃x(Fx & Gx)	Pr
(2)	∀x(Hx → ~Gx)	Pr
(3)	~~SHOW~~: ∃x(Fx & ~Hx)	DD
(4)	Fa & Ga	1,∃O
(5)	Ha → ~Ga	2,∀O
(6)	Ga	4,&O
(7)	~~Ga	6,DN
(8)	~Ha	5,7,→O
(9)	Fa	4,&O
(10)	Fa & ~Ha	8,9,&I
(11)	∃x(Fx & ~Hx)	10,∃I

#33:

(1)	∀x(Fx → Gx)	Pr
(2)	∀x(Gx → Hx)	Pr
(3)	∃x~Hx	Pr
(4)	~~SHOW~~: ∃x~Fx	DD
(5)	~Ha	3,∃O
(6)	Ga → Ha	2,∀O
(7)	~Ga	5,6,→O
(8)	Fa → Ga	1,∀O
(9)	~Fa	7,8,→O
(10)	∃x~Fx	9,∃I

#34:

(1)	∀x(Fx → ~Gx)	Pr
(2)	~~SHOW~~: ~∃x(Fx & Gx)	ID
(3)	∃x(Fx & Gx)	As
(4)	~~SHOW~~: ✕	DD
(5)	Fa & Ga	3,∃O
(6)	Fa	5,&O
(7)	Fa → ~Ga	1,∀O
(8)	~Ga	6,7,→O
(9)	Ga	5,&O
(10)	✕	8,9,✕I

#35:

(1)	∃x(Fx & ~Gx)	Pr
(2)	~~SHOW~~: ~∀x(Fx → Gx)	ID
(3)	∀x(Fx → Gx)	As
(4)	~~SHOW~~: ✕	DD
(5)	Fa & ~Ga	1,∃O
(6)	Fa	5,&O
(7)	Fa → Ga	3,∀O
(8)	Ga	6,7,→O
(9)	~Ga	5,&O
(10)	✕	8,9,✕I

#36:

(1)	∀x(Fx → Gx)	Pr
(2)	∀x(Gx → ~Hx)	Pr
(3)	~~SHOW~~: ~∃x(Fx & Hx)	ID
(4)	∃x(Fx & Hx)	As
(5)	~~SHOW~~: ✕	DD
(6)	Fa & Ha	4,∃O
(7)	Fa	6,&O
(8)	Fa → Ga	1,∀O
(9)	Ga	7,8,→O
(10)	Ga → ~Ha	2,∀O
(11)	~Ha	9,10,→O
(12)	Ha	6,&O
(13)	✕	11,12,✕I

#37:

(1)	∀x(Gx → Hx)	Pr
(2)	∃x(Ix & ~Hx)	Pr
(3)	∀x(~Fx ∨ Gx)	Pr
(4)	~~SHOW~~: ∃x(Ix & ~Fx)	DD
(5)	│ Ia & ~Ha	2,∃O
(6)	│ ~Ha	5,&O
(7)	│ Ga → Ha	1,∀O
(8)	│ ~Ga	6,7,→O
(9)	│ ~Fa ∨ Ga	3,∀O
(10)	│ ~Fa	8,9,∨O
(11)	│ Ia	5,&O
(12)	│ Ia & ~Fa	10,11,&I
(13)	│ ∃x(Ix & ~Fx)	12,∃I

#38:

(1)	∃xFx ∨ ∃xGx	Pr
(2)	∀x~Fx	Pr
(3)	~~SHOW~~: ∃xGx	ID
(4)	│ ~∃xGx	As
(5)	│ ~~SHOW~~: ✕	DD
(6)	│ │ ∃xFx	1,4,∨O
(7)	│ │ Fa	6,∃O
(8)	│ │ ~Fa	2,∀O
(9)	│ │ ✕	7,8,✕I

#39:

(1)	∀x(Fx → Gx)	Pr
(2)	~~SHOW~~: ∃xFx → ∃xGx	CD
(3)	│ ∃xFx	As
(4)	│ ~~SHOW~~: ∃xGx	DD
(5)	│ │ Fa	3,∃O
(6)	│ │ Fa → Ga	1,∀O
(7)	│ │ Ga	5,6,→O
(8)	│ │ ∃xGx	7,∃I

#40:

(1)	∀x[Fx → (Gx → Hx)]	Pr
(2)	~~SHOW~~: ∃x(Fx&Gx)→∃x(Fx&Hx)	CD
(3)	│ ∃x(Fx & Gx)	As
(4)	│ ~~SHOW~~: ∃x(Fx & Hx)	DD
(5)	│ │ Fa & Ga	3,∃O
(6)	│ │ Fa	5,&O
(7)	│ │ Fa → (Ga → Ha)	1,∀O
(8)	│ │ Ga → Ha	6,7,→O
(9)	│ │ Ga	5,&O
(10)	│ │ Ha	8,9,→O
(11)	│ │ Fa & Ha	6,10,&I
(12)	│ │ ∃x(Fx & Hx)	11,∃I

#41:

(1)	~∀x(Fx → Gx)	Pr
(2)	~~SHOW~~: ∃x(Fx & ~Gx)	ID
(3)	│ ~∃x(Fx & ~Gx)	As
(4)	│ ~~SHOW~~: ✕	DD
(5)	│ │ ∃x~(Fx → Gx)	1,~∀O
(6)	│ │ ~(Fa → Ga)	5,∃O
(7)	│ │ Fa & ~Ga	6,~→O
(8)	│ │ ∀x~(Fx & ~Gx)	3,~∃O
(9)	│ │ ~(Fa & ~Ga)	8,∀O
(10)	│ │ ✕	7,9,✕I

#42:

(1)	~∀xFx	Pr
(2)	~~SHOW~~: ∃x(Fx → Gx)	ID
(3)	│ ~∃x(Fx → Gx)	As
(4)	│ ~~SHOW~~: ✕	DD
(5)	│ │ ∃x~Fx	1,~∀O
(6)	│ │ ~Fa	5,∃O
(7)	│ │ ∀x~(Fx → Gx)	3,~∃O
(8)	│ │ ~(Fa → Ga)	7,∀O
(9)	│ │ Fa & ~Ga	8,~→O
(10)	│ │ Fa	9,&O
(11)	│ │ ✕	6,10,✕I

#43:

(1)	∀x(Gx → Hx)	Pr
(2)	∀x(Fx → Gx)	Pr
(3)	~~SHOW~~: ~∀xHx → ∃x~Fx	CD
(4)	│ ~∀xHx	As
(5)	│ ~~SHOW~~: ∃x~Fx	DD
(6)	│ │ ∃x~Hx	4,~∀O
(7)	│ │ ~Ha	6,∃O
(8)	│ │ Ga → Ha	1,∀O
(9)	│ │ ~Ga	7,8,→O
(10)	│ │ Fa → Ga	2,∀O
(11)	│ │ ~Fa	9,10,→O
(12)	│ │ ∃x~Fx	11,∃I

#44:

(1)	∃x(Fx ∨ Gx)	Pr
(2)	~~SHOW~~: ∃xFx ∨ ∃xGx	ID
(3)	│ ~(∃xFx ∨ ∃xGx)	As
(4)	│ ~~SHOW~~: ✕	DD
(5)	│ │ ~∃xFx	3,~∨O
(6)	│ │ ~∃xGx	3,~∨O
(7)	│ │ Fa ∨ Ga	1,∃O
(8)	│ │ ∀x~Fx	5,~∃O
(9)	│ │ ~Fa	8,∀O
(10)	│ │ Ga	7,9,∨O
(11)	│ │ ∀x~Gx	6,~∃O
(12)	│ │ ~Ga	11,∀O
(13)	│ │ ✕	10,12,✕I

#45:

(1)	∃x(Fx → Gx)	Pr
(2)	SHOW: ∃x~Fx ∨ ∃xGx	ID
(3)	~(∃x~Fx ∨ ∃xGx)	As
(4)	SHOW: ✕	DD
(5)	~∃x~Fx	3,~∨O
(6)	~∃xGx	3,~∨O
(7)	Fa → Ga	1,∃O
(8)	∀x~~Fx	5,~∃O
(9)	~~Fa	8,∀O
(10)	Fa	9,DN
(11)	Ga	7,10,→O
(12)	∀x~Gx	6,~∃O
(13)	~Ga	12,∀O
(14)	✕	11,13,✕I

#46:

(1)	∃xFx → ∀xFx	Pr
(2)	SHOW: ∀xFx ∨ ∀x~Fx	ID
(3)	~(∀xFx ∨ ∀x~Fx)	As
(4)	SHOW: ✕	DD
(5)	~∀xFx	3,~∨O
(6)	~∀x~Fx	3,~∨O
(7)	~∃xFx	1,5,→O
(8)	∀x~Fx	7,~∃O
(9)	✕	6,8,✕I

#47:

(1)	∀x(Fx → Gx)	Pr
(2)	~∃x(Gx & Hx)	Pr
(3)	SHOW: ~∃x(Fx & Hx)	ID
(4)	∃x(Fx & Hx)	As
(5)	SHOW: ✕	DD
(6)	Fa & Ha	4,∃O
(7)	Fa	6,&O
(8)	Fa → Ga	1,∀O
(9)	Ga	7,8,→O
(10)	∀x~(Gx & Hx)	2,~∃O
(11)	~(Ga & Ha)	10,∀O
(12)	Ga → ~Ha	11,~&O
(13)	~Ha	9,12,→O
(14)	Ha	6,&O
(15)	✕	13,14,✕I

#48:

(1)	∃xFx ∨ ∃xGx	Pr
(2)	SHOW: ∃x(Fx ∨ Gx)	ID
(3)	~∃x(Fx ∨ Gx)	As
(4)	SHOW: ✕	DD
(5)	∀x~(Fx ∨ Gx)	3,~∃O
(6)	SHOW: ~∃xFx	ID
(7)	∃xFx	As
(8)	SHOW: ✕	DD
(9)	Fa	7,∃O
(10)	~(Fa ∨ Ga)	5,∀O
(11)	~Fa	10,~∨O
(12)	✕	9,11,✕I
(13)	∃xGx	1,6,∨O
(14)	Gb	13,∃O
(15)	~(Fb ∨ Gb)	5,∀O
(16)	~Gb	15,~∨O
(17)	✕	14,16,✕I

#49:

(1)	∃x~Fx ∨ ∃xGx	Pr
(2)	SHOW: ∃x(Fx → Gx)	ID
(3)	~∃x(Fx → Gx)	As
(4)	SHOW: ✕	DD
(5)	∀x~(Fx → Gx)	3,~∃O
(6)	SHOW: ~∃x~Fx	ID
(7)	∃x~Fx	As
(8)	SHOW: ✕	DD
(9)	~Fa	7,∃O
(10)	~(Fa → Ga)	5,∀O
(11)	Fa & ~Ga	10,~→O
(12)	Fa	11,&O
(13)	✕	9,12,✕I
(14)	∃xGx	1,6,∨O
(15)	Gb	14,∃O
(16)	~(Fb → Gb)	5,∀O
(17)	Fb & ~Gb	16,~→O
(18)	~Gb	17,&O
(19)	✕	15,18,✕I

#50:

(1)	∀x(Fx → Gx)	Pr
(2)	∀x[(Fx & Gx) → ~Hx]	Pr
(3)	∃xHx	Pr
(4)	~~SHOW~~: ∃x(Hx & ~Fx)	ID
(5)	~∃x(Hx & ~Fx)	As
(6)	~~SHOW~~: ✕	DD
(7)	Ha	3,∃O
(8)	∀x~(Hx & ~Fx)	5,~∃O
(9)	~(Ha & ~Fa)	8,∀O
(10)	Ha → ~~Fa	9,~&O
(11)	~~Fa	7,10,→O
(12)	Fa	11,DN
(13)	Fa → Ga	1,∀O
(14)	Ga	12,13,→O
(15)	Fa & Ga	12,14,&I
(16)	(Fa & Ga) → ~Ha	2,∀O
(17)	~Ha	15,16,→O
(18)	✕	7,17,✕I

#51:

(1)	∀x(Fx → Gx)	Pr
(2)	~~SHOW~~: ∀x(Fx → ∃yGy)	UD
(3)	~~SHOW~~: Fa → ∃yGy	CD
(4)	Fa	As
(5)	~~SHOW~~: ∃yGy	DD
(6)	Fa → Ga	1,∀O
(7)	Ga	4,6,→O
(8)	∃yGy	7,∃I

#52:

(1)	∀x(Fx → ∀yGy)	Pr
(2)	~~SHOW~~: ∃xFx → ∀xGx	CD
(3)	∃xFx	As
(4)	~~SHOW~~: ∀xGx	UD
(5)	~~SHOW~~: Ga	DD
(6)	Fb	3,∃O
(7)	Fb → ∀yGy	1,∀O
(8)	∀yGy	6,7,→O
(9)	Ga	8,∀O

#53:

(1)	∃xFx → ∀xGx	Pr
(2)	~~SHOW~~: ∀x(Fx → ∀yGy)	UD
(3)	~~SHOW~~: Fa → ∀yGy	CD
(4)	Fa	As
(5)	~~SHOW~~: ∀yGy	UD
(6)	~~SHOW~~: Gb	DD
(7)	∃xFx	4,∃I
(8)	∀xGx	1,7,→O
(9)	Gb	8,∀O

#54:

(1)	∃xFx → ∀xGx	Pr
(2)	~~SHOW~~: ∀x∀y(Fx → Gy)	UD
(3)	~~SHOW~~: ∀y(Fa → Gy)	UD
(4)	~~SHOW~~: Fa → Gb	CD
(5)	Fa	As
(6)	~~SHOW~~: Gb	DD
(7)	∃xFx	5,∃I
(8)	∀xGx	1,7,→O
(9)	Gb	8,∀O

#55:

(1)	∀x∀y(Fx → Gy)	Pr
(2)	~~SHOW~~: ~∀xGx → ~∃xFx	CD
(3)	~∀xGx	As
(4)	~~SHOW~~: ~∃xFx	ID
(5)	∃xFx	As
(6)	~~SHOW~~: ✕	DD
(7)	∃x~Gx	3,~∀O
(8)	~Ga	7,∃O
(9)	Fb	5,∃O
(10)	∀y(Fb → Gy)	1,∀O
(11)	Fb → Ga	10,∀O
(12)	~Fb	8,11,→O
(13)	✕	9,12,✕I

#56:

(1)	∃xFx → ∃x~Gx	Pr
(2)	~~SHOW~~: ∀x(Fx → ~∀yGy)	UD
(3)	~~SHOW~~: Fa → ~∀yGy	CD
(4)	Fa	As
(5)	~~SHOW~~: ~∀yGy	ID
(6)	∀yGy	As
(7)	~~SHOW~~: ✕	DD
(8)	∃xFx	4,∃I
(9)	∃x~Gx	1,8,→O
(10)	~Gb	9,∃O
(11)	Gb	6,∀O
(12)	✕	10,11,✕I

#57:

(1)	∃xFx → ∀x~Gx	Pr
(2)	~~SHOW~~: ∀x(Fx → ~∃yGy)	UD
(3)	~~SHOW~~: Fa → ~∃yGy	CD
(4)	Fa	As
(5)	~~SHOW~~: ~∃yGy	ID
(6)	∃yGy	As
(7)	~~SHOW~~: ✕	DD
(8)	∃xFx	4,∃I
(9)	∀x~Gx	1,8,→O
(10)	Gb	6,∃O
(11)	~Gb	9,∀O
(12)	✕	10,11,✕I

#58:

(1)	∀x(Fx → ~∃yGy)	Pr
(2)	~~SHOW~~: ∃xFx → ∀x~Gx	CD
(3)	∃xFx	As
(4)	~~SHOW~~: ∀x~Gx	UD
(5)	~~SHOW~~: ~Ga	ID
(6)	Ga	As
(7)	~~SHOW~~: ✕	DD
(8)	Fb	3,∃O
(9)	Fb → ~∃yGy	1,∀O
(10)	~∃yGy	8,9,→O
(11)	∀y~Gy	10,~∃O
(12)	~Ga	11,∀O
(13)	✕	6,12,✕I

#59:

(1)	∀x(∃yFy → Gx)	Pr
(2)	~~SHOW~~: ∀x∀y(Fx → Gy)	UD
(3)	~~SHOW~~: ∀y(Fa → Gy)	UD
(4)	~~SHOW~~: Fa → Gb	CD
(5)	Fa	As
(6)	~~SHOW~~: Gb	DD
(7)	∃yFy → Gb	1,∀O
(8)	∃yFy	5,∃I
(9)	Gb	7,8,→O

#60:

(1)	∃xFx → ∀xFx	Pr
(2)	~~SHOW~~: ∀x∀y(Fx ↔ Fy)	UD
(3)	~~SHOW~~: ∀y(Fa ↔ Fy)	UD
(4)	~~SHOW~~: Fa ↔ Fb	DD
(5)	~~SHOW~~: Fa → Fb	CD
(6)	Fa	As
(7)	~~SHOW~~: Fb	DD
(8)	∃xFx	6,∃I
(9)	∀xFx	1,8,→O
(10)	Fb	9,∀O
(11)	~~SHOW~~: Fb → Fa	CD
(12)	Fb	As
(13)	~~SHOW~~: Fa	DD
(14)	∃xFx	12,∃I
(15)	∀xFx	1,14,→O
(16)	Fa	15,∀O
(17)	Fa ↔ Fb	5,11,↔I

#61:

(1)	∀x∀yRxy	Pr
(2)	~~SHOW~~: ∀x∀yRyx	UD
(3)	~~SHOW~~: ∀yRya	UD
(4)	~~SHOW~~: Rba	DD
(5)	∀yRby	1,∀O
(6)	Rba	5,∀O

#62:

(1)	∃xRxx	Pr
(2)	~~SHOW~~: ∃x∃yRxy	DD
(3)	Raa	1,∃O
(4)	∃yRay	3,∃I
(5)	∃x∃yRxy	4,∃I

#63:

(1)	∃x∃yRxy	Pr
(2)	~~SHOW~~: ∃x∃yRyx	DD
(3)	∃yRay	1,∃O
(4)	Rab	3,∃O
(5)	∃yRyb	4,∃I
(6)	∃x∃yRyx	5,∃I

#64:

(1)	∃x∀yRxy	Pr
(2)	~~SHOW~~: ∀x∃yRyx	UD
(3)	~~SHOW~~: ∃yRya	DD
(4)	∀yRby	1,∃O
(5)	Rba	4,∀O
(6)	∃yRya	5,∃I

#65:

(1)	∃x~∃yRxy	Pr
(2)	~~SHOW~~: ∀x∃y~Ryx	UD
(3)	~~SHOW~~: ∃y~Rya	DD
(4)	~∃yRby	1,∃O
(5)	∀y~Rby	4,~∃O
(6)	~Rba	5,∀O
(7)	∃y~Rya	6,∃I

#66:

(1)	∃x~∃y(Fy & Rxy)	Pr
(2)	~~SHOW~~: ∀x(Fx → ∃y~Ryx)	UD
(3)	~~SHOW~~: Fa → ∃y~Rya	CD
(4)	Fa	As
(5)	~~SHOW~~: ∃y~Rya	DD
(6)	~∃y(Fy & Rby)	1,∃O
(7)	∀y~(Fy & Rby)	6,~∃O
(8)	~(Fa & Rba)	7,∀O
(9)	Fa → ~Rba	8,~&O
(10)	~Rba	4,9,→O
(11)	∃y~Rya	10∃I

#67:

(1)	∀x(Fx → ∃y~Kxy)	Pr
(2)	∃x(Gx & ∀yKxy)	Pr
(3)	SHOW: ∃x(Gx & ~Fx)	ID
(4)	~∃x(Gx & ~Fx)	As
(5)	SHOW: ✕	DD
(6)	∀x~(Gx & ~Fx)	4,~∃O
(7)	Ga & ∀yKay	2,∃O
(8)	Ga	7,&O
(9)	~(Ga & ~Fa)	6,∀O
(10)	Ga → ~~Fa	9,~&O
(11)	~~Fa	8,10,→O
(12)	Fa	11,DN
(13)	Fa → ∃y~Kay	1,∀O
(14)	∃y~Kay	12,13,→O
(15)	~Kab	14,∃O
(16)	∀yKay	7,&O
(17)	Kab	16,∀O
(18)	✕	15,17,✕I

#68:

(1)	∃x[Fx & ~∃y(Gy & Rxy)]	Pr
(2)	SHOW: ∀x[Gx → ∃y(Fy & ~Ryx)]	UD
(3)	SHOW: Ga → ∃y(Fy & ~Rya)	CD
(4)	Ga	As
(5)	SHOW: ∃y(Fy & ~Rya)	DD
(6)	Fb & ~∃y(Gy & Rby)	1,∃O
(7)	Fb	6,&O
(8)	~∃y(Gy & Rby)	6,&O
(9)	∀y~(Gy & Rby)	8,~∃O
(10)	~(Ga & Rba)	9,∀O
(11)	Ga → ~Rba	10,~&O
(12)	~Rba	4,11,→O
(13)	Fb & ~Rba	7,12,&I
(14)	∃y(Fy & ~Rya)	13,∃I

#69:

(1)	∃x[Fx & ∀y(Gy → Rxy)]	Pr
(2)	SHOW: ∀x[Gx → ∃y(Fy & Ryx)]	UD
(3)	SHOW: Ga → ∃y(Fy & Rya)	CD
(4)	Ga	As
(5)	SHOW: ∃y(Fy & Rya)	DD
(6)	Fb & ∀y(Gy → Rby)	1,∃O
(7)	∀y(Gy → Rby)	6,&O
(8)	Ga → Rba	7,∀O
(9)	Rba	4,8,→O
(10)	Fb	6,&O
(11)	Fb & Rba	9,10,&I
(12)	∃y(Fy & Rya)	11,∃I

#70:

(1)	~∃x(Kxa & Lxb)	Pr
(2)	∀x[Kxa → (~Fx → Lxb)]	Pr
(3)	SHOW: Kba → Fb	CD
(4)	Kba	As
(5)	SHOW: Fb	DD
(6)	Kba → (~Fb → Lbb)	2,∀O
(7)	~Fb → Lbb	4,6,→O
(8)	∀x~(Kxa & Lxb)	1,~∃O
(9)	~(Kba & Lbb)	8,∀O
(10)	Kba → ~Lbb	9,~&O
(11)	~Lbb	4,10,→O
(12)	~~Fb	7,11,→O
(13)	Fb	12,DN

#71:

(1)	∀x∃yRxy	Pr
(2)	∀x(∃yRxy → Rxx)	Pr
(3)	∀x(Rxx → ∀yRyx)	Pr
(4)	SHOW: ∀x∀yRxy	UD
(5)	SHOW: ∀yRay	UD
(6)	SHOW: Rab	DD
(7)	∃yRby	1,∀O
(8)	∃yRby → Rbb	2,∀O
(9)	Rbb	7,8,→O
(10)	Rbb → ∀yRyb	3,∀O
(11)	∀yRyb	9,10,→O
(12)	Rab	11,∀O

#72:

(1)	∀x∃yRxy	Pr
(2)	∀x∀y(Rxy → ∃zRzx)	Pr
(3)	∀x∀y(Ryx → ∀zRxz)	Pr
(4)	SHOW: ∀x∀yRxy	UD
(5)	SHOW: ∀yRay	UD
(6)	SHOW: Rab	DD
(7)	∃yRay	1,∀O
(8)	Rac	7,∃O
(9)	∀y(Ray → ∃zRza)	2,∀O
(10)	Rac → ∃zRza	9,∀O
(11)	∃zRza	8,10,→O
(12)	Rda	11,∃O
(13)	∀y(Rya → ∀zRaz)	3,∀O
(14)	Rda → ∀zRaz	13,∀O
(15)	∀zRaz	12,14,→O
(16)	Rab	15,∀O

#73:

(1)	∀x∃yRxy	Pr
(2)	∀x∀y(Rxy → Ryx)	Pr
(3)	∀x(∃yRyx → ∀yRyx)	Pr
(4)	~~SHOW~~: ∀x∀yRxy	UD
(5)	│ ~~SHOW~~: ∀yRay	UD
(6)	│ │ ~~SHOW~~: Rab	DD
(7)	│ │ │ ∃yRby	1,∀O
(8)	│ │ │ Rbc	7,∃O
(9)	│ │ │ ∀y(Rby → Ryb)	2,∀O
(10)	│ │ │ Rbc → Rcb	9,∀O
(11)	│ │ │ Rcb	8,10,→O
(12)	│ │ │ ∃yRyb → ∀yRyb	3,∀O
(13)	│ │ │ ∃yRyb	11,∃I
(14)	│ │ │ ∀yRyb	12,14,→O
(15)	│ │ │ Rab	14,∀O

#74:

(1)	∃x∃yRxy	Pr
(2)	∀x∀y(Rxy → ∀zRxz)	Pr
(3)	∀x(∀zRxz → ∀yRyx)	Pr
(4)	~~SHOW~~: ∀x∀yRxy	UD
(5)	│ ~~SHOW~~: ∀yRay	UD
(6)	│ │ ~~SHOW~~: Rab	DD
(7)	│ │ │ ∃yRcy	1,∃O
(8)	│ │ │ Rcd	7,∃O
(9)	│ │ │ ∀y(Rcy → ∀zRcz)	2,∀O
(10)	│ │ │ Rcd → ∀zRcz	9,∀O
(11)	│ │ │ ∀zRcz	8,10,→O
(12)	│ │ │ ∀zRcz → ∀yRyc	3,∀O
(13)	│ │ │ ∀yRyc	11,12,→O
(14)	│ │ │ Rac	13,∀O
(15)	│ │ │ ∀y(Ray → ∀zRaz)	2,∀O
(16)	│ │ │ Rac → ∀zRaz	15,∀O
(17)	│ │ │ ∀zRaz	14,16,→O
(18)	│ │ │ Rab	17,∀O

#75:

(1)	∃x∃yRxy	Pr
(2)	∀x(∃yRxy → ∀yRyx)	Pr
(3)	~~SHOW~~: ∀x∀yRxy	UD
(4)	│ ~~SHOW~~: ∀yRay	UD
(5)	│ │ ~~SHOW~~: Rab	DD
(6)	│ │ │ ∃yRcy	1,∃O
(7)	│ │ │ ∃yRcy → ∀yRyc	2,∀O
(8)	│ │ │ ∀yRyc	6,7,→O
(9)	│ │ │ Rbc	8,∀O
(10)	│ │ │ ∃yRby	9,∃I
(11)	│ │ │ ∃yRby → ∀yRyb	2,∀O
(12)	│ │ │ ∀yRyb	10,11,→O
(13)	│ │ │ Rab	12,∀O

#76:

(1)	∀x[Kxa → ∀y(Kyb → Rxy)]	Pr
(2)	∀x(Fx → Kxb)	Pr
(3)	∃x[Kxa & ∃y(Fy & ~Rxy)]	Pr
(4)	~~SHOW~~: ∃xGx	ID
(5)	│ ~∃xGx	As
(6)	│ ~~SHOW~~: ✕	DD
(7)	│ │ Kca & ∃y(Fy & ~Rcy)	3,∃O
(8)	│ │ ∃y(Fy & ~Rcy)	7,&O
(9)	│ │ Fd & ~Rcd	8,∃O
(10)	│ │ Fd	9,&O
(11)	│ │ Kca → ∀y(Kyb → Rcy)	1,∀O
(12)	│ │ Kca	7,&O
(13)	│ │ ∀y(Kyb → Rcy)	11,12,→O
(14)	│ │ Kdb → Rcd	13,∀O
(15)	│ │ Fd → Kdb	2,∀O
(16)	│ │ Kdb	10,15,→O
(17)	│ │ Rcd	14,16,→O
(18)	│ │ ~Rcd	9,&O
(19)	│ │ ✕	17,18,✕I

#77:

(1)	∃xFx	Pr
(2)	∀x[Fx → ∃y(Fy & Ryx)]	Pr
(3)	∀x∀y(Rxy → Ryx)	Pr
(4)	~~SHOW~~: ∃x∃y(Rxy & Ryx)	DD
(5)	│ Fa	1,∃O
(6)	│ Fa → ∃y(Fy & Rya)	2,∀O
(7)	│ ∃y(Fy & Rya)	5,6,→O
(8)	│ Fb & Rba	7,∃O
(9)	│ Rba	8,&O
(10)	│ ∀y(Rby → Ryb)	3,∀O
(11)	│ Rba → Rab	10,∀O
(12)	│ Rab	9,11,→O
(13)	│ Rab & Rba	9,12,&I
(14)	│ ∃y(Ray & Rya)	13,∃I
(15)	│ ∃x∃y(Rxy & Ryx)	14,∃I

#78:

(1)	∃x(Fx & Kxa)	Pr
(2)	∃x[Fx & ∀y(Kya → ~Rxy)]	Pr
(3)	~~SHOW~~: ∃x[Fx & ∃y(Fy & ~Ryx)]	DD
(4)	│ Fb & Kba	1,∃O
(5)	│ Fc & ∀y(Kya → ~Rcy)	2,∃O
(6)	│ ∀y(Kya → ~Rcy)	5,&O
(7)	│ Kba → ~Rcb	6,∀O
(8)	│ Kba	4,&O
(9)	│ ~Rcb	7,8,→O
(10)	│ Fc	5,&O
(11)	│ Fc & ~Rcb	9,10,&I
(12)	│ ∃y(Fy & ~Ryb)	11,∃I
(13)	│ Fb	4,&O
(14)	│ Fb & ∃y(Fy & ~Ryb)	12,13,&I
(15)	│ ∃x[Fx & ∃y(Fy & ~Ryx)]	14,∃I

#79:

(1)	∃x[Fx & ∀y(Gy → Rxy)]	Pr
(2)	~∃x[Fx & ∃y(Hy & Rxy)]	Pr
(3)	SHOW: ~∃x(Gx & Hx)	ID
(4)	∃x(Gx & Hx)	As
(5)	SHOW: ✕	DD
(6)	Fa & ∀y(Gy → Ray)	1,∃O
(7)	∀x~[Fx & ∃y(Hy & Rxy)]	2,~∃O
(8)	~[Fa & ∃y(Hy & Ray)]	7,∀O
(9)	Fa → ~∃y(Hy & Ray)	8,~&O
(10)	Fa	6,&O
(11)	~∃y(Hy & Ray)	9,10,→O
(12)	∀y~(Hy & Ray)	11,~∃O
(13)	Gb & Hb	4,∃O
(14)	~(Hb & Rab)	12,∀O
(15)	Hb → ~Rab	14,~&O
(16)	Hb	13,&O
(17)	~Rab	15,16,→O
(18)	∀y(Gy → Ray)	6,&O
(19)	Gb → Rab	18,∀O
(20)	Gb	13,&O
(21)	Rab	19,20,→O
(22)	✕	17,21,✕I

#80:

(1)	∀x(Fx → Kxa)	Pr
(2)	∃x[Gx & ~∃y(Kya & Rxy)]	Pr
(3)	SHOW: ∃x[Gx & ~∃y(Fy & Rxy)]	ID
(4)	~∃x[Gx & ~∃y(Fy & Rxy)]	As
(5)	SHOW: ✕	DD
(6)	Gb & ~∃y(Kya & Rby)	2,∃O
(7)	Gb	6,&O
(8)	∀x~[Gx & ~∃y(Fy & Rxy)]	4,~∃O
(9)	~[Gb & ~∃y(Fy & Rby)]	8,∀O
(10)	Gb → ~~∃y(Fy & Rby)	9,~&O
(11)	~~∃y(Fy & Rby)	7,10,→O
(12)	∃y(Fy & Rby)	11,DN
(13)	Fc & Rbc	12,∃O
(14)	Fc	13,&O
(15)	Fc → Kca	1,∀O
(16)	Kca	14,15,→O
(17)	~∃y(Kya & Rby)	6,&O
(18)	∀y~(Kya & Rby)	17,~∃O
(19)	~(Kca & Rbc)	18,∀O
(20)	Kca → ~Rbc	19,~&O
(21)	~Rbc	16,20,→O
(22)	Rbc	13,&O
(23)	✕	21,22,✕I

PICTORIAL SUMMARY OF RULES OF DERIVATION

SENTENTIAL LOGIC

&I

$$\frac{\begin{array}{c}\mathcal{A}\\ \mathcal{B}\end{array}}{\mathcal{A}\ \&\ \mathcal{B}} \qquad \frac{\begin{array}{c}\mathcal{A}\\ \mathcal{B}\end{array}}{\mathcal{B}\ \&\ \mathcal{A}}$$

&O

$$\frac{\mathcal{A}\ \&\ \mathcal{B}}{\mathcal{A}} \qquad \frac{\mathcal{A}\ \&\ \mathcal{B}}{\mathcal{B}}$$

~&O

$$\frac{\sim(\mathcal{A}\ \&\ \mathcal{B})}{\mathcal{A} \rightarrow \sim\mathcal{B}}$$

∨I

$$\frac{\mathcal{A}}{\mathcal{A} \vee \mathcal{B}} \qquad \frac{\mathcal{A}}{\mathcal{B} \vee \mathcal{A}}$$

∨O

$$\frac{\begin{array}{c}\mathcal{A} \vee \mathcal{B}\\ \sim\mathcal{A}\end{array}}{\mathcal{B}} \qquad \frac{\begin{array}{c}\mathcal{A} \vee \mathcal{B}\\ \sim\mathcal{B}\end{array}}{\mathcal{A}}$$

~∨O

$$\frac{\sim(\mathcal{A} \vee \mathcal{B})}{\begin{array}{c}\sim\mathcal{A}\\ \sim\mathcal{B}\end{array}}$$

↔I

$$\frac{\begin{array}{c}\mathcal{A} \rightarrow \mathcal{B}\\ \mathcal{B} \rightarrow \mathcal{A}\end{array}}{\mathcal{A} \leftrightarrow \mathcal{B}} \qquad \frac{\begin{array}{c}\mathcal{A} \rightarrow \mathcal{B}\\ \mathcal{B} \rightarrow \mathcal{A}\end{array}}{\mathcal{B} \leftrightarrow \mathcal{A}}$$

↔O

$$\frac{\mathcal{A} \leftrightarrow \mathcal{B}}{\mathcal{A} \rightarrow \mathcal{B}} \qquad \frac{\mathcal{A} \leftrightarrow \mathcal{B}}{\mathcal{B} \rightarrow \mathcal{A}}$$

~↔O

$$\frac{\sim(\mathcal{A} \leftrightarrow \mathcal{B})}{\sim\mathcal{A} \leftrightarrow \mathcal{B}}$$

DN

$$\frac{\mathcal{A}}{\sim\sim\mathcal{A}} \qquad \frac{\sim\sim\mathcal{A}}{\mathcal{A}}$$

→O

$$\frac{\begin{array}{c}\mathcal{A} \rightarrow \mathcal{C}\\ \mathcal{A}\end{array}}{\mathcal{C}} \qquad \frac{\begin{array}{c}\mathcal{A} \rightarrow \mathcal{C}\\ \sim\mathcal{C}\end{array}}{\sim\mathcal{A}}$$

~→O

$$\frac{\sim(\mathcal{A} \rightarrow \mathcal{C})}{\mathcal{A}\ \&\ \sim\mathcal{C}}$$

×I

$$\frac{\begin{array}{c}\mathcal{A}\\ \sim\mathcal{A}\end{array}}{\times}$$

×O

$$\frac{\times}{\mathcal{A}}$$

R

$$\frac{\mathcal{A}}{\mathcal{A}}$$

DD

~~SHOW~~: $\mathcal{A}$
|
| $\mathcal{A}$

CD

~~SHOW~~: $\mathcal{A} \rightarrow \mathcal{C}$
| $\mathcal{A}$
| ~~SHOW~~: $\mathcal{C}$
| |

ID

~~SHOW~~: $\sim\mathcal{A}$
| $\mathcal{A}$
| ~~SHOW~~: ×
| |

ID

~~SHOW~~: $\mathcal{A}$
| $\sim\mathcal{A}$
| ~~SHOW~~: ×
| |

PREDICATE LOGIC

In the following, v is any variable, a and n are names, and $\mathbb{F}$ is a formula. Furthermore, $\mathbb{F}$[a/v] is the formula that results when a is substituted for v at all its *free* occurrences in $\mathbb{F}$, and similarly, $\mathbb{F}$[n/v] is the formula that results when n is so substituted.

∀O

$$\frac{\forall v\mathbb{F}}{\mathbb{F}[a/v]}$$

a can be **any** name

∃I

$$\frac{\mathbb{F}[a/v]}{\exists v\mathbb{F}}$$

a can be **any** name

∃O

$$\frac{\exists v\mathbb{F}}{\mathbb{F}[n/v]}$$

n must be a **new** name

UD

~~SHOW~~: ∀v$\mathbb{F}$
| ~~SHOW~~: $\mathbb{F}$[n/v] n must be a **new** name

~∀O

$$\frac{\sim\forall v\mathbb{F}}{\exists v\sim\mathbb{F}}$$

~∃O

$$\frac{\sim\exists v\mathbb{F}}{\forall v\sim\mathbb{F}}$$